T0280501

Software Engineering

Software Engineering
A Methodical Approach

Second Edition

Elvis C. Foster

CRC Press
Taylor & Francis Group
Boca Raton London New York

CRC Press is an imprint of the
Taylor & Francis Group, an **informa** business

AN AUERBACH BOOK

Second edition published 2022
by CRC Press
6000 Broken Sound Parkway NW, Suite 300, Boca Raton, FL 33487-2742

and by CRC Press
2 Park Square, Milton Park, Abingdon, Oxon, OX14 4RN

First edition published by Apress 2014

CRC Press is an imprint of Taylor & Francis Group, LLC

ISBN: 978-0-367-76943-7 (hbk)
ISBN: 978-0-367-74601-8 (pbk)
ISBN: 978-0-367-74602-5 (ebk)

DOI: 10.1201/9780367746025

Typeset in Garamond
by Straive, India

Brief Contents

Contents

Preface

This text provides a comprehensive, but concise introduction to software engineering. It adopts a methodical approach to solving software engineering problems, proven over several years of teaching, with outstanding results. The book covers concepts, principles, design, construction, implementation, and management issues of software engineering. Each chapter is organized systematically into brief, reader-friendly sections, with itemization of the important points to be remembered. Diagrams and illustrations also sum up the salient points to enhance learning. Additionally, the book includes a number of the author's original methodologies that add clarity and creativity to the software engineering experience, while making a novel contribution to the discipline.

New in the second edition are the following: additional clarifications to various concepts covered in the first edition; revised and new illustrations all in color; seven new chapters (4, 5, 9, 17, 24, 25, and 26) covering various methodologies for software modeling and design, management support systems, software engineering frameworks, and new frontiers in software engineering.

Special Features: Upholding my overarching goals of succinctness, comprehensive coverage, and relevance, my practical and methodical discussion style aims straight to the salient issues, avoids unnecessary topics, and minimizes superfluous theoretical coverage. After reading this book, readers will acquire the following benefits:

- Working knowledge of how software systems are conceptualized, researched, modeled, designed, constructed, implemented, and managed
- Mastery of the salient phases and activities of the software development life cycle (SDLC)
- Solid grounding in various software engineering methodologies
- Appreciation of the role, scope, expanse, and challenges of software engineering

Software Engineering—A Methodical Approach (2nd edition) makes a useful addition to the literature, because of its unique approach to the subject. Many of the books that have been written on the subject exhibit assorted combinations of the following shortcomings: trivial illustrations that do not easily translate to real-to-life problems; overkill of theoretical discussions; coding obsession in the absence of explanation of the relevant principles; bulkiness due to content overload. In Software Engineering—A Methodical Approach (2nd edition), these problems are avoided by focusing on the essential details of each subject area, expressing each detail concisely without much fluff or fanfare, providing useful illustrations as well as step-by-step guidelines on how to solve various generic problems. From start to finish, the text is written as a friendly conversation with students—the way a typical classroom session would be organized. In just over 500 pages, the text covers the essential details, while providing the additional benefits mentioned earlier.

Target Audience: The book is best suited for undergraduate students who are pursuing a course in software engineering. However, graduate students with a similar need may also find it useful; practicing software engineers may also use it for quick referencing of specific methodologies.

Main Selling Features: Users of the text should find the following features quite reader friendly:

- Comprehensive but concise coverage of the discipline of software engineering
- Overview and summary of each chapter
- Short paragraphs that express the salient aspects of the subject matter being discussed
- Attractive original diagrams and illustrations in color
- Bullet points or itemization of important things to be remembered
- Inclusion of a few original software engineering methodologies, for example, the Information Topology Chart (ITC), the User Interface Topology Chart (UITC), the Object/Entity Specification Grid (O/ESG), object naming convention (ONC), and the Extended Operation Specification (EOS)

Organization of the Text

The text is organized into 26 chapters and 3 appendices. The chapters are placed into six divisions, as summarized below:

Part A: Fundamentals
Chapter 1: Introduction to Software Engineering
Chapter 2: The Role of the Software Engineer

Part B: Software Investigation & Analysis
Chapter 3: Project Selection & Initial System Requirement
Chapter 4: Fundamentals of Object-Oriented Methodologies
Chapter 5: Object-Oriented Information Engineering
Chapter 6: The Requirements Specification
Chapter 7: Information Gathering
Chapter 8: Communication via Diagrams
Chapter 9: More Diagrams—Modeling Objects and their Behavior
Chapter 10: Decision Models for System Logic
Chapter 11: Project Management Aids

Part C: Software Design
Chapter 12: Overview of Software Design
Chapter 13: Database Design
Chapter 14: User Interface Design

Chapter Summaries

Chapter 1 commences by providing the reader with a rationale for the course, followed by a discussion of the nature and scope of software engineering. The chapter proceeds with the following captions: Definitions and Concepts; The Organization as a System; Information Levels in the Organization; Software Life Cycle; Categories of Software; Alternate Software Acquisition Approaches; Software Engineering Paradigms; Desirable Features of Computer Software; Summary and Concluding Remarks.

Chapter 2 discusses the role of the software engineer under the following themes: Historical Role; Modern Role of the Software Engineer; Job Description of the Software Engineer; Tools Used by the Software Engineer; Management Issues with Which the Software Engineer Must be Familiar; Summary and Concluding Remarks.

Chapter 3 introduces the first major deliverable in a software engineering project—the *initial system requirement* (also called the *project proposal*). The chapter also discusses other related preliminary issues under the following captions: Project Selection; Problem Definition; Proposed Solution; Scope and Objectives of the System; System Justification; Feasibility Analysis Report; Alternate Approach to Feasibility Analysis; Summary of System Inputs and Outputs; Initial Project Schedule; Project Team; Summary and Concluding Remarks.

Chapter 4 provides an overview of fundamental object-oriented methodologies under the following captions: Rationale for Object-Oriented Methodologies (OOM); Characteristics of OOM; Benefits of OOM; Objects and Object Types and Classes; Operations and Methods; Encapsulation and Classes; Inheritance and Amalgamation; Object-Object Communication; Polymorphism and Reusability; Interfaces; Late Binding; Multithreading; Perception versus Reality; Object-Oriented Software Engineering and the SDLC; Summary and Concluding Remarks.

Chapter 5 introduces the principles of object-oriented information engineering (OOIE)—a set of techniques that reinforces an object-oriented approach to the early phases of the SDLC. The discussion advances under the following captions: Introduction; Engineering the Infrastructure; Related Diagramming Techniques; Enterprise Planning; Business Area Analysis; Software System Design; Software System Construction; Summary and Concluding Remarks.

Chapter 6 discusses the *requirements specification* as the second major deliverable of a software engineering project. The chapter includes Introduction; Contents of the Requirements Specification; Documenting the Requirements; Requirements Validation; How to Proceed; Presenting the Requirements Specification; The Agile Approach; Summary and Concluding Remarks.

Chapter 7 introduces various information-gathering strategies that the software engineer may employ in the pursuit of related information about software systems to be constructed. The chapter includes the following captions: Rationale for Information Gathering; Interviews; Questionnaires and Surveys; Sampling and Experimenting; Observation and Document Review; Prototyping; Brainstorming and Mathematical Proof; Object Identification; End-User Involvement; Summary and Concluding Remarks.

Chapter 8 covers various diagramming techniques that are used in software engineering. The discussion proceeds based on the following captions: Introduction; Traditional System Flow Charts; Procedure Analysis Chart; Innovation: Topology Charts; Data Flow Diagrams; Object Flow Diagram; Other Contemporary Diagramming Techniques; Program Flow Chart; Summary and Concluding Remarks.

Chapter 9 continues the discussion of diagramming with emphasis on object classification and object behavior: Overview of the Unified Modeling Language; Identifying Object Relationships; Fern Diagram; Object Relationship Diagrams; Representing Details about Object Types; Avoiding Multiple Inheritance Relationships; Top-Down versus Bottom-Up; Use-cases; States and State Transitions; Finite State Machines; Event Diagrams; Triggers; Activity Diagrams; Sequence Diagrams and Collaboration Diagrams; Summary and Concluding Remarks.

Chapter 10 introduces various models for system logic. It discusses Structured Language; Decision Tables; Decision Trees; Which Technique to Use; Decision Techniques versus Flowcharts; System Rules; Summary and Concluding Remarks.

Chapter 11 introduces project management aids that are commonly used in software engineering: PERT & CPM; Ghantt Chart; Project Management Software; Summary & Concluding Remarks.

Chapter 12 provides an overview of the software design phase, including a discussion of The Software Design Process; Design Strategies; Architectural Design; Interface Design; Software Design and Development Standards; The Design Specification; Summary and Concluding Remarks.

Chapter 13 introduces the reader to database design as an important aspect of software engineering. The chapter presents its discussion under the following subheadings: Introductory notes (outlining the rationale for and objectives of database design); Approaches to Database Design; Summary of File Organization; Summary and Concluding Remarks.

Chapter 14 discusses user interface design as another important component of software design. It includes the following captions: Introduction; Types of User Interfaces; Steps in User Interface Design; Output Design; Output Methods Versus Content & Technology; Guidelines for Designing Output; Guidelines for Designing Input; Summary and Concluding Remarks.

Chapter 15 discusses operations design, another important aspect of software design. It includes Introduction; Categorization of Operations; Essentials of Operation Design; Informal Methods of Specification; Formal Methods of Specification; Summary and Concluding Remarks.

Chapter 16 looks at other design considerations not discussed in the previous chapters: System Catalog; Product Documentation; User Message Management; Design for Real-time Software; Design for Reuse; System Security; The Agile Effect; Summary and Concluding Remarks.

Chapter 17 crystalizes the salient points from earlier chapters to emphasize the essential steps involved in designing software systems. The chapter also identifies several prospective software engineering projects for students to explore.

Chapter 18 directs the reader to important software development issues to be managed during a software engineering project. It includes Introduction; Standards and Quality Assurance; Management of Targets and Financial Resources; Leadership and Motivation; Planning of the Implementation Strategy; The Agile Effect; Summary and Concluding Remarks.

Chapter 19 provides an overview of human resource management from a software engineering perspective. The chapter proceeds under the following captions: Management Responsibilities; Management Styles; Developing the Job Description; Hiring; Maintaining the Desired Environment; Preserving Accountability; Grooming and Succession Planning; Summary and Concluding Remarks.

Chapter 20 discusses software economics—development costs, marketing cost, and other related issues. The chapter proceeds under the following captions: Software Cost versus Software Price; Software Value; Assessing Software Productivity; Estimation Techniques for Engineering Cost; Summary and Concluding Remarks.

Chapter 21 discusses software implementation issues that the software engineer should be familiar with. It includes the following subheadings: Introduction; Operating Environment; Installation of the System; Code Conversion; Change Over; Training; Marketing of the Software; Summary and Concluding Remarks.

Chapter 22 discusses software maintenance under the captions: Introduction; Software Maintenance; Legacy Systems; Software Integration; Software Re-engineering; The Agile Effect; Summary and Concluding Remarks.

Chapter 23 discusses the challenge of organizing software engineering teams for effective work. The discussion covers the following captions: Introduction; Functional Organization; Parallel Organization; Hybrid Organization; Organization of Software Engineering Firms; Summary and Concluding Remarks.

Chapter 24 focuses on management support systems as an area of increased importance in contemporary software engineering. The discussion advances through the following captions: Overview of Management Support Systems (MSS); Building System Security through Database Design; Case Study on a Dynamic Menu Interface Designer; Some MSS Project Ideas; Summary and Concluding Remarks.

Chapter 25 discusses software engineering frameworks as a significant building block for the design and construction of contemporary software systems. The chapter includes Software Architecture Tools; Software Frameworks; The Model-View-Controller Framework; Software Patterns; Summary and Concluding Remarks.

Chapter 26 examines new frontiers in software engineering and their implications for the future. Among the issues addressed are the following: Empirical Software Engineering; Data Science; Bioinformatics; Machine Learning; Game Design; Augmented and Virtual Reality; Internet of Things (IoT); Cloud Computing; Summary and Concluding Remarks.

Appendices A–C pulls some of the methodologies discussed in the text into three deliverables for a generic inventory management system: the project proposal, requirements specification, and design specification.

Text Usage

The text could be used as a one-semester or two-semester course in software engineering, augmented with an appropriate CASE or RAD tool. Below are two suggested schedules for using the text; one assumes a one-semester course; the other assumes a two-semester course. The schedule for a one-semester course is a very aggressive one that assumes adequate preparation on the part of the participants. The schedule for a two-semester course gives the participants more time to absorb the material, and gain mastery in the various methodologies discussed by engaging in a meaningful project. This obviously, is the preferred scenario.

One-Semester Schedule:

Week	Topic
01	Chapter 01
02	Chapter 02
03	Chapter 03
Chapter 04 may require only a cursory treatment	
04	Chapters 05 & 06
05	Chapter 07
06	Chapters 08 - 10
Chapter 09 may require only a cursory treatment	
07	Chapter 11
08	Chapter 12
09	Chapter 13
10	Chapters 14 & 15
11	Chapters 16 & 17
12	Chapters 18 & 19
13	Chapter 20
14	Chapters 21 & 22
15	Chapter 23
16	Review
Chapters 24 – 26 are not covered	

Two-Semester Schedule:

Week	Topic
01	Chapter 01
02	Chapter 02
03	Chapter 03
Chapter 04 may require only a cursory treatment	
04	Chapters 05 & 06
05	Chapter 07
06	Chapters 08 - 10
Chapter 09 may require only a cursory treatment	
07	Chapter 11
08	Chapter 12
09	Chapter 13
10	Chapters 14 & 15
11	Chapters 16 & 17
12	Chapters 18 & 19
13	Chapter 20
14	Chapters 21 & 22
15	Chapter 23
16	Review
17	Chapters 24 - 26
18-32	Course Project

Approach

Throughout the text, I have adopted a practical, methodical approach to software engineering, avoiding an overkill of theoretical calculations where possible (these can be obtained elsewhere). The primary objective is to help the reader to gain a good grasp of the activities in the software development life cycle (SDLC). At the end of the course, the participants should feel confident about taking on a new software engineering project.

Feedback & Support

It is hoped that you will have as much fun using this book as I had preparing it. Additional support information can be obtained from the publisher's website (https://www.routledge.com) or my website (https://www.elcfos.com). Also, your comments will be appreciated.

Acknowledgements

My profound gratitude is owed to my wife, Jacqueline, for putting up with me during the periods of preparation of this text. Also, I must recognize several of my past and current students (from four different institutions and several countries) who at various stages have encouraged me to publish my notes, and have helped to make it happen. In this regard, I would like to make special mention of Dionne Jackson, Kerron Hislop, Brigid Winter, Sheldon Kennedy, Ruth Del Rosario, Brian Yap, Rossyl Lashley, Jesse Schmidt, and Georgeana Hill.

Two silent but significant influencers of my work are former mentors Ezra Mugissa and Han Reichgelt. Indeed, their work in shaping my career in computer science is enduring.

My colleague and friend Brad Towle deserves special mention; he made significant contributions to chapters 25 and 26 of the text, while also assisting with the general editing.

The editorial and production teams at CRC Press (of the Taylor & Francis Group) deserve mention for their work in facilitating the publication of this volume. Thanks to everyone, with special mention of John Wyzalek, Glenon Butler, and Thivya Vasudevan.

Finally, I should also make mention of reviewers Marlon Moncrieffe, Siegwart Mayr, and Matthew Taylor, each a practicing software engineer, or computer science professor who has taken time to review the manuscript and provide useful feedback. Thanks, gentlemen.

Elvis C. Foster, PhD
Keene State College
Keene, New Hampshire, USA

FUNDAMENTALS

This preliminary division of the course is designed to cover some fundamentals. The objectives are as follows:

- To define and provide a rationale for software engineering as a discipline
- To provide you with a wide perspective of computer software and its varied usefulness and applications
- To discuss different approaches to software acquisition
- To define the job of the software engineer in the organization
- To discuss various tools used by the software engineer

The division consists of two chapters:

- Chapter 1: Introduction to Software Engineering
- Chapter 2: The Role of the Software Engineer

DOI: 10.1201/9780367746025-1

Chapter 1

Introduction to Software Engineering

Welcome and congratulations on your entry to this course in software engineering. The fact that you are in this course means that you have covered several fundamental topics in programming, data structures, and perhaps user interface. You have been writing computer programs to solve basic and intermediate-level problems. Now you want to take your learning experience to another level—you want to learn how to design, develop, and manage complex software systems (small, medium sized, and large), which may consist of tens or hundreds of programs all seamlessly integrated into a coherent whole. You will learn all of these and more in this course, but first, we must start at the beginning. This chapter introduces you to the discipline of software engineering. Topics covered include the following:

- Definitions and Concepts
- The Organization as a System
- Information Levels in the Organization
- Software Life Cycle
- Categories of Software
- Alternate Software Acquisition Approaches
- Software Engineering Paradigms
- Desirable Features of Computer Software
- The Software Engineering Dilemma
- Summary and Concluding Remarks

1.1 Definitions and Concepts

Computer software affects almost all aspects of human life. A study of the process of software construction and management is therefore integral to any degree in computer science, computer information systems, multimedia technology, or any other related field. Software systems are not created, and do not exist in a vacuum; rather they are typically created by individuals and/or organizations, for use by individuals and/or organizations. We will therefore start by defining a system, defining software, identifying the relationship between the two, and then showing how they both relate to the organization.

1.1.1 System

A system is a set of interacting, interrelated, interdependent components that function as a whole to achieve specific objectives. An effective system must be synergistic. The system usually operates in an environment external to itself. A system may also be defined as the combination of personnel, materials, facilities, and equipment working together to convert the input into meaningful and needed outputs.

Following are some fundamental principles about systems:

- The components of a system are interrelated and interdependent.
- The system is usually viewed as a whole.
- Every system has specific goals.
- There must be some type of inputs and outputs.
- Processes prescribe the transformation of inputs to outputs.
- Systems exhibit entropy i.e. tendency to become disorganized.
- Systems must be regulated (planning, feedback, and control).
- Every system has subsystems.
- Systems exhibit a tendency to a final state.

As a personal exercise, you are encouraged to identify examples of systems in areas with which you are familiar. A good place to start is the human body.

1.1.2 Software and Software Engineering

Software is the combination of program(s), database(s), and documentation in a systemic suite, and with the sole purpose of solving specific system problems and meeting predetermined objectives. Software adds value to the hardware components of a computer system. In fact, without software, a computer is reduced to nothing more than an electronic box of no specific use to most human beings. Also, it should not surprise you that computer software is a special kind of system. This course will teach you how to design, construct, and manage such systems.

Software engineering is the discipline concerned with the *research, evaluation, analysis, design, construction, implementation, and management of software systems.* It also includes the re-engineering of existing software systems with a view to improving their role, function, and performance. The ultimate objective is the provision or improvement of desirable conveniences and the enhancement of productivity within the related problem domain.

Each term in the above definition is significant, involving its own set of related methodologies; this will become clearer as you proceed through the course. For instance, software design involves the crafting of various models and algorithms via assorted techniques and methodologies; and construction involves the automation of those algorithms via various other techniques and methodologies.

System transformation may take various paths, some of which may be:

- Improving the internal workings of the system
- Modifying inputs and outputs
- Modifying the goals and objectives of the system
- Redesigning the system
- Designing and developing a new system based on existing problems

Before embarking on any major work in software engineering, a process of research and analysis takes place. This process may be summarized in the following steps:

1. Define the problem.
2. Understand the problem (system)—the interrelationships and interdependencies; have a picture of the variables at work within the system; define the extent of the system (problem).
3. Identify alternate solutions.
4. Examine alternate solutions.
5. Choose the best alternative.
6. Pursue the chosen alternative.
7. Evaluate the impact of the (new/modified) system.

1.2 The Organization as a System

Traditionally, many software systems were created by organizations for use in organizations. To a large extent, this scenario still holds true. For the moment, let us therefore take a look at the organization, in the context of this approach. An organization is a collection of individuals that employ available facilities, resources, and equipment to work in pursuit of a predetermined set of objectives. This concept of the organization is consistent with many scholars of organization theory (including [Morgan 2006]). However, the focus here is software engineering, not organization theory. The organization qualifies as a system since it has all the ingredients in the definition of a system: people, facilities, equipment, materials, and methods of work.

Every organization has certain functional areas (called divisions, departments, sections, etc.). Typical areas include finance and planning, human resource management (HRM), marketing, production and operations (or the equivalent), technical/manufacturing (or the equivalent). These are usually further divided into departments, units, and subunits. Traditionally, a data processing department/unit would be enlisted as a subunit of finance. However, more enlightened organizations are now enlisting information technology (IT) as a functional area at the senior management level, servicing all other areas.

Figure 1.1 shows what a highly summarized organizational chart for a modern organization might look like. In tiny or small organizations, each unit that appears

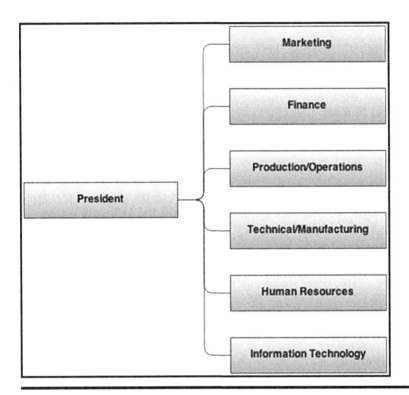

Figure 1.1 Typical Organizational Chart

under the president or chief executive officer (CEO) may be a department. In medium-sized and large organizations, each unit under the president or CEO is typically a division, consisting of several departments and/or sections. It should also be noted that you are unlikely to find IT (or the equivalent) at the senior level in many traditional organizations, as managers still struggle with appreciating the scope and role of IT in the organization. However, in more progressive and forward-thinking organizations, you will find that IT is correctly and appropriately positioned at the senior level of the management hierarchy.

Discussion:
Why should IT (or its equivalent) be positioned at the senior level in the organization. If the answer to this question is not immediately obvious to you, it will be by the time you complete this course.

Here are three additional insights about the functional units of the organization chart:

- Every organization has a set of functional unit(s) representing the essence of the organization. These are represented as Production/Operations and technical/ Manufacturing in the figure. Depending on the organization, the nomenclature may be different. For instance, in a college/university, these two units would typically be replaced by an Academic Administration unit consisting of various schools/divisions and/or academic departments.

- Again, depending on the type of organization, the Marketing unit may go by a different nomenclature. For example, some institutions prefer the name Public Relations.
- In some instances, the name Information Technology may be replaced by Information Services or some other appropriate alternative.

1.3 Information Levels in the Organization

Figure 1.2 shows the information levels in an organization. Information flows vertically as well as horizontally. Vertically, channels are very important. The chart also summarizes the characteristics and activities at each level. Let us examine these a bit closer:

1.3.1 Top Management

Activities are strategic and may include:

- Goal setting
- Long term planning
- Expansion or contraction or consolidation
- Merging
- Other strategic issues

For the individuals that operate at this level, there is always an external perspective in the interest of organizational image and interest.

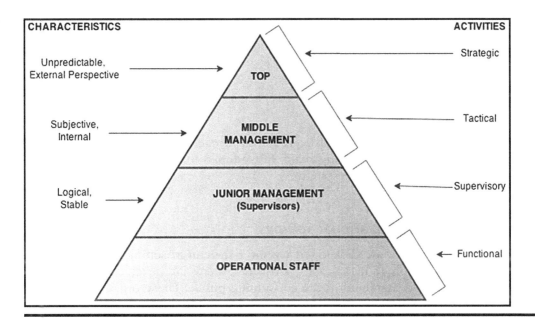

Figure 1.2   Information Levels in the Organization

1.3.2 Middle Management

Activities are of a tactical nature and may include:

- Allocation and control of resources
- Medium range planning
- Delegation
- Performance measurement and analysis

For the individuals that operate at this level, subjectivity is rather high—one's own style is brought into play. There is an internal perspective.

1.3.3 Junior Management and Operational Staff

At the junior management level, one is mainly concerned with:

- Job scheduling
- Checking operational results
- Maintaining discipline and order

Operations at the junior management level are very logical and predictable. Here is where you are likely to find supervisors who oversee routine activities by the operational staff.

The lowest level is the operational staff. The individuals who work at this level carry out the daily routine activities that keep the organization functioning.

1.3.4 Importance of Information Levels in Software Engineering

The information levels are of importance to the software engineer for two reasons:

- Information gathering and analysis must span the entire organization; therefore, communication at the appropriate level is important.
- The information needs vary with each level. An effective software system must meet the need at each level.

Discussion:
What kind of information would be required at each level? Propose an integrated software system (showing main subsystems) for a university, a variety store, and a hardware store, respectively.

1.3.5 Alternate Organizational Settings

Not all software systems are developed within a structured setting as portrayed in the foregoing discussion. In many cases, software systems are developed by software engineering firms and marketed to the consuming public. These organizations tend to be more flexible in the way they are structured, often favoring more highly skilled employees, and a more flattened structure. This approach is more fully discussed in Chapter 23.

Software systems may also be constructed by amorphously structured organizations. This classification includes individuals operating independently, or as collaborative groups. The open-source community is an excellent example of this kind of amorphous operation.

Whatever the circumstance, each software system is only relevant if it fulfills a need that people recognize. It must help solve a problem, and it typically has a period of relevance.

1.4 Software Life Cycle

Every software system has a life cycle—a period over which it is investigated/conceived, designed, developed, and remains applicable or needed. Various life cycle models have been proposed; we shall examine seven:

- Waterfall Model
- Phased Prototype Model
- Iterative Development Model
- Rapid Prototype Model
- Formal Transformation Model
- Component-Based Model
- Agile Development Model
- V-shaped Model

Irrespective of the model is used, however, a software system passes through the five phases, as depicted in Figure 1.3 (related deliverables also shown). These phases constitute the *software development life cycle* (SDLC).

SDLC Phase	Related Deliverable(s)
Investigation & Analysis	Initial System Requirements; Requirements Specification
Design (Modeling)	Design Specification
Development (Construction)	Actual Software System; Product Documentation
Implementation	Actual Software System; Product Documentation
Management	Enhanced Software System; Revised Documentation

Figure 1.3   SDLC Phases and Related Deliverables

1.4.1 Waterfall Model

The *waterfall model* is the traditional approach to software engineering; it may be summarized in the following points:

- It assumes that total knowledge of the requirements of a system can be obtained before its development.
- Each phase of the system life cycle is signed off with the users before advancing to the next phase.

- The process is irreversible—retracting is not allowed until the system is completed.

The model has the following advantages:

- It ensures a comprehensive, functional, well-integrated system.
- It minimizes the level user complaints.
- It ensures user participation (since the user must sign off on each phase).
- It is likely to result in a well-documented system.

The model is not void of major disadvantages:

- System development is likely to take a long time; users may become impatient.
- The requirements of the system may change before the system is completed.
- One may therefore have a well-designed, well-documented system that is not being used, due to its irrelevance.

1.4.2 Phased Prototype Model

This model (also referred to as the *evolutionary development* model) may be summarized in the following steps:

1. Investigate and fully define the system, identifying the major components.
2. Take a component and model it, then develop and implement it.
3. Obtain user feedback.
4. Revise the model if necessary.
5. If the system is not completed, go back to step 2.

The advantages of phased prototype model are:

- The user gets a feel of the system before its completion.
- Improved user participation over the waterfall model.
- The likelihood of producing an acceptable system is enhanced.

The disadvantages of the phased prototype model include:

- The increased likelihood of a poorly documented system.
- The system may be poorly integrated.
- The system will therefore be more difficult to maintain.

1.4.3 Iterative Development Model

The iterative development model is in some respects a refinement of the phased prototype model. In this approach, the entire life cycle is composed of several iterations. Each iteration is a mini-project in and of itself, consisting of the various lifecycle phases (investigation and analysis, design, construction, implementation, and management). The iterations may be in series or in parallel but are eventually integrated into a release of the project. The final iteration results in a release of the complete software product.

The advantages of the iterative development model are identical to those of the phased prototype model. However, due to the precautions inherent in the approach, disadvantages (of the phased prototype model) are minimized.

In many respects, the iterative development approach to software construction has been immortalized by the *Rational Unified Process* (RUP). RUP is an iterative development lifecycle framework that has become a very popular in the software engineering industry. RUP was first introduced by the pioneers of Rational Software—Grady Booch, James Rumbaugh, and Ivar Jacobson. Since 2003, the company has been acquired by International Business Machines (IBM). IBM currently markets the Rational product line as one of its prime product lines.

1.4.4 Rapid Prototype Model

Rapid prototyping is a commonly used (perhaps overused) term in software engineering. It refers to the rapid development of software via the use of sophisticated software development tools, typically found in CASE (*computer-aided software engineering*) tools and DBMS (*database management system*) suites. Another term that keeps flying around is *rapid application development* (RAD).

The line of distinction of a RAD tool from a CASE tool is not always very clear: in both cases, we are talking about software systems that facilitate the development of other software systems by providing among others, features such as:

* Automatic generation of code
* Convenient, user-friendly (and typically graphical) user interface
* Executable diagrams

To further blur the distinction, contemporary DBMS suites provide those features also. These tools will be further discussed in Chapter 2.

The rapid prototype model may be summarized in the following steps:

* Obtain an idea of the system requirements from user.
* Develop a prototype (possibly under the observation of the user).
* Obtain user feedback.
* Revise if necessary.

One point of clarification: RAD tools, CASE tools, DBMS suites and the like may be employed in any software engineering project, irrespective of the model being followed. Rapid prototyping describes a process, not the tools used.

Rapid prototyping provides us with two significant advantages:

* The system developed is obtained quickly if the first prototype is correct.
* The approach is useful in the design of *expert systems* as well as small end-user applications.

The main disadvantages of rapid prototyping are the following:

- The system may be poorly documented.
- The system may be difficult to maintain.
- System development could take long if the prototypes are wrong.

1.4.5 Formal Transformation Model

The formal transformation model produces software from mathematical system specifications. These transformations are "correctness preserving" and therefore ensure software quality. A number of formal specification languages have been proposed. However, much to the chagrin of its proponents, software development via this method is not as popular as hoped. Formal methods will be further discussed in Chapter 15.

The main advantage of the formal transformation model is the production of *provable software* i.e. software generated based on sound mathematical principles. This means that the reliability and quality of the software is high.

The model suffers from three haunting disadvantages:

- The approach (of formal methods) is not always relevant to the problem domain.
- The approach uses abstract specifications with which the software engineer must become familiar.
- Due to the use of quite abstract notations, end-user participation is not likely to be high, thus violating an essential requirement for software acceptance.

1.4.6 Components-Based Model

The component-based approach produces software by combining tested and proven components from other software products. As the discipline of software engineering becomes more established and more software standards are established, this approach has is expected to be more widely used. The approach is commonly called *component-based software engineering* (CBSE).

The main advantages of CBSE are the following:

- Improvement in the quality and reliability of software
- Software construction can be faster

The main disadvantage of the approach is that like the formal transformation model, it is not always relevant to the problem domain. The reason for this is that software engineering, being a relatively new discipline, has not established enough standards for solving the many and varied software needs that are faced by the world.

1.4.7 Agile Development Model

The agile development model pulls ideas from phased prototyping, iterative development, and rapid prototyping into a model that champions the idea of emphasizing the results of the software engineering effort over the process of getting to the results. The traditional approach of investigation and analysis, design, development, implementation, and management, is deemphasized. In contrast, the agile approach emphasizes construction and delivery.

Agile development calls for small, highly talented, highly responsive teams that construct software in small increments, focusing on the essential requirements. The chief architects of the methodology have articulated 12 principles that govern the agile development approach [Beck 2001]. They are paraphrased below:

1. Place the highest priority on customer satisfaction through early and continuous delivery of valuable software systems.
2. Welcome changing requirements that enhance the customer's competitive, irrespective of the stage in the development.
3. Deliver working software frequently and within a short timeframe.
4. Get the business people and the software developers to work together on a consistent basis throughout the project.
5. Build projects around motivated individuals. Give them the required resources and trust them to get the job done.
6. The most efficient and effective method of communication within a software engineering team is face-to-face conversation.
7. The primary measure of progress and success is a working software system or component.
8. The agile development process promotes sustainable development. The sponsors, developers, and users should be able to maintain a constant pace indefinitely.
9. There should be continuous attention to technical excellence and good design.
10. There should be great emphasis on simplicity as an essential means of maximizing the amount of work not done.
11. The best architectures, requirements, and designs emerge from self-organizing teams.
12. The software engineering team should periodically reflect on how to become more effective, and then refine its behavior accordingly.

The advantages of agile development are similar to those of phased prototyping:

- The user gets a version of the required software system in the shortest possible time.
- By merging business personnel and software developer in a single project, there is improved user participation.
- The likelihood of producing an acceptable system is enhanced.

Like the advantages, the disadvantages of the agile development model are comparable to those of the phased prototyping model:

- There is an increased likelihood of a poorly documented system. In fact, some extreme proponents of agile development go as far as to deemphasize the importance of software documentation.
- The system may be poorly integrated.
- A system that is poorly designed, documented, and poorly integrated is likely to be more difficult to maintain, thus increasing the said cost that agile development seeks to control.

The agile life cycle has emerged as the preferred approach for many contemporary software engineering enterprises. One possible reason for this is that the approach easily aligns to the business aspects of software engineering, while responding to the need for nimbleness in meeting end-user demands.

1.4.8 V-shaped Model

The V-shaped model is so named because the main activities are arranged in a V-shape as illustrated in Figure 1.4. The left side of the V-shape represents project definition and clarification. Advancing through stages of project conceptualization, requirements analysis, and detailed design. The base of the V-shape represents project construction. The right side of the V-shape represents project testing and integration, advancing through stages of unit testing, integration testing, system testing, and acceptance testing.

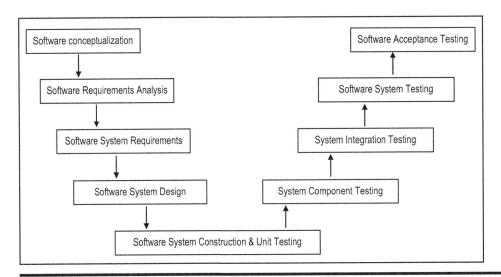

Figure 1.4   The V-shaped Model

The V-shaped model provides several advantages:

- Implicitly engraved in the model is the importance of software quality, evaluated and ascertained through testing.

- Through various levels of rigorous testing and evaluation, end-users are provided some assurance that the final product will likely meet some established benchmarks of quality.
- The model underscores the importance of software planning.
- The various deliverables of the model are subject to evaluation and testing.
- The model forces some minimum level of software documentation.

Like most things in software engineering, the V-shaped model has several flaws and/or potential disadvantages. Like the waterfall model:

- Due to the various checks and balances, system development is likely to take a long time; users may become impatient.
- The model is not very friendly to dynamic scenarios where the system requirements may change before the system construction is completed.
- One may therefore have a well-designed, thoroughly tested software system that is not being used, due to its irrelevance.

1.5 Categories of Software

Software engineering is a very wide, variegated field, constrained only by one's own imaginations and creativity. There are, however, some observable categories of software. Figure 1.5 provides a list of common software categories. Most of the software products that you are likely to use or be involved with fall into one or more of these categories.

1.6 Alternate Software Acquisition Approaches

Software may be acquired by any of the following approaches, each with its advantages and challenges:

- Traditional waterfall approach (in-house or via contracted work)
- Prototyping (phased or rapid, in-house or via contracted work)
- Iterative development (in-house or via contracted work)
- Assembly from re-usable components (in-house or via contracted work)
- Formal transformation (in-house or via contracted work)
- Agile development (in-house or via contracted work)
- V-shaped approach (in-house or via contracted work)
- Customizing an application software package
- End-user development
- Outsourcing

Whatever the acquisition approach that is employed, a software engineering objective of paramount importance is the production of software that has a significantly greater value than its cost. This is achieved by packing quality in the product from the outset—a principle that is emphasized throughout the text.

Discussion:
As a personal exercise, determine what scenario(s) would warrant each approach.

Operating System: A set of programs that provide certain desirable and necessary features for users of a computer system.
Compiler: A program that allows users (programmers) to code instructions to a computer system in a high-level language (HLL). The compiler converts the instructions from source code to object (machine) code.
Interpreter: An interpreter is similar to a compiler. However, it operates in an interactive mode, whereas the compiler operates in batch mode.
Assembler: A special compiler that works on lower level (assembly language) programs, converting them to object code.
Database Management System (DBMS): A set of programs that facilitate the creation and management of a database. A database is a collection of related records. A database consists of at least one file containing data, but typically includes several files.
Network Protocol: A software system that facilitates electronic communication on a computer network, according to a prescribed set of rules and standards.
Desktop Applications: Describe all generic computer software applications that run on microcomputers and notebook computers. They include subcategories such as word games, multimedia applications, and web browsers.
Management Support Systems: Refer to a family of software systems that are geared towards the promotion and facilitation of efficient and effective management and decision-making in the organization. Included among MSSs are the following categories: *strategic information systems* (SISs), *decision support systems* (DSSs), *executive information systems* (EISs), *expert systems* (ESs), *knowledge management systems* (KMSs), *business intelligence systems* (BISs), and *enterprise resource planning systems* (ERPSs).
Data Warehouse: An integrated, subject-oriented, time-variant, nonvolatile, consistent database, constructed from multiple source databases, and made available (in the form of read-only access) to support decision making in a business context.
Business Applications: Describe software applications that solve specific problems in a business. They include but are not confined to desktop applications and some information systems. Business applications therefore include accounting packages, library management systems, manufacturing systems, desktop applications, college/University administration systems, inventory management systems, point of sale systems, airline reservation systems.
Artificial Intelligence (AI) Systems: A system that causes the computer to exhibit humanlike intelligence. Popular branches include neural networks, natural language processing and expert systems.
Expert System (ES): A special case AI system that emulates a human expert in a particular problem domain, e.g. medical diagnosis and robotics.
Hypermedia System: A special desktop application that facilitates the creation and maintenance of multi-media-based systems. This includes geographic information systems (GIS), documentaries, documentation systems, etc.
Computer Aided Design (CAD) System: Special business/desktop application used in manufacturing and architecture to design blueprints.
Computer Aided Manufacturing (CAM) System: Used in manufacturing environments.
Computer Integrated Manufacturing (CIM) System: A combination of CAD and CAM.
Computer Aided Software Engineering (CASE) Tool: A sophisticated software product that is used to automate design and construction of other software products.
Rapid Application Development (RAD) Tool: A brand of CASE tool that facilitates the rapid design and construction of other software applications.
Software Development Kit (SDK): A conglomeration of software products bundled together for the purpose of software development.
Documentation System: These are multimedia-based knowledge management systems (KMSs) that provide information. While many documentation systems relate to, and are typically integrated into other software systems, such software systems could explore other aspects of life, and are usually designed to be independent systems.

Figure 1.5   Common Software Categories

1.7 Software Engineering Paradigms

There are two competing paradigms of software construction: the (traditional) *function-oriented* (FO) approach and the (more contemporary) *object-oriented* (OO) approach. The two approaches, though sometimes divergent, are not mutually exclusive—an experienced software engineer can design and construct software systems using aspects of both approaches. This leads to a third alternative—the *hybrid* approach—which ideally borrows the strong points from both approaches while avoiding the vulnerable points in either (see [Foster 2016]).

This course pursues a balance between the object-oriented paradigm and the function-oriented paradigm, with a strong focus on the fundamentals, and an evident bias to the object-oriented paradigm. A full treatment of the object-oriented approach is (best) treated in another course—*object-oriented software engineering* (OOSE) or *object-oriented methodologies* (OOM). However, in keeping with the stated bias, the course incorporates all the essential principles and concepts of OOM.

One of the distinguishing features of the software engineering discipline is that most of the principles and methodologies are ubiquitous. Once you have learned these principles and methodologies, you can apply them to problem-solving in any geographic location. And if the job is done right, the software system produced can be used in a range of related scenarios anywhere in the world. This feature accounts for the huge impact that software engineering has made on our world, impacting contemporary lifestyle in a way, and to a magnitude that no other discipline has.

Notwithstanding the ubiquitous nature of software engineering, the discipline encourages creativity and innovation. This innovative heritage leads to new frontiers being formed and explored on an ongoing basis.

1.8 Desirable Features of Computer Software

The following are some desirable features of computer software:

- **Maintainability:** How easily maintained is the software? This will depend on the quality of the design as well as the documentation.
- **Documentation:** How well documented is the system?
- **Efficiency:** How efficiently are the core operations carried out? Of importance are the response time and the system throughput.
- **User Friendliness:** How well designed is the user interface? How easy is it to learn and use the system?
- **Compatibility** with existing software products.
- **Security:** Are there mechanisms to protect and promote confidentiality and proper access control?
- **Integrity:** What is the quality of data validation methods?
- **Reliability:** Will the software perform according to requirements at all times?
- **Growth potential:** What is the storage capacity? What is the capacity for growth in data storage?
- **Functionality and Flexibility:** Does the software provide the essential functional features required for the problem domain? Are there alternate ways of doing things?
- **Differentiation:** What is unique about the software?
- **Adaptability:** How well are unanticipated situations handled?
- **Productivity:** How will productivity be affected by the software system?
- **Comprehensive Coverage:** Is the problem comprehensively and appropriately addressed?

These features are often referred to as *software quality factors* and for good reasons, since they affect the quality of the software products. In this course, you will learn how to research, plan, design, and construct software of a high quality. You will do so in a deliberately methodical manner. As such, we will revisit these quality factors later, and show how they can be used to guide the software engineering process.

1.9 The Software Engineering Dilemma

The software engineering industry is characterized by a multifaceted challenge, which this author has consistently called the *software engineering dilemma*. This dilemma is stated as a question, simple but profound: *How do we consistently produce and maintain software systems of the highest quality?*

This course guides you through the principles and methodologies that will enable you to produce and maintain such software systems. However, there are various challenges along the journey toward that objective. Following are some of the challenges:

Nonbinding Standards: While most software engineering methodologies are ubiquitous, they are not legislatively binding. Practicing software engineers have the liberty of embracing or ignoring any combination of methodologies that they choose to. There is an ongoing debate within the software engineering community on whether and how universal standards may be enforced (for example, see [SOD 2017]). This debate has taken place within professional associations such as the Association for Computing Machinery (ACM).

Bogus Claims: There is no universally accepted set of benchmarks for software engineers. Anyone can make claim to the profession. This looseness typically leads to the proliferation of inferior solutions by individuals falsely wearing the software engineering badge, thus leaving a vexing trail of mess for properly trained software engineers to clean up.

Consumer Demands: The onus is on the software engineering team to produce software systems that meet the demands of the consumers while conforming of universally accepted standards. Success in the industry often involves meeting these two criteria as well as the software quality factors (covered in the previous section).

Market Impatience: Software engineering has come to the stage where the quality factors are expected by the consuming public within a much tighter timeframe than in the past. Indeed, software consumers have become quite impatient with what may be perceived to be inferior or inadequate, and wary about persisting with products that do not meet their expectations. In short, software engineers are expected to deliver more sophisticated software systems and within a shorter timeframe than in the past.

So, the answer to the software engineering dilemma is to observe high software engineering standards and employ sound methodologies. However, as you practice the discipline, you will find that due to the aforementioned challenges, this is easier said than done. Moreover, more often than not, you are likely to be operating against a headwind sourced to any combination of those challenges.

1.10 Summary and Concluding Remarks

Let us summarize what has been covered in this chapter:

- A system is a set of interacting, interrelated, interdependent components that function together as a whole to achieve specific objectives.
- Software is the combination of program(s), database(s), and documentation in a systemic suite, with the sole purpose of solving specific system problems and meeting predetermined objectives.

- Software engineering is the process by which an information system or software is investigated, planned, modeled, developed, implemented, and managed.
- The organization is a complex system consisting of personnel, facilities, equipment, materials, and methods
- Software systems are created specifically for organizational usage and benefits.
- Every software system has a life cycle—a period over which it is investigated/conceived, designed, developed, and remains applicable or needed. Various life cycle models have been proposed; this chapter examined the following seven models: waterfall, phased prototyping, iterative development, rapid prototyping, formal transformation, component-based development, and agile development.
- Irrespective of the life cycle model employed, computer software goes through the phases of investigation and analysis, design, development, implementation, and management. These phases are referred to as the software development life cycle (SDLC).
- Two paradigms of software construction are the function-oriented (FO) approach and the object-oriented (OO) approach.
- Among the desirable features of computer software are the following: maintainability, documentation, efficiency, user-friendliness, compatibility, security, integrity, reliability, growth potential, functionality and flexibility, differentiation, productivity, and adaptability.
- The software engineering dilemma describes the challenge to sound software engineering practices due to the mitigating forces of nonbinding standards, bogus claims, consumer demands, and market impatience.

So now you have academic knowledge of what is meant by computer software engineering. As you will soon discover, this is not enough. To excel in this field, you will need experiential knowledge as well. But you have made an important start in exploring this exciting field. Much more could be said in this introductory chapter. The next chapter will discuss the role of the software engineer in the organization.

1.11 Review Questions

1. Give definitions of:
   - A system
   - Computer software
   - Software engineering

2. Explain how software engineering relates to the management of an organization.

3. Develop an organization chart for an organization that you are familiar with. Analyze the chart and propose a set of interrelated software systems that may be used in helping the organization to achieve its objectives.

4. Why should a software engineer be cognizant of the information levels in an organization?

5. What is the software development life cycle? Explain its significance.

6. Discuss the seven life cycle models covered in the chapter. For each model:
   - Describe the basic approach
   - Identify the advantages
   - Identify the disadvantages
   - Describe a scenario that would warrant the use of this approach

7. Identify ten major categories of computer software. For each category, identify a scenario that would warrant application of such a category of software.

8. Discuss some desirable features of computer software.

9. Write a paper on the importance of software engineering and your expectations for the future.

10. Discuss the challenges to development and maintaining software systems of the highest quality.

References and Recommended Readings

[Beck 2001] Beck, Kent et al. 2001. Manifesto for Agile Software Development. Accessed June 2009. http://www.agilemanifesto.org/

[Bruegge 2010] Bruegge, Bernd and Allen H. Dutoit. 2010. *Object-Oriented Software Engineering*, 3rd ed. Boston, MA: Pearson. See chapter 1.

[Foster 2016] Foster, Elvis C. with Shripad Godbole. 2016. *Database Systems: A Pragmatic Approach*, 2nd ed. New York, NY: Apress. See chapter 23.

[Kendall 2014] Kendall, Kenneth E., and Julia E. Kendall. 2014. *Systems Analysis and Design*, 9th ed. Boston, MA: Pearson. See chapters 1 and 2.

[Maciaszek 2005] Maciaszek, Leszek A. and Bruce Lee Long. 2005. *Practical Software Engineering*. Boston, MA: Addison-Wesley. See chapter 1.

[Mili & Fchier 2015] Mili, Ali and Fairouz Tchier. 2015. *Software Testing: Concepts and Operations*. Hoboken, NJ: John Wiley. See chapters 3–5.

[Morgan 2006] Morgan, Gareth. *Images of Organization*, updated ed. 2006. Thousand Oaks, CA: Sage Publications.

[Peters 2000] Peters, James F. and Witold Pedrycz. 2000. *Software Engineering: An Engineering Approach*. New York, NY: John Wiley & Sons. See chapters 1 and 2.

[Pfleeger 2006] Pfleeger, Shari Lawrence. 2006. *Software Engineering Theory and Practice*, 3rd ed. Upper Saddle River, NJ: Prentice Hall. See chapters 1 and 2.

[Pressman 2015] Pressman, Roger. 2015. *Software Engineering: A Practitioner's Approach*, 8th ed. New York, NY: McGraw-Hill. See chapters 1–6.

[Schach 2011] Schach, Stephen R. 2011. *Object-Oriented & Classical Software Engineering*, 8th ed. New York, NY: McGraw-Hill. See chapters 1–3.

[SOD 2017] Office of the Deputy Assistant Secretary of Defense for Systems Engineering. 2017. Best Practices for Using Systems Engineering Standards. Accessed June 2018. https://www.acq.osd.mil/se/docs/15288-Guide-2017.pdf.

[Sommerville 2016] Sommerville, Ian. 2016. *Software Engineering*, 10th ed. Boston: Pearson. See chapters 1–3.

[Spence 2004] Spence, Ian and Kurt Bittner. 2004. Managing Iterative Software Development with Use Cases. IBM. Accessed December 2009. http://www.ibm.com/developerworks/rational/library/5093.html.

[Van Vliet 2008] Van Vliet, Hans. *Software Engineering*, 3rd ed. 2008. New York, NY: John Wiley & Sons. See chapters 1–3.

Chapter 2

The Role of the Software Engineer

This second chapter will examine the role, responsibilities, functions, and characteristics of the software engineer in the organization. The following will be discussed:

- Historical Role
- Modern Role of the Software Engineer
- Job Description of the Software Engineer
- Tools Used by the Software Engineer
- Management Issues with Which the Software Engineer Must be Familiar
- Summary and Concluding Remarks

2.1 Historical Role

Historically, most systems were manual and coordinated by a system and procedure analyst with responsibilities such as:

- Forms design and record management
- Analysis, design, and management of manual systems
- Report distribution analysis
- Work measurement and specification
- Development and maintenance of procedure manuals

Traditionally, the systems and procedure analyst position was attached to the Finance Department. Most systems were accounting oriented. This created an imbalance. A second difficulty was the collation of information from various departments.

The arrival of computer technology triggered several changes and brought several advantages:

DOI: 10.1201/9780367746025-3

- A vast amount of data could be collected and stored.
- Inherent validation checks could ensure data accuracy.
- The response was fast.
- There was significant reduction of manual effort, redundancies, and data security problems.

2.2 Modern Role of the Software Engineer

In addition to the historical roles, the software engineer (SE) takes on additional responsibilities, some of which are mentioned here:

- The software engineer acts as a *consultant* to the organization.
- The software engineer is the *supporting expert* on system design, development, implementation, and maintenance.
- The software is the *change agent* in the organization, lobbying for and effecting system enhancements (manual and automated; strategic and functional).
- The software engineer acts as a *project leader* on the development or modification of systems.
- The software engineer acts as a *software designer and developer* (especially in developing countries or other environments where resources are scarce).

The contemporary trend reflects a bias to the title of *software engineer* instead of the traditional systems analyst (SA). Software engineer is more accurate as it spans the whole software development life cycle. However, many organizations tend to stick to the traditional. Where both titles are used in the same organization, the SA usually operates upper phases of SDLC, while the software engineer operates in the lower phases; some overlap typically takes place in the area of software design.

2.3 Job Description of the Software Engineer

If you are interviewed for, or employed as a software engineer, one of the first pieces of document you will receive is a job description. Each organization has its own standards as to how job descriptions are written. However, generally speaking, the job description will have the following salient components:

- **Heading:** An organizational heading followed by a departmental heading.
- **Summary:** A summary of what the job entails.
- **Core Functions:** A list of core functions and/or responsibilities of the job. This is further elucidated below.
- **Desirable Qualities:** A section describing the desirable qualities of the incumbent. This is further elucidated below.
- **Authority:** A section that outlines the authority associated with the job.
- **Job Specification:** A section that describes reporting relationship and other related information about the job.

Figure 2.1 provides an illustration of a job description of a software engineer. It includes all the essential ingredients mentioned above. The next two subsections further provide content guidelines for the core functions of a software engineer, and the desirable qualities of the incumbent.

LAMBERT COX COLLEGE JOB DESCRIPTION
Department of Computer Science

JOB TITLE of Software Engineer

SUMMARY
The incumbent works with the software engineering team of the college to develop and maintain quality software that will be relevant to the college in particular and the information technology industry in general. The incumbent also participates in academic matters as the need arises.

CORE FUNCTIONS
01. Investigates and determines the software system needs of user departments as assigned by the Chair of Computer Science Department.
02. Prepares detailed requirements specification and design of software systems to be developed.
03. Discusses and confirms the project concept with Chair of Computer Science and Director of Computer Services.
04. Develops system design specifications of various software systems for review and approval.
05. Participates in the establishment of software development standards for the institution.
06. Develops software systems according to approved and agreed specifications and standards.
07. Maintains software systems previously developed as required.
08. Prepares and submits relevant reports and documentation on systems developed.
09. Conducts training of end-users of various software systems, lab assistants and other members of staff as required.
10. Participates in consulting ventures on behalf of the college.
11. Participates in teaching of no more than one course per quarter if and when required.
12. Keeps abreast of current developments and trends in information technology and ensures that the college is likewise kept up-to-date.
13. Provides leadership and guidance to students and/or trainees who may from time to time work in the Software Engineering Center.
14. Performs any other related duties consistent with the nature, functions and objectives of the job.

AUTHORITY TO
01. Participate in budget preparation and review exercises.
02. Design and develop resource materials.
03. Take necessary action to ensure security and proper use of resources.
04. Make recommendations for the selection of resources.
05. Apply disciplinary measures to students and/or trainees working in the Software Engineering Center.

JOB SPECIFICATION
Required Qualification: Post graduate degree in computer science or related field

Alternate Qualification: Bachelor of Science in computer science or related field; bachelor of science (or graduate diploma) in management studies

Required Experience: A minimum of five years in systems development, implementation and management is required. In the absence of such experience, the incumbent will be subject to one year of intense on-the-job training and guidance.

Special Requirement: Flexible working hours; sometimes required to work outside of normal working hours

Required Expertise
- Excellent communication skills
- Conversant with various computer platforms, information systems methodologies and approaches
- Excellent human resource management skills
- Excellent software development skills

Reports to: Chair, Computer Science Department

Liaises with: Director of Computer Services, employees at all levels of operation of the college, external organizations in the field of information technology, and other companies as required.

Figure 2.1   Sample Job Description for a Software Engineer

2.3.1 Core Functions of the Software Engineer

Some core functions of the software engineer are as follows:

- Investigates and defines software system problems and makes proposals for their solution.
- Carries out detailed feasibility analysis and system specification.
- Plays a critical role in the design and development of software systems in the organization.
- Participates in the process of hardware/software selection for the organization.
- Participates in the definition of software engineering standards for the organization.
- Performs as project leader according to assignment from his superior.
- Conducts system enhancements as required.
- Keeps abreast of current technological developments and trends and ensures that the organization is strategically positioned.
- Plays a crucial role in the development and maintenance of system and user documentation.
- Plays a crucial role in the training of staff.
- Ensures the safety of all information technology (IT)-related resources.

2.3.2 Desirable Qualities of the Software Engineer

Among the required qualities of the software engineer are the following:

- Excellent communication skills
- Pleasant personality
- Tolerant and businesslike
- Experienced in various computer platforms, applications, and software
- Wide experience in (appreciation of) business administration
- Able to effectively and efficiently assimilate large volumes of information
- Imaginative and perceptive
- Able to understand complex problems
- A good salesperson of ideas

2.4 Tools Used by the Software Engineer

Traditionally, the tools the software engineer uses in performing include:

- Coding Systems
- Forms Design
- Data Analysis Charts
- Decision Tables and Decision Trees
- Flow Charts and Diagrams
- Other Creative Technical Documents and Modeling Techniques
- Software Planning and Development Tools

Additionally, *object-oriented software engineering* (OOSE), also referred to as *object-oriented methodologies* (OOM), provides other standards, conventions, and tools. This course covers the main OO design techniques and methodologies; however, a more detailed treatment can be obtained from a course in OOSE or OOM (see the appendices). Flow charts, diagrams, and decision tables/trees are covered in Chapters 5 and 6, respectively. The other mentioned tools will be briefly discussed here.

2.4.1 Coding Systems

A code is a group of characters used to identify an item of data and show its relationship to other similar items. It is a compact way of defining a specific item or entity. Coding systems are particularly useful during forms design and database design (to be discussed in Chapter 10). Benefits of a proper coding system include:

- Reduction of storage information
- Easy identification and recollection
- Easy classification of data

2.4.1.1 Desirable Features of a Coding System

A coding system must exhibit some of the following features:

- **Uniqueness:** The code must give specificity to different items.
- **Purpose:** The code must serve a useful purpose (e.g. comparison or location).
- **Compactness**: The code must not be too bulky.
- **Meaningful:** The code must be meaningful. The code can relate to shape, size, location, type, or other features of items represented.
- **Self-checking:** The code may be self-checking, utilizing a mathematically calculated check digit as part of the code to ensure validity.
- **Expandable:** The code must be expandable; it must be easy to add new blocks of items.

2.4.1.2 Types of Coding Systems

There are six common types of coding systems as summarized below. Bear in mind, however, that in many practical cases, coding types are combined to form the desired coding system.

1. Simple Sequence
   - Numbers are assigned consecutively, e.g.1,2,3,...
   - The main advantage of the approach is that an unlimited number of items can be stored.
   - The main drawback is that little or no information conveyed about items.
2. Block Sequence
   - Blocks of consecutive numbers identify groups of similar items.
   - The main advantage of this approach is that more information is conveyed about the items.

- The main drawback is that it is not always possible to maintain sequencing in each block.
3. Group Classification
   - This approach involves a major, intermediate, and minor classification of items by ordering or digits. A good example is the US zip code.
   - The main advantage of the approach is that data can be subdivided.
   - One possible drawback is that groups can become quite large.
4. Significant Digit
   - For this type of code, a digit is assigned to denote a physical characteristic of the item.
   - The main advantage of the approach is that it aids physical identification of the items.
   - Except for a few simple scenarios, the approach is inadequate to be used on its own. For this reason, it is often combined with another type of coding.
5. Alphabetic Codes
   - This coding system involves the use of letters to identify items. The coding system may be mnemonic (e.g. H_2O represents water) or alphabetic according to a set rule (e.g. NY represents New York).
   - The main advantage of the approach is that it facilitates quick reference to items.
   - One drawback is that if not controlled, the code used can be too long and confusing.
6. Self-checking Code
   - In this coding system, a number is mathematically calculated and applied to the code.
   - This approach is particularly useful in situations where data validation is required after transmission over telecommunication channels.
   - The approach is redundant in situations where electronic transmission is not required.

2.4.2 Forms Design

A significant proportion of data input to a business information system will be done with the use of forms. For instance, most universities and colleges require applicants to fill out application forms. Completed application forms are then used to key data into some database for subsequent processing. When an applicant is accepted, it is this same data from the application form that is used as the student's personal information.

Forms design provides the following advantages:

- Elimination of redundant operations
- Reduction of the organization's operational cost
- Facilitation of filing and retrieval
- Enabling of easy data capture
- Facilitation of interdepartmental flow
- Facilitation of control e.g. assigning a form number or color code

A form has three parts: a heading, a body, and a footing. The heading should contain the important company information (for example the company name), along with a descriptive title for the form. The body contains the essential details that the form contains. The footing contains instructional guidelines and is optional (some forms are self-explanatory and therefore do not need footnotes). Figure 2.2 indicates the three styles for providing/requesting information on a form—open style, box style, or a mixture of both.

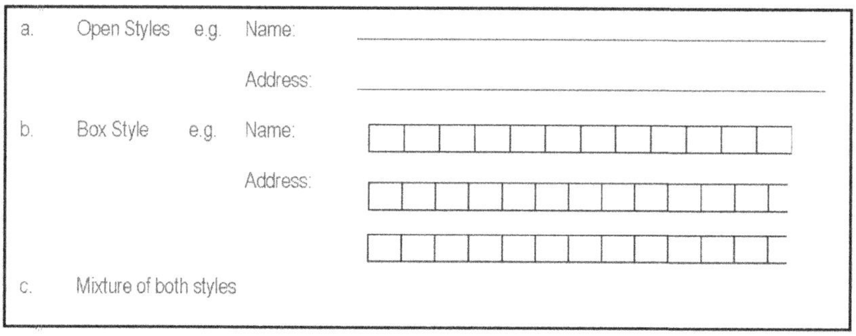

Figure 2.2   Form Fill-in Styles Used

There are three broad categories of forms, as described below:

- **Cut Sheet Forms:** These are usually made in pads; sheets can be detached. This is the most common type of form.
- **Specialty Forms:** There are four basic types of specialty forms:
  o Bound in books e.g. receipts
  o Detachable stub e.g. a checkbook
  o Mailer
  o Continuous e.g. machine-printed report
- **Electronic Forms:** These are forms for which there may or may not be printed versions. They serve as either data collection instruments for system input, or information dissemination instruments for system output.

In designing forms, the following guiding principles are important:

- Captions must be unambiguous; instructions must be clear, copies must be numbered or color-coded and/or numbered.
- The form should require minimum writing by the person filling it in, collecting only necessary data.
- Arrangement of information must be logical and subservient to efficiency.
- The form must be easy to fill out.
- The completed form should be easy to use in the system.
- The form layout must be consistent with input screen design (see Chapter 14).
- The form must have good legibility features. The Le Courier's legibility chart (Figure 2.3), though dated, still provides a useful frame of reference.

	Print Color	Background Color
1	Black	Yellow
2	Green	White
3	Red	White
4	Blue	White
5	White	Blue
6	Black	White
7	Yellow	Black
8	White	Red
9	White	Green
10	White	Black
11	Red	Yellow
12	Green	Red
13	Red	Green

Note: Since this chart was proposed, computer graphics has significantly improved; today, the list of possible colors is much richer. The chart should therefore be construed as a basic guideline.

Figure 2.3   Le Courier's Legibility Chart

2.4.3 *Data Analysis Charts*

Data analysis charts are used to present information in a clear, concise manner. Uses of charts include:

*   Analysis of trends
*   Forecasting of future events
*   Comparison of data
*   Conveying technical information about the requirements of a system

Four commonly used types of charts (illustrated in Figure 2.4) are:

*   Line Chart, as used in regression analysis and break-even analysis
*   Pie Chart, used to show percentage distributions
*   Bar Chart, used to show movements and relationships

2.4.4 *Technical Documents and Modeling Techniques*

In carrying the responsibilities of the job, the software engineer prepares several technical documents. There are no set rules or formats for these; the software engineer may be as innovative as he/she wishes provided that the following guidelines are observed:

*   The document must be concise, but comprehensive in its treatment of the related subject.
*   Where relevant, established tools and techniques must be employed.

The following are some examples of technical documents (typically referred to as *deliverables*) that the software engineer might be responsible for preparing:

*   Project Proposal

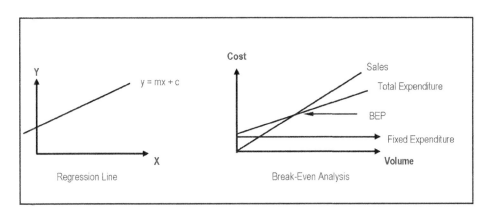

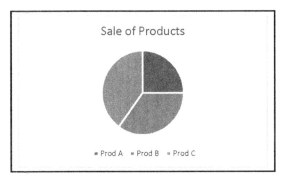

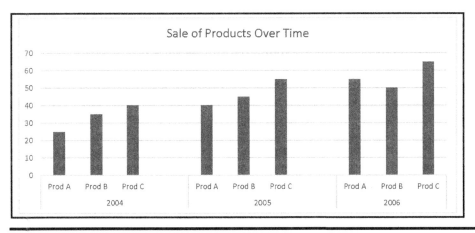

Figure 2.4 (a) Sample Line Charts (b) Sample Pie Chart for Sale of Products (c) Sample Bar Charts for Sale over Time

- Initial System Requirement
- Feasibility Report
- Requirements Specification
- Design Specification
- Help System

- User's Guide
- System Manuals

With respect to modeling techniques, this course, along with a course in OOM, will cover most of the established techniques that are at the disposal of the software engineer. Additionally, the experienced software engineer will from time to time, devise innovative variations of standard techniques, commensurate with the problem domain under consideration. He/she may even propose new techniques. To think about it for a moment, all the methodologies and techniques that are covered in this course are approaches that were developed by software engineers, in their attempts to solve problems.

2.4.5 *Software Planning and Development Tools*

Software planning and development tools were mentioned in the previous chapter. They may be loosely classified as modeling and design tools, database planning tools, software development tools, and DBMS suites. Some of the popular products have been included in Figure 2.5 (in alphabetic order). Please note that this is by no means a comprehensive list. As you view it, please note the following:

- Software engineering is a challenging, exciting, and rewarding field that is supported by a rich reservoir of excellent resources to help you learn and gain mastery in the discipline.
- Terms such as *CASE tool* and *RAD tool* describe a wide range of software systems that include any combination of the above-mentioned classifications. One's choice of CASE/RAD tool(s) will depend on the prevailing circumstances and the desired objectives for the software engineering project.

Given such a wide plethora of software engineering tools and resources, the matter of software evaluation and selection becomes pertinent to the success of a software engineering venture. Evaluation and selection should be conducted based on clearly defined criteria. This process will be further discussed in the upcoming chapter.

You have no doubt noticed that many of the tools mentioned in Figure 2.5 relate to the field of database systems. This happens to be a very important field of software engineering that will be summarized in Chapter 13, but which you will (or should) study in a separate course. A database management system (DBMS) is the software that is used to develop and manage a database. Three main categories of DBMS suites (see [Date 2004] and [Foster 2016]):

- **Relational:** A relational DBMS supports the relational model for database design.
- **Object-Oriented:** An object-oriented DBMS (OODBMS) is a DBMS that has been designed based on OO methodologies.
- **Universal:** A universal DBMS supports the relational model as well as the object-oriented model for database design.

Product	Parent Company	Comment
Model and Design Tools		
ConceptDraw	CS Odessa	Supports UML diagrams, GUI designs, flowcharts, ERD, and project planning charts.
Enterprise Architect	Sparx Systems	Facilitates UML diagrams that support the entire software development life cycle (SDLC). Includes support of business modeling, systems engineering, and enterprise architecture. Supports reverse engineering as well.
MagicDraw	No Magic	A relatively new product that has just been introduced to the market. Appears to be similar to Enterprise Architect.
Power Designer	Sybase	Supports UML, business process modeling, and data modeling. Integrates with development tools such as .NET, Power Builder (a Sybase Product), Java, and Eclipse. Also integrates with the major DBMS suites.
SmartDraw	SmartDraw	A graphics software that facilitates modeling in the related disciplines of business enterprise planning, software engineering, database modeling, and information engineering (IE). Provides over 100 different templates (based on different methodologies) that you can choose from. Supported methodologies include UML, Chen Notation IE notation, etc.
TogetherSoft	Borland	Provides UML-based visual modeling for various aspects of the software development life cycle (SDLC). Allows generation of DDL scripts from the data model. Also supports forward and reverse engineering for Java and C++ code.
Toolkit for Conceptual Modeling (TCM)	University of Twente, (of Holland)	Includes various resources for traditional software engineering methodologies as well as object-oriented methodologies based on the UML standards.
UML Diagrammer	Pacestar	Facilitates easy creation of UML diagrams for software systems.
Visio	Microsoft	Facilitates modeling in support of business enterprise planning, software engineering, and database management.
Visual Thought	CERN	Similar to Visio but is free.

Product	Parent Company	Comment
Database Model and Design Tools		
DataArchitect	theKompany.com	Supports logical and physical data modeling. Interfaces with ODBC and DBMSs such as MySQL, PostgreSQL, DB2, MS SQL Server, Gupta SQLBase, and Oracle. Runs on Linux, Windows, Mac OS X, HP-UX, and Sparc Solaris platforms.
Database Design Studio	Chili Source	Allows modeling via ERD, data structure diagrams, and data definition language (DDL) scripts. Three products are marketed: DDS-Pro is ideal for large databases; DDD-Lite is recommended for small and medium-sized databases; SQL-Console is a GUI-based tool that connects with any database that supports ODBC.
Database Design Tool (DDT)	Open Source	A basic tool that allows database modeling that can import or export SQL.
DBDesigner 4 and MySQL Workbench	fabFORCE.net	This original product was developed for the MySQL database. The replacement version, MySQL Workspace is targeted for any database environment, and is currently available for the Windows and Linux platforms.
DeZign	Datanamic	Facilitates easy development of ERDs and generation of corresponding SQL code. Supports DBMSs including Oracle, MS SQL Server, MySQL, IBM DB2, Firebird, InterBase, MS Access, PostgreSQL, Paradox, dBase, Pervasive, Informix, Clipper, Foxpro, Sybase, SQLite, ElevateDB, NexusDB, DBISAM.
ER Creator	Model Creator Software	Allows for the creation of ERDs, and the generation of SQL and the generation of corresponding DDL scripts. Also facilitates reverse engineering from databases that support ODBC.
ER Diagrammer	Embarcadero	Similar to ER Creator.
ERWin Data Modeler	Computer Associates	Facilitates creation and maintenance o data structures for databases, data warehouses, and enterprise data resources. Runs on the Windows platform. Compatible with heterogeneous DBMSs.
Oracle Designer	Oracle	Supports design for Oracle databases.
Oracle JDeveloper	Oracle	Supports UML diagramming.
xCase	Resolution Software	A database modeling tool that supports all aspects of the database development life cycle (DDLC): it supports ERD design, documentation, SQL code generation, logical and physical migration across multiple DBMS platforms, and data analysis.

Figure 2.5 Some Popular Software Planning and Development Tools

(Continued)

Product	Parent Company	Comment
Software Development Tools		
Delphi, C++ Builder, JBuilder, and Kylix	Formerly by Borland; Now marketed by Embarcadero Technologies	A group of RAD tools that facilitates the development of various categories of software systems (including database connectivity). Code generated in Object Pascal, C++, or Java (depending on user's choice). Runs on various operating system (OS) platforms.
LiveModel	Intellicorp	An OO-ICASE tool that facilitates modeling and development of various software systems.
LANSA	LANSA Corporation (of Australia)	An integrated CASE tool for OS/400 and Windows environments.
MS Visual Studio	Microsoft	A conglomerate of software development environments including Visual Basic, Visual C++ and Visual FoxPro
NetBeans	Sun Microsystems	A Java development environment that runs on various OS platforms.
Object Domain Standard	Object Domain Systems	Facilitates forward and reverse engineering for C++, Java, Python, SVG, and Python Scripting.
Rational Rose	IBM	An OO-CASE tool for various OS platforms including UNIX and Windows
SAP	SAP AG (of Germany)	A suite of products that are used in *enterprise resource planning* (ERP). The English equivalent of the German acronym SAP means Systems, Applications and Products.
Synon	Computer Associates	A CASE tool for OS/400 and Unix environments.
Software through Pictures UML	Aonix	Facilitates forward and reverse engineering for C, C++, Java, and Ada.
Team Developer	Formerly by Gupta Technologies, now marketed by Unify Corporation	An OO-CASE tool that facilitates the development of various categories of software systems (including database connectivity) on the Windows platform.
UML Studio	Pragsoft Corporation	Facilitates forward and reverse engineering for C++, Java, and IDL.
Visual UML Standard Edition	Visual Object Modelers	Facilitates forward and reverse engineering for C++, C#, and Java. Also supports data modeling.
Database Management Systems (DBMSs)		
DB2	IBM	A universal DBMS (UDBMS) suite for various OS platforms including OS-400, Windows, UNIX and RISC.
Informix	IBM	A DBMS suite for various OS platforms including UNIX, Linux, Windows and RISC.
MySQL	Formerly MySQL AB (of Sweden); now Marketed by Oracle	An open source DBMS that is gaining in credibility. Runs on major OS platforms including Windows, Linux, and UNIX.
Oracle	Oracle	A universal DBMS (UDBMS) suite for various OS platforms, including UNIX, Linux, and Windows.
SQL Server	Microsoft	A universal DBMS suite for the Windows platform.
Sybase	Sybase	A DBMS suite for various OS platforms including UNIX, Windows, and Solaris.

Figure 2.5   (Continued)

CASE tools may be placed into three broad categories, each with three sub-categories [Martin 1993]:

- **Traditional:** A traditional CASE tool is a CASE tool that is based on the (traditional) function-oriented approach to software development. It may be a front-end system (concentrated on software investigation, analysis, and specification), a back-end system (concentrated on software development), or an integrated (concentrated on the entire SDLC).
- **Object-oriented:** An OO-CASE tool is a CASE tool that is based on the OO paradigm for software construction. It may be front-ended, back-ended, or integrated.

• **Hybrid:** A hybrid CASE tool is a CASE tool that supports both the FO paradigm and the OO paradigm for software construction. It may be front-ended, back-ended, or integrated.

Figure 2.6 illustrates the categorizations of CASE tools. In examining software development tools for subsequent development, products will fit (some vaguely) in various cells of the matrix. Prudent software evaluation will aid the selection process.

As illustrated in Figure 2.6, the conventional wisdom is that OO-CASE tools are more desirable than other types of CASE tools. However, given the abundance of legacy software systems that still abound (this will be further discussed in Chapter 22), there is reasonable explanation (if not justification) for the variety. Figure 2.7 provides some significant benefits of OO-CASE tools.

	Non-OO	Traditional with OO Flavor	Hybrid	Pure OO
Fragmented CASE	Front-end (upper)			
	Back-end (lower)			
I-CASE (but not IE)				
IE-CASE				
			Ideal Choice	

Figure 2.6   Categories of CASE Tools

Logical support for OOSE: As mentioned earlier, OO-CASE is a perfect facilitator for OOSE.

Simplified SDLC: The use of OO-CASE tools facilitates a significant change in system life cycle from several disjoint phases to two integrated, interwoven stages relating to the structure and behavior of objects.

Faster System Development: By using O-CASE we oxporionco a cignificant ovorall roduction in cyctom dovolopment (by a factor of 40-80 percent). In fact, OO-CASE promotes RAD, so that most RAD tools are object oriented to some degree.

Powerful Integrated Development: OO-CASE tools facilitate a change from line-by-line system development to chunk-by-chunk (component-by-component) system development, with the facility to integrate previously built components from different systems. The CASE toolset typically includes a repository where resources (components, classes, and methods) to be used in different systems can be stored.

More Reliable Systems: With OO-CASE tools that generate code automatically, more reliable code is produced, leading to more reliable systems.

More Maintainable Systems: Systems developed using OO-CASE tools are much easier to maintain than systems developed by other means.

Figure 2.7   Significant Benefits of OO-CASE Tools

The matter of integrated software system development warrants a bit more clarification. With the use of OO-CASE tools, the software development life cycle (SDLC) is reduced from several disjoint phases to two integrated, interwoven stages—modeling and code generation—as illustrated in Figure 2.8.

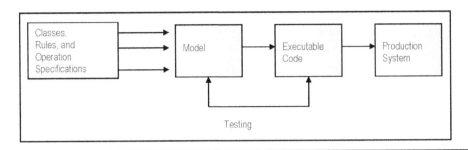

Figure 2.8   OO Modeling and Code Generation

Like CASE tools, most RAD tools tend to be object-oriented. Moreover, the distinction between CASE and RAD is not always well defined. However, what is not in question is spectrum of significant benefits that these software development tools bring to the software engineering discipline. These benefits include (but are not confined to):

- Automated project management techniques (see Chapter 11)
- System modeling (including database and user interface)
- Conveniences of graphical user interface
- Prototype modeling and testing
- Sophisticated diagramming techniques, including executable diagrams
- Fourth-generation language (4GL) interface to aid in application development
- Automatic code generation

2.4.6 *Object-Oriented Programming Languages*

An *object-oriented programming language* (OOPL) is a high-level language that supports object-oriented programming. The language must support the fundamental principles of OO programming such as classes, inheritance, polymorphism, encapsulation, etc. (as discussed in Chapter 2) in order to qualify as an OOPL. The important features to consider when choosing an OOPL are summarized in Figure 2.9.

There are two categories of OOPLs: pure OOPLs and hybrid languages. Pure OOPLs are languages which from the outset, were designed to be object oriented; they do not support procedural programming at all. Examples of pure OOPLs include SmallTalk, Actor, Eiffel, Java, Python, Ruby, and Scalable Application Language (SAL).

Hybrid OOPLs are languages that have resulted from the upgrade of traditionally procedural languages. They support both procedural and object-oriented programming. Examples include C++ (from C), Objective C (from C), Visual Basic (from Basic), Object Pascal (from Pascal), Object COBOL (from COBOL), and CLOS (from LISP).

1. The OOPL should support fundamental principles of object technology as discussed in chapter 4 (classes, inheritance, polymorphism, encapsulation, etc.).

2. Is the OOPL interpretive or a compiling language?
   - An interpretive language employs an *interpreter* to convert source code to object code. Translation is done on a line-by-line or command-by-command basis. This is very convenient for the developer, but is not as efficient as compiling languages. The reason for this is that the source code is translated each time the program is executed.
   - A compiling language employs a *compiler* to convert source code to object code. The object code is saved as an executable file for subsequent execution of the program. A compiling language does not provide the convenience that an interpretive language provides because the compiling process is a batch process (which produces a result after the entire program is analyzed).
   - *Dynamic compilation* is a compromise between the previous two approaches. Only modified methods (of an application) are recompiled. Also, if the user (programmer) tries to run a program, but there are methods that have been modified since the last compilation, a new compilation is automatically generated (without user intervention) and then (assuming no errors) the program is executed.

3. The OOPL should support pointers or dynamic arrays. Object technology is predicated on the use of pointers; without them, the whole idea breaks down.

4. The OOPL should support *late (dynamic) binding* as opposed to static binding. The process of determining which object another object references (points to) is called binding. Binding identifies the receiver of a request; it can be done at compilation time, but preferably at execution time. Compilation time binding is also called early binding or static binding. Execution time binding is also called late binding or dynamic binding.

Figure 2.9   Desirable Features of OOPLs

2.5 Management Issues with Which the Software Engineer Must Be Familiar

Whether your organization is a software engineering firm, or a typical business that specializes in some (traditional or nontraditional) field of interest, there is a virtual inevitability that it will need the services of software engineers. Once the enterprise relies on, uses, or markets software, such services will be required. Software engineering has therefore become a pervasive, ubiquitous discipline.

The software engineer must be acquainted with plans that will significantly affect the organization. The effects may be in any of a number of areas, for instance:

- Plans for expansion, or contraction, or reorganization might affect the physical infrastructure of the organization.
- Relocation will affect the physical infrastructure also.
- A merger or contraction will affect the scope of the organization's information infrastructure.
- Changes in the methods of internal work (e.g. budget preparation) or mode of operation (global or in specific areas), might require adjustment to underlying information system operations.
- Changes in the business focus of the enterprise may also impact on organization's information infrastructure.

In many cases, these changes directly impact the software systems on which the organization relies; in other instances, the effect is indirect. Whether the effects are direct or indirect, the underlying systems may require modification, and the software engineering team will be required to implement these modifications.

2.6 Summary and Concluding Remarks

Here is a summary of what we have covered in this chapter:

- The software engineer is responsible for the investigation, analysis, design, construction, implementation, and management of software systems in the organization. This requires a highly skilled individual with certain desirable qualities.
- The software engineer uses a number of important tools in carrying out the functions of the job. These include coding systems, forms design, charts, diagrams, technical documents, and software development tools.
- Coding systems allow for data representation, particularly during database design and forms design.
- Forms design relates to data collection instruments for input to as well as output from the system.
- Data analysis charts are used for data representation in a concise, unambiguous manner. Line charts, pie charts, bar charts, and step charts are commonly used.
- Technical documents convey useful information about the work of the software engineer. These include but are not confined to project proposal, initial system requirement, feasibility report, requirements specification, design specification, help specification, user's guide, and system manual. We will discuss these documents as we proceed through the course.
- Flow charts and other diagrams will be discussed later in the course.
- Software planning and development tools are available in the form CASE/RAD tools and SDKs in categories such as model and design tools, database planning tools, software development tools, programming languages, and DBMS suites.
- The software engineer must be aware of any planned organizational changes that will have direct or indirect implications for the software systems of the organization.

This completes the introductory comments about software engineering. The next few chapters will discuss various issues relating to software requirements investigation and analysis.

2.7 Review Questions

1. Prepare a job description for a software engineer in a software engineering firm.

2. Propose coding systems for the following:
   - Animals in a zoo
   - Employees in an organization

- Students at a college
- Resources in a college library
- Airline flights at an international airport

3. Imagine that you were employed as a software engineer for a firm. Your team is in the process of overhauling the company's information infrastructure. You have been asked to propose a front-end software application that will automate the job application process. Do the following:
   - Propose a job application form for the company.
   - Identify the steps that you envisage that a job application would pass through.
   - After an applicant has been employed, what (internal) software system(s) do you suppose your application processing system would feed its output into? Why?

4. Give examples of technical documents that software engineers have to prepare in the course of their job.

5. Give examples of some contemporary software planning and development tools that are available in the marketplace.

6. Why do software engineers need to concern themselves with management issues in an organization?

References and Recommended Readings

[CDP 2006] CDP. 2006. CDP Print Management. Le Couriers Table of Legibility. Accessed July 2009. http://www.cdp.co.uk/lecourierstable.shtml.

[Date 2004] Date, Christopher J. 2004. *An Introduction to Database Systems*, 8th ed. Reading, MA: Addison-Wesley.

[Foster 2016] Foster, Elvis C. with Shripad Godbole. 2016. *Database Systems: A Pragmatic Approach*, 2nd ed. New York, NY: Apress Publishing.

[Martin 1993] Martin, James. 1993. *Principles of Object-Oriented Analysis and Design*. Eaglewood Cliffs, NJ: Prentice Hall.

[Pfleeger 2006] Pfleeger, Shari Lawrence. 2006. *Software Engineering Theory and Practice*, 3rd ed. Upper Saddle River, NJ: Prentice Hall. See chapter 2.

[Schach 2011] Schach, Stephen R. 2011. *Object-Oriented & Classical Software Engineering*, 8th ed. New York, NY: McGraw-Hill. See chapter 5.

SOFTWARE INVESTIGATION AND ANALYSIS

B

The next nine chapters will focus on the Investigation and Analysis Phase of the SDLC. The objectives of this phase are

- To clearly identify, define, and describe software system problems
- To thoroughly analyze system problems
- To identify and evaluate alternate solutions to identified software system problems. Such solutions should improve efficiency and productivity of the system(s) concerned
- To provide adequate documentation of the requirements of software systems

Two deliverables will emerge from this phase of the software development life cycle (SDLC): the initial software/system requirement (ISR) and the requirements specification (RS).

Chapters to be covered include:

- Chapter 3: Project Selection & Initial System Requirement
- Chapter 4: Fundamentals of Object-Oriented Methodologies
- Chapter 5: Object-Oriented Information Engineering
- Chapter 6: The Requirements Specification
- Chapter 7: Information Gathering
- Chapter 8: Communication via Diagrams
- Chapter 9: More Diagrams—Modeling Objects and their Behavior
- Chapter 10: Decision Models for System Logic
- Chapter 11: Project Management Aids

DOI: 10.1201/9780367746025-4

Chapter 3

Project Selection and the Initial System Requirements

This chapter covers a number of activities that are typical of the early stage of software requirements investigation. These activities converge into the first deliverable—the *initial system requirement* (ISR), also called the *project proposal*. The ISR contains a number of important components, which will be discussed. The chapter proceeds under the following subheadings:

- Project Selection
- Problem Definition
- Proposed Solution
- Scope and Objectives of the System
- System Justification
- Feasibility Analysis Report
- Alternate Approach to Feasibility Analysis
- Summary of System Inputs and Outputs
- Initial Project Schedule
- Project Team
- Summary and Concluding Remarks

3.1 Project Selection

For obvious reasons, let us assume that an organization stands in need of various software systems. The directors of the organization may or may not know the intricate details of their need, but hopefully are smart enough to recognize that something is lacking. At this early stage, management would invite the attention of one or more IT and/or software systems professionals (from the host organization, or a contracted software engineering firm) to conduct a needs assessment and make

DOI: 10.1201/9780367746025-5

appropriate recommendations. The needs assessment should clearly identify all candidate software systems projects for the organizational scenario.

Very early in this software needs assessment exercise, the matter of project selection would become pertinent: It is often common that an organization stands in need of several software systems. Since it is impractical to attempt to acquire them all at the same time, some determination is normally made as to which projects will be pursued first. The intricacies of such determination are beyond the scope of this course; however, you should be aware of the salient factors that will likely affect such decisions. They are:

- Backing from management
- Appropriate timing
- Alignment to corporate goals
- Required resources and related cost
- Internal constraints of the organization
- External constraints of the environment/society

If the software product is to be marketed to the public, there are some additional factors to be considered: anticipated marketability of the product; expected impact on the industry; and expected market position to be attained by introducing this new software system.

Consideration of these factors against each prospective software engineering project will lead to selection on one or more projects to be pursued. Once a project is selected, the next target is preparation of the ISR. The rest of this chapter discusses various components of the ISR, starting with the problem definition.

3.2 Problem Definition

Once the software engineering project has been selected, the first activity that is performed is that of problem definition. It must therefore be done thoroughly, accurately, and persuasively. The following are some basic guidelines:

- Define what aspect(s) of the organization is affected (recall the functional areas of an organization).
- Describe (in convincing, precise language) the effects and possible causes. Show how problems may be interrelated.
- Describe the worker's attitude to the problem.
- Cite the impact the problem has on productivity and revenue.
- Identify the operating constraints.

3.2.1 Constraints of a System

In defining a system, there are certain constraints that the software engineer must be aware of. These include the following:

- External Constraints

o Customers of the organization
o Government policies
o Suppliers of the organization
o State of the economy
o State of the industry in which the organization operates
o Competitors of the organization
• Internal constraints include
o Management support
o Organizational policies
o Human resource
o Financial constraints
o Employee acceptance
o Information requirements
o Organizational politics

3.2.2 Aid in Identifying System Problems

How does the software engineer identify system problems? The following pointers should prove useful:

1. Check output against performance criteria. Look for the following:
   o Frequency and rate of system errors
   o Efficiency and throughput of work done via the software system
   o Accuracy and/or adequacy of work done via the software system
2. Observe behavior of employees. Look for the following:
   o High absenteeism and/or job turnover
   o High job dissatisfaction
3. Examine source documents. Look for the following:
   o Adequacy of the forms design for input as well as output
   o Redundancy on the input and/or output forms
4. Examine the workflow and outcomes in the organization. Look for the following:
   o Inadequate/poor work management
   o Redundancies and/or omissions
5. Listen to external feedback from vendors, customers, and suppliers. Be on the alert for complaints, suggestions for improvement, or loss of business.

3.2.3 Identifying the System Void

If the organization in question is a software engineering firm, or an organization that lacks adequate software systems support, then what might be the issue is not necessarily identifying problems in an existing system, but identifying the need for introducing a new software system.

If the proposal is for a new software system for internal use, the proposal should follow a thorough internal research that involves prospective users of the product. The research should clearly outline the areas of user complaint. Here are a few points to note:

- The research might be conducted formally, involving the use of questionnaires (see Chapter 7), user forums/workshops, expert review, and other information collection instruments.
- Information could also be gathered via informal/nonconventional means, for instance, brain-storming and observation (see Chapter 7).

If the proposal is for a new software system to be marketed to the public, identification of the need could arise from any combination of a brain-storming session, a market research, or industry observation.

3.3 The Proposed Solution

The software engineer must propose a well thought out solution to the problem as outlined. This typically includes a system name, operating platform, and required resources. Following are a few points about the proposed solution:

- The proposed solution may be to modify the existing software system or to replace it.
- The proposal must show clearly how the problems specified will be addressed.
- It should briefly mention what areas of the organization will be impacted and how.
- The proposal must mention the chosen system platform (do not include details here as this will be required in the feasibility report).
- The proposal should mention the estimated cost of the software system and project duration.

3.4 Scope and Objectives of the System

In specifying the scope of the proposed solution, the software engineer should be guided by the following:

- Identify main components (subsystems and/or modules) to your best knowledge (which is limited at this early stage), as well as their interconnections.
- Identify the principal user(s) of the system.
- Cite the organizational areas to be impacted by the system.
- State for each functional area, whether user participation will be data input, processing of data, or access to data.

An itemized list of objectives serves to sharpen focus on the essentials of the system. Basic guidelines for stating system objectives are:

- State system objectives clearly and concisely.
- Identify primary objectives and secondary objectives where applicable.

- Ensure that the needs of end users are addressed, especially those of the principal users.
- Align system objectives to corporate goals.
- Use corporate language in stating the objectives.

3.5 System Justification

It is imperative that the proposal underscores the benefits that the new software system brings to the table. These benefits go a considerable way to justifying the project in the eyes of management. Basic guidelines for stating software system benefits are as follows:

- Show how the system will help to improve productivity and efficiency.
- Mention advantages in the functional areas.
- Identify the benefits that the proposed software system brings to organization as a whole.
- Align the stated advantages to corporate goals of the organization.
- Use the corporate language in stating the advantages.
- If possible, do a payback analysis or break-even analysis to show how soon the organization could expect to recover the cost of the system. An amortization schedule would be useful here.
- Briefly mention risks or drawbacks (these will be discussed in the feasibility analysis).

3.6 Feasibility Analysis Report

A feasibility study is a thorough analysis of a prospective project to determine the likelihood of success of the venture. Traditionally, the analysis is done in three areas: *the technical feasibility, the economic feasibility, and the operational feasibility.* Additionally, the alternate solutions to the defined problem must be carefully examined, and recommendation made in favor of the most prudent alternative.

3.6.1 Technical Feasibility

The technical feasibility analysis addresses the basic question of whether the project is technically feasible. The analysis must be done with respect to four main considerations. Once these four areas are addressed, the question of technical feasibility can be answered with reasonable accuracy. The areas are

- Availability of hardware required
- Availability of software required
- Availability of knowledge and expertise
- Availability of other technology that may be required e.g. telecommunication, new software development tool, etc.

3.6.2 Economic Feasibility

The economic feasibility analysis is aimed at measuring the economic results that would materialize if the project is pursued. Important considerations are:

- Costs versus benefits expressed in terms of payback period, cash flow, and return on investment (ROI)
- Development time and the financial implications
- Economic life—for how long the system will be economically advantageous
- Risk analysis—examining the associated risks

The matter of risk is very important and therefore needs some elucidation. Every business venture has associated with it, some inherent risks; software engineering projects are not excluded from this reality. The software engineer should be aware of the following:

- Generally speaking (from a financial perspective), the higher the risks are, the greater is the possible return on investment (and vice versa). However, there are exceptions.
- In many instances, taking prudent precautionary steps can reduce software engineering project risks. The software engineer must evaluate each risk and determine whether it can be reduced.

Two commonly used business techniques for evaluating and analyzing risks are *payback analysis* and *return on investment* (ROI). ROI is often facilitated by what is called a *net present value* (NPV) analysis. Figure 3.1 provides the formulas used in payback analysis and ROI, respectively.

The payback analysis is contingent on the assumption that interest rates will remain stable over the economic life of the project. However, even in stable economies (such as exist for developed countries), this assumption does not coincide with reality; and in more volatile economies (such as in many developing countries), this assumption does not come close to reality. For these reasons, payback analysis is considered as a crude estimate, just to provide a rough idea of the implications of the financial investment.

PB = Investment / [Average Cash flow per Annum]

$NPV = -A_0 + A_1 / (1+r_1) + A_2 / (1+r_2)^2 + \ldots + A_n / (1+r_n)^n$

Key
PB = Pay-back
NPV = Net Present-value
A_0 = Initial investment
A_i = Annual cash flow in the form of savings and/or additional income
r_i = Rate of return (may be constant)

Figure 3.1   Formulas for Payback and NPV

The NPV analysis is a more pragmatic approach that attempts to recognize what is called the *time value of money* (TMV)—that money today may be worthless (more likely) or more several years in the future, depending on economic factors at play. The NPV analysis tries to anticipate interest rates in the future and uses this to conduct an up-front analysis on the effectiveness of a financial investment. The more accurate the future interest rate can be estimated, the more precise the NPV calculation will be. However, since it is virtually impossible to predict future interest rate with 100% certainty, an estimate is often made, based on the advice (conventional wisdom) of economic experts. The anticipated average interest rate (r) over the related period is typically used.

The annual cash inflow from the investment may come in the form of increased savings on the cost of operation or increased revenue. Similar to interest rates, one cannot predict with 100% certainty, what the future cash inflow is going to be, so typically, an average estimate (A) over the economic life of the project is taken.

Once the anticipated average interest rate (r) and average cash inflow (A) have been determined, these are then fed into the NPV formula of Figure 3.1 to yield the following:

$$\begin{aligned} NPV &= -A_0 + A/(1+r) + A/(1+r)^2 + \cdots + A/(1+r)^n \\ NPV &= -A_0 + A\left[1/(1+r) + 1/(1+r)^2 + \cdots + 1/(1+r)^n \right] \end{aligned} \tag{3.1}$$

From your knowledge of Discrete Mathematics and/or Data Structures and Algorithms, the expression in square brackets constitutes a geometric series whose first term (denoted by a) is $1/(1+r)$ and its common ratio (denoted R) is $1/(1+r)$. The general sum of the first *n* terms of a geometric series is determined by the following formula:

$$\begin{aligned} \text{Sum of first } n \text{ terms}: S_n &= a\left(R^n - 1\right)/(R-1) \text{ if } R > 1 \\ \text{Sum of first } n \text{ terms}: S_n &= a\left(1 - R^n\right)/(1-R) \text{ if } R < 1 \end{aligned} \tag{3.2}$$

Applying Equation 3.2 to the bracketed term of Equation 3.1, we obtain the following

$$\begin{aligned} S_n &= a\left(1 - R^n\right)/(1-R) = 1/(1+r)\left[1 - (1/1+r)^n \right]/\left[1 - (1+r) \right] \\ &= 1/r\left[1 - (1+r)^n \right] \\ &= 1/r * \left[(1+r)^n - 1\right)/(1+r)^n \right] \end{aligned}$$

Therefore, by applying Equation 3.2 to Equation 3.1, we obtain the following:

$$NPV = -A_0 + \left[A/r\right]\left[(1+r)^n - 1\right)/(1+r)^n \right] \tag{3.3}$$

This final equation can be used to calculate the NPV for a project with initial financial investment of A_0, anticipated average cash inflow A, and estimated average interest rate r over a period of n years. Alternately, you may set up a spreadsheet to do the calculation (left as an exercise for you). If the calculation produces a positive result, that the financial investment is financially sound. A negative result indicates that the financial investment is risky; other alternatives and/factors should be considered.

3.6.3 Operational Feasibility

The operational feasibility speaks to issues such as the likely user/market response to the system, as well as the required changes on the part of operational staff. Generally speaking, user involvement in a project tends to positively impact user acceptance. The factors to be considered in assessing the operational feasibility of a project include the following:

- User attitude to (likely) changes
- Likelihood of changes in organizational policies
- Changes in methods and operations that may be necessary
- Timetable for implementation (long implementation time implies low feasibility)

3.6.4 Evaluation of System Alternatives

Evaluation of system alternatives should be based on software quality factors and feasibility factors. In the previous sub-section, we looked at the feasibility factors. Let us now examine the quality factors, explore the cost factors a bit further, and see how the evaluation is done.

3.6.4.1 Evaluation Based on Quality

The quality factors of Chapter 1 are relevant here. In the interest of clarity, they are repeated:

- **Maintainability:** How easily maintained is the software? This will depend on the quality of the design as well as the documentation.
- **Documentation:** How well documented is the system?
- **Efficiency:** How efficiently are certain core operations carried out? Of importance are the response time and the system throughput.
- **User Friendliness:** How well designed is the user interface? How easy is it to learn and use the system?
- **Compatibility** with existing software products.
- **Security:** Are there mechanisms to protect and promote confidentiality and proper access control?
- **Integrity:** What is the quality of data validation methods?
- **Reliability:** Will the software perform according to requirements at all times?
- **Growth potential:** What is the storage capacity? What is the capacity for growth in data storage?

- **Functionality and Flexibility:** Does the software provide the essential functional features required for the problem domain? Are there alternate ways of doing things?
- **Adaptability:** How well are unanticipated situations handled?
- **Differentiation:** What is unique about the software?
- **Productivity:** How will productivity be affected by the software?
- **Comprehensive Coverage:** Is the problem comprehensively and appropriately addressed?

Since these factors are qualitative, weights usually are applied, and each alternative is assessed on each factor. The total grade for each alternative is then noted. The result may be neatly presented in tabular form as in Figure 3.2. In this illustration, alternative-B would be the best option.

Quality Factors		System Alternatives		
		Alternative-A	Alternative-B	Alternative-C
Maintainability	[Max 10]	8	9	9
Efficiency	[Max 10]	9	9	6
User Friendliness	[Max 10]	7	9	6
Documentation	[Max 10]	5	8	10
Compatibility	[Max 10]	4	8	9
Security	[Max 10]	4	9	9
Reliability	[Max 10]	5	8	9
Flexibility & Functionality	[Max 10]	8	9	9
Adaptability	[Max 10]	8	9	10
Growth Potential	[Max 10]	6	9	7
Productivity	[Max 10]	6	9	7
Comprehensive Coverage	[Max 10]	6	9	7
Overall Evaluation Score	**[Max 120]**	**76**	**105**	**92**
Recommended alternative is Alternative-B.				

Figure 3.2 Quality Evaluation Grid Showing Comparison of System Alternatives

3.6.4.2 Evaluation Based on Cost

The main cost components to be considered are development cost, operational cost, equipment cost, and facilities cost. Following is a breakout of each.
   Engineering costs include

- Investigation, Analysis, and Design cost
- Development cost
- Implementation cost
- Training cost

Operational costs include

- Cost of staffing
- Cost for data capture and preparation

- Supplies and stationery cost
- Maintenance cost

Equipment costs include

- Cost of hardware
- Cost of software
- Cost of transporting equipment
- Depreciation of equipment over the acquisition and usage period

Facilities costs are sometimes bundled with operational cost. If there is a distinction, then facilities costs would typically include

- Computer installation cost
- Electrical and cooling cost
- Security Cost

A similar analysis may be constructed for the feasibility factors (including cost factors). When the alternatives are evaluated against these feasibility factors, an evaluation matrix similar to that shown in Figure 3.3 may be constructed. In this illustration, alternative-C would be the best option. Of course, the two figures (Figures 3.2 and 3.3) may be merged.

Feasibility Factors		Alt-A	Alt-B	Alt-C
Technical Feasibility		68	60	78
Availability of Hardware (bigger means better)	[Max 20]	20	15	20
Availability of Software (bigger means better)	[Max 20]	18	15	20
Availability of Expertise (bigger means better)	[Max 20]	14	16	20
Availability of Technology (bigger means better)	[Max 20]	16	12	18
Economic Feasibility		106	107	108
Engineering Cost (bigger means lower cost)	[Max 20]	18	17	17
Equipment Cost (bigger means lower cost)	[Max 20]	17	17	17
Operational Cost (bigger means lower cost)	[Max 20]	19	19	19
Facilities Cost (bigger means lower cost)	[Max 20]	19	19	19
Development Time (bigger means shorter time)	[Max 20]	18	18	19
Economic Life (bigger means longer time)	[Max 20]	16	18	18
Risk (bigger means lower risk)	[Max 20]	18	18	18
Operational Feasibility		50	52	56
User Attitude to Likely Changes (bigger means better)	[Max 20]	20	16	18
Likelihood of Organizational Policy Changes (bigger means fewer changes)	[Max 20]	16	19	18
Implementation Time (bigger means shorter time)	[Max 20]	14	17	20
Overall Evaluation Score	[Max 280]	243	238	261

Figure 3.3   Feasibility Evaluation Grid Showing Comparison of System Alternatives

3.6.4.3 Putting the Pieces Together

A final decision should be taken as to the most prudent alternative, after examination of each alternative with respect to the quality factors and the feasibility factors. Figure 3.4 provides an illustrative evaluation grid that may be used. The rule of thumb is to recommend the alternative with the highest score. However, other constraints such as financial constraints, focus of the organization, etc., may mitigate against that option. Given the overall organizational situation, the most prudent and realistic alternative should be recommended by the software engineer.

Feasibility Factors		Alt-A	Alt-B	Alt-C
Technical Feasibility				
Availability of Hardware (bigger means better)	[Max 20]			
Availability of Software (bigger means better)	[Max 20]			
Availability of Expertise (bigger means better	[Max 20]			
Availability of Technology (bigger means better)	[Max 20]			
Economic Feasibility				
Engineering Cost (bigger means lower cost)	[Max 20]			
Equipment Cost (bigger means lower cost)	[Max 20]			
Operational Cost (bigger means lower cost)	[Max 20]			
Facilities Cost (bigger means lower cost)	[Max 20]			
Development Time (bigger means shorter time)	[Max 20]			
Economic Life (bigger means longer time)	[Max 20]			
Risk (bigger means lower risk)	[Max 20]			
Operational Feasibility				
User Attitude to Likely Changes (bigger means better)	[Max 20]			
Likelihood of Organizational Policy Changes (bigger means fewer changes)	[Max 20]			
Implementation Time (bigger means shorter time)	[Max 20]			
Software Quality				
Maintainability (bigger means better)	[Max 20]			
Efficiency (bigger means better)	[Max 20]			
User Friendliness (bigger means better)	[Max 20]			
Documentation (bigger means better)	[Max 20]			
Compatibility (bigger means better)	[Max 20]			
Security (bigger means better)	[Max 20]			
Reliability (bigger means better)	[Max 20]			
Flexibility & Functionality (bigger means better)	[Max 20]			
Adaptability (bigger means better)	[Max 20]			
Growth Potential (bigger means better)	[Max 20]			
Productivity (bigger means better)	[Max 20]			
Comprehensive Coverage (bigger means better)	[Max 20]			
Overall Evaluation Score	**[Max 520]**			

Figure 3.4 Overall Feasibility Evaluation Grid for Project Alternatives

3.7 Alternate Approach to the Feasibility Analysis

A popular alternate approach to conducting the feasibility analysis involves the consideration of the feasibility factors, the quality factors, and the cost factors in a comprehensive way. The evaluation criteria are grouped in the following categories:

- TELOS Feasibility Factors: Technical, Economic, Legal, Operational, Schedule feasibility factors
- PDM Strategic factors: Productivity, Differentiation, and Management factors
- MURRE Design Factors: Maintainability, Usability, Reusability, Reliability, Extendibility factors

The principle of weights and grades described in the previous section is also applicable here. Figure 3.5 illustrates a sample rating worksheet that could be employed.

TELOS Feasibility Factors:		Alt-A	Alt-B	Alt-C
Technical Feasibility	[Max 20]			
Economic Feasibility	[Max 20]			
Legal Feasibility	[Max 20]			
Operational Feasibility	[Max 20]			
Schedule Feasibility	[Max 20]			
Subtotal	**[Max 100]**			

PDM Feasibility Factors:		Alt-A	Alt-B	Alt-C
Productivity	[Max 20]			
Differentiation	[Max 20]			
Management	[Max 20]			
Subtotal	**[Max 60]**			

MURRE Feasibility Factors:		Alt-A	Alt-B	Alt-C
Maintainability	[Max 20]			
Usability	[Max 20]			
Reusability	[Max 20]			
Reliability	[Max 20]			
Extendibility	[Max 20]			
Subtotal	**[Max 100]**			
Overall Evaluation Score	**[Max 260]**			

Figure 3.5   Alternate Evaluation Matrix for Project Alternatives

3.8 Summary of System Inputs and Outputs

It is sometimes useful to provide a summary of the main inputs to a software system. Depending on the category and complexity of the software, the inputs may vary from a small set of electronic files to a large set involving various input media. Input and output design will be discussed in more detail in Chapter 14. For now, and at this point in the project, a simple list of system inputs will suffice.

Additionally, a brief indication of what to expect from the system from a user viewpoint is sometimes useful in giving the potential end user a chance to start identifying with the product even before its development has begun. Usually, a summary of possible system outputs will suffice.

In situations where visual aid is required, a nonoperational prototype (see Chapter 7) might be useful. For a real-time system (see Chapter 16), a simulation may be required.

3.9 Initial Project Schedule

A preliminary project schedule outlining the estimated duration of project activities (from that point) is usually included in the ISR. Please note:

- The initial schedule may be detailed or highly summarized depending on information available at the time of preparation. As such, subsequent revision may be necessary (during the design specification).
- A PERT diagram, Gantt chart (see Chapter 11), or simple table will suffice.
- A cost schedule is also provided showing projected costs for various activities. This may also be subject to subsequent revision.

Figure 3.6 illustrates a sample tabular representation of a project schedule. As illustrated in the figure, it is a good idea to include in the schedule, an estimate of the number of professionals who will be working on the project. As more details become available, it will be necessary to revise this schedule; but for now, a tabular representation (as shown in the figure) will suffice. Chapter 11 discusses project scheduling in more detail.

Activity	Activity Description	Estimated Time (Days)	Estimated Cost ($)
A	Design System Architecture	30	72,000
B	Design Operation Specifications	12	28,800
C	Design Control Operations	8	19,200
D	Design Modification Operations	15	36,000
E	Design Inquiry/Report Operations	7	16,800
F	Code Control Operations	2	4,800
G	Prepare System Users Guide	5	12,000
H	Test Control Operations	2	4,800
I	Code Modification Operations	6	14,400
J	Test Modification Operations	4	9,600
K	Code Inquiry Operations	2	4,800
L	Code Report Operations	3	7,200
M	Test Inquiry/Report Operations	1	2,400
N	Integration Test	4	9,600
Total estimated time and cost		91	242,400
Estimated Number of Fulltime Professionals is 3			

Figure 3.6   Sample Initial Project Schedule

3.10 Project Team

The final component of your ISR is a proposed project team. The team must be well balanced; typically, it should include the following job roles:

- Project Manager
- Secretary and/or record keeper
- Software Engineer(s) and Developer(s)
- Software Documenter(s) (if different from software engineers)
- Data Analyst(s)
- User Interface Officer(s)

The number of personnel and the chosen organization chart will depend on the size and complexity of the project. Additionally, a user workgroup or steering committee is normally set up to ensure that the stated objectives are met. This may include the following personnel:

- Project Manager
- Director of Information Technology (if different from Project Manager)
- Representation from the principal users
- Other managers whose expertise may be required from time to time (on invitation)
- Software Engineer(s) (on invitation)
- Managing Director (depending on nature of the project)

In some instances, the project team forms a sub-committee of the steering committee; in small organizations, the two teams may be merged. Figure 3.7 shows three possible traditional configurations of the project team. As an exercise, you are encouraged to identify the pros and cons of each project team configuration. It must be

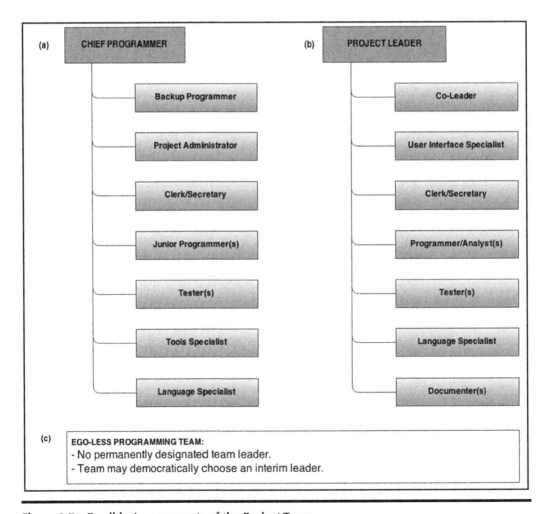

Figure 3.7   Possible Arrangements of the Project Team

pointed out, however, that there may be various other configurations of a software engineering team. For instance, an agile development team tends to have a different configuration, consisting of role players such as user representatives, product owners, designers, developers, quality assurance engineers, and analysts. Chapter 23 provides more insights on organizing software engineering teams.

3.11 Summary and Concluding Remarks

It's now time to summarize what we have covered in this chapter:

- Selection of a software engineering project depends on the corporate mission and objectives of the organization.
- The ISR is the first deliverable of the software engineering project. It consists of components such as problem definition, proposed solution, scope and objectives, system justification, feasibility analysis report, summary of system outputs, initial project schedule, and the project team.
- Problem definition sets the tone for the project. It must therefore be done thoroughly and accurately.
- The proposed solution provides a summary of how the problem will be addressed.
- The system scope defines the extent and boundaries of the software system; system objectives are also clearly specified.
- The system justification provides the rationale and main benefits of the software system.
- The feasibility analysis report analyzes and reports on the likelihood of success of the project. The feasibility study examines the technical, economic and operational feasibility, as well as other related factors such as associated risks, maintainability, usability, reusability, reliability, extendibility, productivity, differentiation, management issues, legal issues, functionality, flexibility, security, etc.
- Summary of system inputs and outputs is self-explanatory.
- The initial project schedule provides a draft schedule of the project activities that will subsequently be updated as more detailed information becomes available.
- The project team indicates the individuals that are or will be involved in the project.

Appendix A provides excerpts from the ISR for an inventory management system. Please take some time to carefully review the document and gain useful insights.

Once the ISR is completed, this document is submitted to the management of the organization for consideration and approval. Typically, the software engineer responsible for its preparation may be required to make a formal presentation to the organization. In this presentation, you sell the project idea to the organization. Once approval is obtained, you then start working on your next deliverable which is the *requirements specification* (RS). The next few chapters will prepare you for this. As you proceed, bear in mind that one primary objective of software engineering is to produce software that has a significantly greater value than its cost. We do this by building value in the areas of the software's quality factors.

3.12 Review Questions

1. What factors are likely to influence the selection of a software engineering project? Briefly discuss these factors and explain why they are important.

2. Discuss the importance of problem definition in a software engineering project.

3. What should the proposed solution entail?

4. What information is provided when you state system scope and system objectives?

5. How do you justify the acquisition of a new software system in an organization?

6. What is a feasibility study? Discuss the main components of a feasibility report.

7. Describe an approach that is commonly used for evaluating system alternatives and determining the most prudent alternative.

8. Identify three approaches to organizing a project team. For each approach, discuss is strengths and weaknesses, and describe a scenario that would warrant such approach.

9. Conduct an examination of an organization that you are familiar with, and do the following:
   • Propose an organization chart for the organization.
   • By examining the organization chart, propose a set of software systems that you would anticipate for this organization.
   • Choose one of those potential systems, and for this system, develop an ISR as your first deliverable. Your ISR must include all the essential ingredients discussed in the chapter.

References and Recommended Readings

[Harris 1995] Harris, David. 1995. *Systems Analysis and Design: A Project Approach*. Fort Worth, TX: Dryden Press. See chapter 3.

[Kendall 2014] Kendall, Kenneth E., and Julia E. Kendall. 2014. *Systems Analysis and Design*, 9th ed. Boston, MA: Pearson. See chapter 3.

[Peters 2000] Peters, James F. and Witold Pedrycz. 2000. *Software Engineering: An Engineering Approach*. New York, NY: John Wiley & Sons. See chapter 4.

[Pfleeger 2006] Pfleeger, Shari Lawrence. 2006. *Software Engineering Theory and Practice*, 3rd ed. Upper Saddle River, NJ: Prentice Hall. See chapter 3.

[Schach 2011] Schach, Stephen R. 2011. *Object-Oriented & Classical Software Engineering*, 8th ed. New York, NY: McGraw-Hill. See chapters 3& 4.

[Sommerville 2016] Sommerville, Ian. 2016. *Software Engineering*, 10th ed. Boston, MA: Pearson. See chapters 4 & 5

[Van Vliet 2008] Van Vliet, Hans. 2008. *Software Engineering*, 3rd ed. New York, NY: John Wiley & Sons. See chapters 2, 5 & 6.

Chapter 4

Overview of Fundamental Object-Oriented Methodologies

This chapter introduces the philosophy and rationale for object-oriented methodologies (OOM). It also covers the characteristics and benefits associated with OOM. Thirdly, the chapter provides clarification on the fundamental concepts in OOM. The chapter proceeds under the following captions:

- Software Revolution and Rationale for Object-Oriented Techniques
- Information Engineering and the Object-Oriented Approach
- Integrating Hi-tech Technologies
- Characteristics of Object-Oriented Methodologies
- Benefits of Object-Oriented Methodologies
- Objects and Object Types
- Operations
- Methods
- Encapsulation and Classes
- Inheritance and Amalgamation
- Requests
- Polymorphism and Reusability
- Interfaces
- Late Binding
- Multithreading
- Perception versus Reality
- Overview of the Object-Oriented Software Engineering Process
- Summary and Concluding Remarks

DOI: 10.1201/9780367746025-6

4.1 Software Revolution and Rationale for Object-Oriented Techniques

One serious anomaly in information technology is the advancement of hardware technology disproportionately to software technology. We are in the sixth generation of computer hardware; we are in the fourth generation (perhaps the fifth depending on one's perspective) of computer software.

Software generations include the following:

- Machine code
- Assembly language
- High-level languages; databases based on hierarchical and network approaches
- Relational systems, 4GLs, CASE tools, and applications
- I-CASE, object-oriented techniques, multi-agent applications, and intelligent systems

One urgent concern in the software industry today is to create more complex software, faster and at lower costs. The industry demands at reduced cost, a quantum leap in:

- Complexity
- Reliability
- Design capability
- Flexibility
- Speed of development
- Ease of change
- Ease of usage

How can the software engineering industry respond to this huge demand? How can software engineers be equipped to produce software of a much higher quality than ever demanded before, at a fraction of the time? A few years ago, companies would be willing to invest millions of dollars into the development of business applications that took three to five years to be fully operational. Today, these companies demand immediate solutions, and they want more sophistication, more flexibility, and more convenience.

As the software engineering industry gropes for a solution to the demands of twenty-first lifestyle, a principle worth remembering is, *Great engineering is simple engineering* (see [Martin 1993]). Object-oriented (OO) techniques simplify the design of complex systems; so far, it is the best promise of the quantum leaps demanded by industry. Through OO techniques, complex software can be designed and constructed block-by-block, component-by-component. Moreover, tested and proven components can be reused multiple times for the construction of other systems, potentially spiraling into a maze of complex, sophisticated software that was once inconceivable to many.

OO methodologies (OOM) involve the use of OO techniques to build computer software. It involves *OO analysis and design* (OOAD), as well as *OO programming languages* (OOPLs). It also involves the application of *object-oriented technology* (OOT or OT) to problem-solving.

4.2 Information Engineering and the Object-Oriented Approach

Traditionally, software systems have been developed through a life cycle model approach of investigation, analysis, design, development, implementation, and maintenance. These activities were separate, often done by different people. Further, systems were developed in a piecemeal fashion—roughly on a department-by-department basis. This traditional approach still prevails in a number of organizations.

Information engineering (IE) is about looking at the entire enterprise as a system of interrelated subsystems, then proposing and engineering a strategic information system that meets the corporate goals and objectives of the enterprise. It is more dynamic, challenging, and rewarding than the traditional approach. As an illustration, the CUAIS (college/university administrative information system) *information topology chart* (ITC) of Figure 4.1 provides excerpts from a comprehensive, integrated information system for a college/university, as opposed to a department-wise automation. The ITC—a technique developed by the current author—provides an overview of the main information entities covered by the software system, and the subsystems that they are commonly managed in. This technique will be revisited later in the course (Chapters 5 and 9).

Whereas the traditional approach is usually managed at the middle management level, enterprise-wide engineering must be managed at the executive level, if it is to be successful. The information technology (IT) professional in charge must be very experienced, well versed in information systems, and able to inspire confidence at all levels of the organization.

We can apply OOM to information engineering in a way that enables users to conceptualize an information system as a set of interacting, inter-related component objects. At the highest level, the entire system is a large object. The component objects encompass encapsulated states (data), methods, and means of communicating with each other as well as with other external objects, all this being transparent to the end user.

When OOM is applied to information engineering, the process becomes more effective and rewarding. We use the term OO information engineering (OOIE) to describe this scenario. In OOIE, the modeling process is more integrated, involving the previously disjointed activities of investigation, design, development, implementation, and maintenance—assuming the use of OO-ICASE tools. Another term that you will hear bandied around is *OO enterprise modeling* (OOEM). Some authors like to parse and dissect and come up with differences between OOIE and OOEM. This course makes no such distinction. OOIE and OOEM essentially describe the same thing. Use of one term over another is a matter of personal preference.

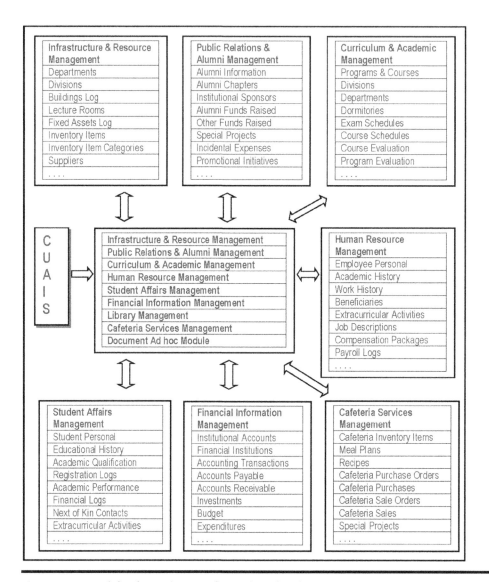

Figure 4.1   Partial Information Topology Chart for the CUAIS Project

4.3 Integrating Hi-tech Technologies

OO analysis and design (OOAD) is not OO programming (in the early days of OOM, the two were mistaken to be the same). Any high-level language can implement an OO design, some more elegantly than others. An OO high-level language (which will be subsequently discussed) facilitates easy implementation of certain of OO concepts, for example: class, inheritance, encapsulation, etc. So do OO-CASE tools.

OOAD facilitates the integration of a number of leading technologies (so called hi-tech) towards the generation of complex systems, with a number of advantages as spin-offs. The technologies include:

- CASE and I-CASE
- Visual Programming via OO programming languages (OOPLs)
- Code Generation
- Repository and Class Libraries
- Repository-Based Methodologies
- Information Engineering
- OO Databases
- Non-procedural languages
- Formal (Mathematically Based) Methods
- Client-Server Technology

According to Martin's prediction, "the sooner OOAD becomes widespread, the sooner software creation will progress from a cottage industry to an engineering discipline" (see [Martin 1993]). We are in the midst of that transformation.

4.4 Characteristics of Object-Oriented Methodologies

Figure 4.2 provides some fundamental characteristics of OO techniques: These characteristics lead to various benefits as mentioned in the upcoming section.

- OO methodologies change the way we conceptualize software systems: We view a system as a set of interacting objects (components) rather than a set of interacting functions. This is more natural than traditional function-oriented approach; it is easier to think about objects and their behavior than functions and their procedures.
- OO methodologies naturally support CBSE: Systems can often be constructed from existing system components that have already been tested and proven. Hence, there is a high degree of reusability, shortened development time, and increased reliability.
- Through OOM, more complex systems can be constructed with less effort.
- OO techniques fit naturally with CASE technology.
- OO techniques promote the use of a CASE repository — an ever-growing library of object types. These object types are designed to be customizable to different system needs.
- OO techniques lead to easier and more likely creation of systems that work correctly. As the saying goes, OO techniques are "more modular than modular programming design."

Figure 4.2   Characteristics of OOM Summarized

4.5 Benefits of Object-Oriented Methodologies

Object technology offers many benefits to the software engineering industry and the business environment. The more prominent ones are mentioned in Figure 4.3. Indeed, OO methodologies have become commonplace in the software engineering industry. To attempt software engineering in a manner that is oblivious to these methodologies would be at best extremely risky.

- **Reusability of Code**: A tested system component can be reused in the design of another component.
- **Stability and Reliability**: Software can be constructed from tested components. Organizations can be assured of guaranteed performance of software.
- **More Sophistication**: Through powerful OO-CASE tools and CBSE, more complex systems can be constructed.
- **Understandability**: Designer and user think in terms of object and behavior rather than low-level functional details. This results in more realistic modeling that is easier to learn and communicate.
- **Faster Design**: Most RAD tools and contemporary CASE tools are object oriented to some degree. Also, code reusability enhances faster development.
- **Higher Quality Design**: New software can be constructed by using tested and proven components.
- **Easier Maintenance**: Since systems are broken down into manageable component objects, isolation of system faults is easy.
- **Dynamic Lifecycle**: I-OO-CASE tools integrate all stages of the software development life cycle (SDLC).
- **Interoperability**: Classes may come from different vendors.
- **Platform Independence**: Classes may be designed to operate and/or communicate across different platforms.
- **Improved Communication** between CS/IT professionals and business people.

Figure 4.3   Benefits of OOM

4.6 Objects and Object Types

An *object* is a concept or thing about which data can be stored, and upon which a set of operations is applicable. The concept of an object is similar to that of an entity in the relational model. Here, however, the definition includes possible actions (operations) which may be performed on the object. In the relational model, this is not the case.

Object may be tangible or intangible (conceptual). Examples of tangible objects are:

- A book
- A building
- A chair
- A motor vehicle engine
- A student
- An employee

Examples of abstract (conceptual) objects are:

- A shape in a program for drawing
- A screen with which the user interacts
- An engineering drawing
- An airline reservation
- An employee's work history

An object may be composed of other objects, which in turn may be composed of other objects and so on. This could result in various levels of object composition.

Example 4.1   Object Composition

- A car consists of a chassis, body, engine, and electrical system. The engine consists of pistons, engine block, input manifold, etc.

- A computer system is made up of various components including processor(s), data buses, control buses, magnetic storage units, I/O devices, memory, etc.
- A software system may be composed of subsystems, each consisting of various objects and facilities.

An *object type* refers to a group or category of (related) objects. The instances of an object type are the actual objects. An object type differs from an entity (in relational model) in three ways:

- The object type involves encapsulation of data structure and operation.
- This means the use of potentially more complex data structures.
- For an object type, there is an emphasis on inheritance.

Example 4.2   Object Type vs Object Instance

If **Employee** is an object type, then the person Bruce Jones could be an instance (object) belonging to object type **Employee**.

It is often the case that the term "object" is used loosely to mean either an instance of an object type, or the object type itself. In such cases, the context should be used to determine what the applicable definition is.

4.7 Operations

Operations are the means of accessing and manipulating the data (instances) of an object (type). Since the operations are defined on the object type, they necessarily apply to each instance of that object type. If for instance, we define the operations **Hire, Modify, Remove, Retire, Fire, Resign, Inquire, Print** on an object type, **Employee**, then for each employee object, these operations are applicable.

Operations of an object type (should) reference only the data structure of that object type. Strict OO design forbids an operation to directly access the data structure of any object type other than its own object type. To use the data structure of another object type, the operation must send a *request* (message) to that object.

In a truly OO software (CASE or programming language) environment, operations and data of an object type are encapsulated in a class (more on this later). However, several quasi OO software development environments exist; they typically facilitate the creation of a relational database, with an OO GUI superimposed.

In the absence of an OO software development tool, operations may be implemented as programs (with singular functions), some of which may manipulate a database. Obviously, this is a less desirable situation.

As you will see in Chapter 15, we may expand the meaning of an operation as described above, to mean a set of related activities. If this is done, then an operation may be implemented as a class consisting of related suboperations.

4.8 Methods

Methods specify the way in which operations are encoded in the OO software. Typically, they are implemented as functions and/or procedures, depending on the development software. The ability to create and maintain methods related to classes is one significant advantage of OO-CASE tools (more on this later).

In OO software, methods are stored within classes, each method having its parent class. The methods are accessible to object instances of that class. *Static methods* are accessible to instances of the host class as well as other classes. In non-OO software, methods translate to procedures and/or functions of application programs.

In environments such as Java, an *interface* is defined as a collection of service declarations with no implementation. Services that are generic (i.e. applicable to objects of various object types) may be implemented as interfaces. Implementation details are left for the specific objects which utilize such interfaces (we will revisit interfaces later on). Some OOPLs and hybrid languages use other terms for methods. For instance, C++ uses *functions* and Pascal uses *procedures*. For the duration of this course, we will continue to use the term method.

OO techniques naturally lead to more organized software systems than do the more traditional techniques. Complex operations are naturally broken down into simple manageable activities.

Example 4.3   Illustrating Inherent Bent of the OO Approach

Consider the class **Employee**, encompassing several aggregate object types:

- **Employee-Personal-Data**
- **Academic-Log**
- **Extracurricular**
- **Employment-History**

The operation, **Create-Employee-Application** would require a number of sub-operations such as:

- **Create-Employee-Personal-Data**
- **Create-Applicant-Academic-Log**
- **Create-Applicant-Extracurricular**
- **Create-Applicant-Employment-History**

Each of these component suboperations would have other constituent methods associated with them. For instance, **Create-Employee-Personal-Data** must allow for data entry and validation, as well as write a record (an object instance) to the data structure, **Employee-Personal-Data**.

4.9 Encapsulation and Classes

What do we mean by *encapsulation*, and what is a *class*? These two concepts are closely related to each other, so that it is difficult to discuss one without reference to the other.

4.9.1 Encapsulation

Encapsulation is the packaging together of data and operations. It is the result (act) of hiding implementation details of an object from its user. The object type (class) hides its data from other object types and allows data to be accessed only via its defined operations—this feature is referred to as *information hiding*.

Encapsulation separates object behavior from object implementation. Object implementations may be modified without any effect on the applications using them.

4.9.2 Class

A *class* is the software implementation of an object type. It has data structure and methods that specify the operations applicable to that data structure.

In an OO software environment, object types are implemented as classes. In a procedural programming environment, object types are implemented as modules.

Example 4.4   Organizing Software Via Classes as Building Blocks

In the CUAIS model, we may have an **Employee** class of an **Employment Subsystem** with options to **Hire, Modify, Fire, Remove, Resign, Retire, Inquire,** or **List** employees. These would be defined as operations on the **Employee** class.

Note that even in the absence of an OO software development environment, OOSE affects the user interface topology of the system. More significantly, it affects the fundamental design of the software being constructed.

4.10 Inheritance and Amalgamation

Inheritance and amalgamation are two critical principles of OO methodologies. They both affect code reusability, and by extension, the productivity of the software design and construction experiences. This section briefly examines each principle.

4.10.1 Inheritance

A class may inherit properties (attributes and operations) of a *parent class*. This is called *inheritance*. The parent class is called the *superclass* or *base class*; the inheriting class is called the *subclass* or *child class*.

A class may qualify as both a subclass and a superclass at the same time. Additionally, a class may inherit properties from more than one superclass. This is called *multiple inheritances*.

Example 4.5   Illustrating Multiple Inheritance

SoftwareEngineer may be designed to be a subclass of the superclass **FacultyMember**, which is in turn a subclass of **Employee**, thus making **FacultyMember** a subclass and superclass at the same time. **Employee** may be designed to inherit from **CollegeMember**, thus making it a subclass and superclass.

Student may be designed to inherit from **CollegeMember. StudentWorker** may be designed to inherit from both Student and **Employee**.

As you will later see, multiple inheritances pose potential problems to software implementation. For this reason, it is not supported in some software development environments. It may therefore be prudent to avoid multiple inheritances, depending on the software limitations.

4.10.2 Amalgamation

An object may be composed of other objects. In OOM, we refer to such an object as an *aggregate* or *composite* object. The act of incorporating other component objects into an object is called *aggregation* or *composition*. Since this is done through the object's class, the class is also called an aggregation (or composition) class. Throughout this course, we shall use the term amalgamation to mean an aggregation and/or composition.

Example 4.6   Illustrating Amalgamation

Following on from Example 4.5, **Employee** may be designed to be composition or aggregation of **EmployeePersonalInfo**, **EmployeeEmploymentHistory**, **EmployeeCompensation**, and **EmployeeAcademicLog**.

4.11 Requests

To make an object do something, we send a *request* to that object. The request is conveyed via a message. The message contains the following:

- Object name (of target object)
- Operation
- Optional parameters
- The type of the value the request (service call) returns

In an OO environment, objects communicate with each other by making requests and sending them via messages. The software is designed to respond to messages. Objects respond to requests by returning values and evoking certain operations intrinsic to such objects.

4.12 Polymorphism and Reusability

An operation or object may appear in different forms; this is referred to as *polymorphism*. Polymorphism may be operational or structural. Operational polymorphism may occur in any of the following ways: overriding an inherited operation, extending an inherited operation, restricting an inherited operation, or overloading an operation.

- **Overriding:** A subclass may override the features of an inherited operation to make it unique for that subclass.
- **Extending:** A subclass may *extend* an inherited operation by adding additional features.
- **Restricting:** A subclass may restrict an inherited operation by inhibiting some of the features of the inherited operation.
- **Overloading:** A class may contain several versions of a given operation. Each version (implemented as a method) will have the same name but different parameters and code. Depending on the argument(s) supplied when the operation is invoked, the appropriate version will run.

Structural polymorphism relates to the structural form an object may take. Generally speaking, an instance of a superclass is likely to be structurally different from an instance of a subclass, though both objects may belong to the same inheritance hierarchy. Moreover, an object belonging to the hierarchy can be instantiated to take on different forms within the hierarchy.

Example 4.7   Illustrating Polymorphic Operations

Suppose that the subclass **Student** inherits from superclass **College Member**. **College Member** defines a PRINT operation which is inherited by **Student**. **Student**'s PRINT operation may be extended to include printing attributes of **Student** which are not attributes of **College Member**.

A **Student** object may be instantiated as a **CollegeMember** object or a **Student** object. However, a **CollegeMember** object may not be instantiated as a **Student** object.

Polymorphism presumes inheritance; polymorphic objects and/or methods can be defined only within the context of an inheritance hierarchy. However, note that the converse does not necessarily hold: inheritance does not necessarily mean that polymorphism is in play.

The most powerful spinoff from inheritance and polymorphism is reusability of code. Reusability may occur in any of the following ways:

- A method (or data structure) is inherited and used by a subclass
- A method is inherited and extended in a subclass
- A method is inherited and restricted in a subclass
- A method is copied and modified in another method

Note: Reusability and polymorphism are not miracles; they must be planned. Reusability of classes may span several systems, not just one. To this end, class libraries are of paramount importance. The OO-CASE tool should therefore support class libraries.

4.13 Interfaces

An interface is a class-like structure with the following exceptions:

- The data items of an interface are constants only.
- The interface contains method signatures only.

The idea of an interface is to promote polymorphism and facilitate a safe alternative to multiple inheritances. Polymorphism is achieved when different classes *implement* (a term used for interfaces instead of inherit) the interface and override its methods (rather, method signatures). Multiple inheritances occur when a particular class implements several interfaces, or inherits a superclass and implements one or more interfaces.

4.14 Late Binding

Late binding is the ability to determine the specific receiver (requester) and the corresponding method to service the request at run time. In traditional programming, a procedure or function call is essentially translated to a branch to a particular location in memory where that section of instructions executes. As part of the compilation process, the necessary pieces of the program are put (bound) together before program execution. This process is referred to as (early) *binding*.

In the object-oriented paradigm, no concern is given to which object will request any given service of another object. Further, objects (more precisely classes) are not designed to anticipate what instances of other classes will require their services. There therefore needs to be late binding between a message (request) and the method that will be used to respond to that request.

4.15 Multithreading

The principle of *multithreading* has been immortalized by the Java programming language. However, software industry leaders have latched on to the concept, so much so that we now talk about multithreading operating systems, as well as multithreading software applications.

Multithreading is the ability of an object to have concurrent execution paths. The effect is to speed up processing and thereby enhance throughput and hence, productivity. Multithreading is one of the reasons object technology offers better machine performance.

4.16 Perception versus Reality

Something is said to be *transparent* if it appears not to exist, when in fact it does. For example, calculations, cross-linking of data, procedures for methods, etc., should be transparent to the user.

Something is said to be *virtual* if it appears to exist, when in fact, it does not. A good example of this concept is the underlying code that facilitates the OO concept. We are making the computer (software) behave like humans do, as opposed to the traditional approach of thinking like the computer.

Methods are not actually stored in each object—this would be wasteful. Instead, the software examines each request that refers to an object and selects the appropriate methods (code) to execute. The method is part of the class (or a higher class in the hierarchy), not part of the object. Yet the software gives the illusion that they are, thus creating a virtual environment for the user.

4.17 Overview of the Object-Oriented Software Engineering Process

Software is constructed using one of two approaches:

* The function-oriented approach
* The object-oriented approach

The function-oriented approach is the traditional approach, based on a life cycle model. The software system passes through several distinct phases, as summarized by the software development life cycle (SDLC): Each of these phases employs different techniques and representations of the software system.

* Investigation
* Analysis
* Design
* Development

- Implementation
- Maintenance

The object-oriented approach is the contemporary approach to software construction. The software system passes through phases which are integrated by a standard set of techniques and representations, as summarized by the following revised SDLC: Each phase employs and builds on standard set of techniques and representations of the software system. The combined set of activities is referred to as modeling.

- Investigation
- Object Structure Analysis (OSA)
- Object Behavior Analysis (OBA)
- Class Structure Design (CSD)
- Operations Design (OD)
- Implementation
- Maintenance

There are two significant advantages of this revised SDLC over the traditional one:

- In the traditional approach, the symbols used in representing the system requirements vary with the different phases, thus presenting the possibility for confusion. In the OO approach, the symbols used are consistent throughout the entire engineering process. The risk for confusion is therefore minimized.
- In the OO approach, there is no need for a distinction between the design phase and development phase. This is particularly true if an OO-ICASE tool is employed. The combination of analysis and design activities in the OO paradigm constitutes system modeling, which often results in the automatic generation of code (typically in an OOPL) which can then be accessed and modified.

4.18 Summary and Concluding Remarks

Let us summarize what we have covered so far:

- There is a disproportional advancement in computer hardware when compared to computer software. Nonetheless, the software engineering industry has made quantum leaps and has achieved an astounding lot over the past five decades.
- To write quality software in the twenty-first century is a very challenging experience: Users are demanding more complexity, flexibility, and functionality for less cost and within a much shorter timeframe. Because of this, software engineering via traditional methods is an early nonstarter. OOM and OOSE present the opportunity to meet the demands of the current era.
- Information engineering (IE) is the act of engineering the information infrastructure of an organization via OO methodologies. The result is a more efficient organization and information infrastructure.
- OOM facilitates the integration of various technologies for the more efficient operation of the organization.

- OOM brings a number of significant advantages to the software engineering industry and the business environment.
- An object type is the term used to describe a family of like objects. The object type defines the data items and operations that can be applied to objects of that family. An object is a concept of thing about which data can be stored and upon which certain operations can be applied. Each object is an instance of the object type that it belongs to.
- An operation is a set of related activities to be applied to an object. Operations are implemented as methods or classes with related methods. A method is a set of related instructions for carrying out an operation.
- A class is the implementation of an object type or an operation. It encapsulates the structure (i.e. data items) and methods for all instances of that class.
- In an OO environment, objects communicate with each other by making requests and sending them via messages.
- An object inherits all the properties (attributes and methods) of its class. A subclass inherits properties from a superclass. A class may qualify as both a subclass and a superclass at the same time. A class may also inherit from multiple superclasses.
- Polymorphism is the act of an operation or object exhibiting a different behavior or form depending on the circumstances. It is implemented via method overriding, method extension, method restriction, and method overloading.
- An interface is a class-like structure that consists of constants and/or method signatures only. It facilitates polymorphism and a safe alternative to multiple inheritances.
- Late binding is the ability to determine the specific receiver (requester) and the corresponding method to service the request at run time.
- Multithreading is the ability of an object to have concurrent execution paths. The effect is to speed up processing and thereby enhance throughput and hence, productivity.
- Something is transparent if it appears not to exist, when in fact it does. Something is virtual if it appears to exist when in fact, it does not.
- With the introduction of OOM, the SDLC has been simplified and more reliable software products can be constructed.

Software construction via the OO paradigm is an integrated, pragmatic approach, which is very exciting and rewarding. It is hoped that you will catch this excitement and be part of the software revolution. The next chapter starts the process of delving deeper into OOM. We will start by looking at the organization as a large complex object type, then work downwards into the details.

4.19 Review Questions

1. Why is contemporary software engineering conducted using object-oriented methodologies? Provide a robust rationale for OOM.

2. Identify the main characteristics of OOM.

3. Identify the main benefits of OOM.

4. Provide the summarized clarifications on the main tenets of object technology.

Sources and Recommended Readings

[Due 2002] Due, Richard T. 2002. *Mentoring Object Technology Projects*. Saddle River, NJ: Prentice Hall. See chapters 1 and 2.

[Lee 2002] Lee, Richard C. and William Tepfenhart. 2002. *Practical Object-Oriented Development with UML and Java*. Upper Saddle River, NJ: Prentice Hall. See chapters 1 and 2.

[Lethbridge 2005] Lethbridge, Timothy C. and Robert Laganiere. 2005. *Object-Oriented Software Engineering*, 2nd ed. Berkshire, England: McGraw-Hill. See chapters 1 and 2.

[Martin 1993] Martin, James and James Odell. 1993. *Principles of Object-Oriented Analysis and Design*. Englewood Cliffs, NJ: Prentice Hall. See chapters 1–3.

[Pressman 2015] Pressman, Roger. 2015. *Software Engineering: A Practitioner's Approach*, 8th ed. New York: McGraw-Hill. See chapters 2 & 3.

[Schach 2011] Schach, Stephen R. 2011. *Object-Oriented & Classical Software Engineering*, 8th ed. New York: McGraw-Hill. See chapter 3.

Chapter 5

Object-Oriented Information Engineering

This chapter discusses object-oriented information engineering (OOIE) as the starting point in developing an information infrastructure for the organization. We will start by looking at the organization as a large complex object type, then work downwards into the details. The chapter includes the following sections:

- Introduction
- Engineering the Infrastructure
- Diagramming Techniques
- Enterprise Planning
- Business Area Analysis
- Software System Design
- Software System Construction
- Summary and Concluding Remarks

5.1 Introduction

Software engineering as we know refers to the discipline of researching, specifying, designing, constructing, implementing, and managing computer software. Information engineering is heavily influenced by software engineering but takes a different perspective: it refers to a set of interrelated disciplines required to build a computerized enterprise based on information systems and managing the resulting information infrastructure. *Information engineering* (IE) concerns itself with the planning, construction, and management of the entire information infrastructure of the organization. It typically pulls several software engineering projects and other related projects together and provides general direction to the organization. Object-oriented information engineering (OOIE) simply means that we conduct IE with an object-oriented focus; and much of OOIE involves the application of OOSE. You might be familiar with the term *business process re-engineering* (BPR). Information

DOI: 10.1201/9780367746025-7

engineering (also referred to as enterprise engineering) includes that concept and much more. Another term that is widely used is OO *enterprise modeling* (OOEM). As mentioned in the previous chapter, for practical reasons, we may consider OOIE and OOEM to be identical, so the rest of the discussion will stick with OOIE.

The following are some broad objectives of IE:

- Support of the management needs of the organization
- Aligning the information infrastructure to the corporate goals of the organization
- Increasing the value and credibility of the computerized systems
- Effective management of the information so that it is available whenever required
- Increasing the speed and ease of software application development
- Reducing the rigors of software systems maintenance
- Achievement of integration of component systems
- Facilitation and promotion of reusable design and code
- Promotion of a higher level of communication and understanding between IS professionals and non-IS professionals in the organization

Desirable characteristics of IE include the following:

- It involves higher user participation.
- It involves the use of sophisticated software planning and development tools (CASE tools, DBMS suites, and RAD tools).
- It embraces CBSE through integration of component systems.
- It leads to an organization that is more progressive, efficient, and technologically aware.

In order to achieve maximum benefit, information engineering should be conducted in an object-oriented manner, hence the acronym OOIE. This means focusing on object first, and behavior last—identify the main subsystems of a software system, then the objects and/or information entities at play within each subsystem, then the behavior of those interacting objects and/or entities.

5.2 Engineering the Infrastructure

In planning an organization's information infrastructure, the following three important ingredients must be determined and clearly specified:

- The structure of the system's required object types
- The behavior of the required system operations
- The underlying technology required to support the organization's mission and operations

James Martin's information system (IS) pyramid (see [Martin 1989a] and [Martin 1993]) proposes four levels of planning that address these ingredients:

- Enterprise Planning
- Business Area Analysis

- System Design
- System Construction

According to Martin, these four strands of information management must relate to the three sides of the pyramid—relating to structure, operations, and technology. These concepts are summarized in Figure 5.1. Upcoming sections of this chapter will discuss each strand of information management, but first, a discussion of diagramming techniques is in order.

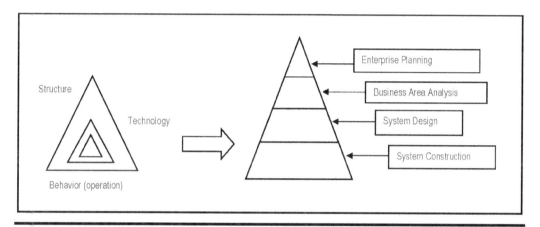

Figure 5.1   Overview of Martin's IS Pyramid

5.3 Diagramming Techniques

If you are going to make any significant progress in designing the information infrastructure, you will first need to put together a number of technical documents. Covering large volumes of information in a succinct, comprehensive manner is critical; hence the importance of diagrams.

Figure 5.2 provides a list of recommended diagramming techniques that are applicable to OOSE. These techniques will be covered as we advance through the course. In the interest of additional clarity, some of them will be revisited, and the new ones discussed in the upcoming sections of this and subsequent chapters. There are other diagramming techniques that are used in OOSE, but these are the ones that this course recommends. At the highest level of planning the information infrastructure of a business enterprise, or architecture of a software system, object flow diagrams and information topology charts are useful. From this point, you then proceed down into the structure of the various object types that comprise the system. Then you address the behavior of instances of each object type.

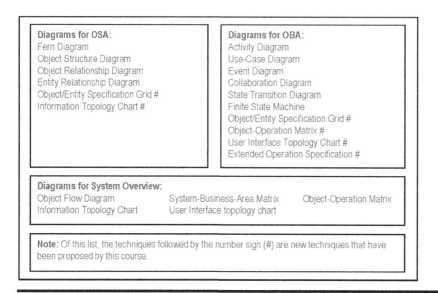

Figure 5.2   Recommended OOM Diagramming Techniques

5.4 Enterprise Planning

A full discussion of enterprise planning is beyond the scope of this course; however, in the interest of comprehensive coverage, a brief summary is provided here. In the broadest sense, enterprise planning relates to planning the essentials of the organization. In OOIE, we concentrate on issues that relate (directly or indirectly) to the information infrastructure of the organization. As such, enterprise planning involves a number of important activities, including the following:

• Planning the information infrastructure of the organization
• Managing that infrastructure in the face of changing user requirements, changing technology, challenges from competitors, etc.
• Planning and pursuing appropriate IT strategies for the organization
• Having a proper backup and recovery policy
• Human and technical resource management
• Project management

Of course, each of these activities involves its own set of principles and methodologies. Three very useful diagramming techniques here are the information topology chart (ITC), the *user interface topology chart* (UITC), and the *object flow diagram* (OFD). The UITC—another innovation from the current author—is an extension of the ITC, focusing on the operational components and their related subsystems. The OFD focuses on the main subsystems and how they interact with each other. Figure 5.3 illustrates a partial ITC for the CUAIS project; Figure 5.4 illustrates a partial UITC for the project; and Figure 5.5 shows an OFD for the project.

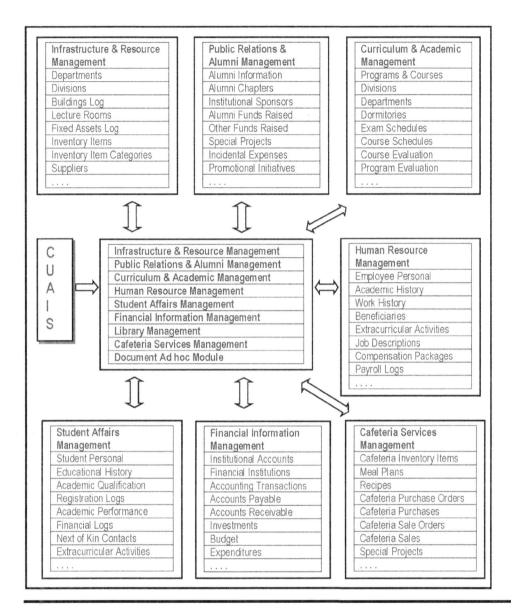

Figure 5.3   Partial Information Topology Chart for the CUAIS Project

CUAIS Main Menu
1. Infrastructure & Resource Management System
2. Curriculum & Academic Management System
3. Financial Information Management System
4. Student Affairs Management System
5. Public Relations & Alumni Management System
6. Human Relations Management System
7. Cafeteria Services Management System
8. Library Information Management System
9. Other Ad Hoc Services

1. Infrastructure & Resource Management System
1.1 Department Definitions
1.1.1 Add Department Definitions
1.1.2 Modify Department Definitions
1.1.3 Delete Department Definitions
1.1.4 Inquire/Report on Department Definitions
1.2 School/Division Definitions
1.2.1 Add School/Division Definitions
1.2.2 Modify School/Division Definitions
1.2.3 Delete School/Division Definitions
1.2.4 Inquire/Report on School/Division Definitions
1.3 Building Definitions
1.3.1 Add Building Definitions
1.3.2 Modify Building Definitions
1.3.3 Delete Building Definitions
1.3.4 Inquire/Report on Building Definitions
1.4 Lecture Room Definitions
1.4.1 Add Lecture Room Definitions
1.4.2 Modify Lecture Room Definitions
1.4.3 Delete Lecture Room Definitions
1.4.4 Inquire/Report on Lecture Room Definitions
1.5 Fixed Asset Logs
1.5.1 Add Fixed Assets
1.5.2 Modify Fixed Assets
1.5.3 Delete Fixed Assets
1.5.4 Inquire/Report on Fixed Assets
1.6 Inventory/Resource Items
1.6.1 Add Inventory/Resource Items
1.6.2 Modify Inventory/Resource Items
1.6.3 Delete Inventory/Resource Items
1.6.4 Inquire/Report on Inventory/Resource Items

1. Infrastructure & Resource Management System
1.7 Supplier Definitions
1.7.1 Add Supplier Definitions
1.7.2 Modify Supplier Definitions
1.7.3 Delete Supplier Definitions
1.7.4 Inquire/Report on Supplier Definitions
1.8 Purchase Order Summaries
1.8.1 Add Purchase Order Summaries
1.8.2 Modify Purchase Order Summaries
1.8.3 Delete Purchase Orders
1.8.4 Inquire/Report on Purchase Order Summaries
1.9 Purchase Order Details
1.9.1 Add Purchase Order Details
1.9.2 Modify Purchase Order Details
1.9.3 Delete Purchase Order Details
1.9.4 Inquire/Report on Purchase Order Details
1.10 Purchase Invoice Summaries
1.10.1 Add Purchase Invoice Summaries
1.10.2 Modify Purchase Invoice Summaries
1.10.3 Delete Purchase Invoices
1.10.4 Inquire/Report on Purchase Invoice Summaries
1.11 Purchase Invoice Details
1.11.1 Add Purchase Invoice Details
1.11.2 Modify Purchase Invoice Details
1.11.3 Delete Purchase Invoice Details
1.11.4 Inquire/Report on Purchase Invoice Details
1.12 Purchase Return Summaries
1.12.1 Add Purchase Return Summaries
1.12.2 Delete Purchase Returns
1.12.3 Inquire/Report on Purchase Return Summaries
1.13 Purchase Return Details
1.13.1 Add Purchase Return Details
1.13.2 Modify Purchase Return Details
1.13.3 Delete Purchase Return Details
1.13.4 Inquire/Report on Purchase Return Details
1.14 Supplier-Resource Mappings
1.14.1 Add Supplier-Resource Mappings
1.14.3 Delete Supplier-Resource Mappings
1.14.4 Inquire/Report on Supplier-Resource Mappings
. . . .

Figure 5.4   Partial User Interface Topology Chart for the CUAIS Project

2. Curriculum & Academic Management System
2.1   Academic Program Definitions
2.1.1 Add Academic Program Definitions
2.1.2 Modify Academic Program Definitions
2.1.3 Delete Academic Program Definitions
2.1.4 Inquire/Report on Academic Program Definitions
2.2   Course Definitions
2.2.1 Add Course Definitions
2.2.2 Modify Course Definitions
2.2.3 Delete Course Definitions
2.2.4 Inquire/Report on Course Definitions
2.3   Academic Department Definitions
2.3.1 Add Academic Department Definitions
2.3.2 Modify Academic Department Definitions
2.3.3 Delete Academic Department Definitions
2.3.4 Inquire/Report on Academic Department Definitions
2.4   Dormitory Definitions
2.4.1 Add Dormitory Definitions
2.4.2 Modify Dormitory Definitions
2.4.3 Delete Dormitory Definitions
2.4.4 Inquire/Report on Dormitory Definitions
2.5   Course Schedules
2.5.1 Add Course Schedules
2.5.2 Modify Course Schedules
2.5.3 Delete Course Schedules
2.5.4 Inquire/Report on Course Schedules
2.6   Examination Schedules
2.6.1 Add Examination Schedules
2.6.2 Modify Examination Schedules
2.6.3 Delete Examination Schedules
2.6.4 Inquire/Report on Examination Schedules
2.7   Course Evaluations
2.7.1 Add Course Evaluations
2.7.2 Modify Course Evaluations
2.7.3 Delete Course Evaluations
2.7.4 Inquire/Report on Course Evaluations
2.8   Academic Program Evaluations
2.8.1 Add Academic Program
2.8.2 Modify Academic Program
2.8.3 Delete Academic Program
2.8.4 Inquire/Report on Academic Program
. . . .

3. Financial Information Management System
3.1   Chart of Accounts
3.1.1 Add Account Definitions
3.1.2 Modify Account Definitions
3.1.3 Delete Account Definitions
3.1.4 Inquire/Report on Account Definitions
3.2   Financial Institutions
3.2.1 Add Financial Institution Definitions
3.2.2 Modify Financial Institution Definitions
3.2.3 Delete Financial Institution Definitions
3.2.4 Inquire/Report on Financial Institution Definitions
3.3   Financial Transactions
3.3.1 Add Financial Transactions
3.3.2 Modify Financial Transactions
3.3.3 Delete Financial Transactions
3.3.4 Inquire/Report on Financial Transactions
3.4   Purchase Orders — Summaries & Details
3.4.1 Add Purchase Orders
3.4.2 Modify Purchase Orders
3.4.3 Delete Purchase Orders
3.4.4 Inquire/Report on Purchase Orders
3.5   Purchase Invoices — Summaries & Details
3.5.1 Add Purchase Invoices
3.5.2 Modify Purchase Invoices
3.5.3 Delete Purchase Invoices
3.5.4 Inquire/Report on Purchase Invoices
3.6   Sale Orders — Summaries & Details
3.6.1 Add Sale Orders
3.6.2 Modify Sale Orders
3.6.3 Delete Sale Orders
3.6.4 Inquire/Report on Sale Orders
3.7   Sale Invoices — Summaries & Details
3.7.1 Add Sale Invoices
3.7.2 Modify Sale Invoices
3.7.3 Delete Sale Invoices
3.7.4 Inquire/Report on Sale Invoices
3.8   Investments
3.8.1 Add Investment Entries
3.8.2 Modify Investment Entries
3.8.3 Delete Investment Entries
3.8.4 Inquire/Report on Investments
. . . .

Figure 5.4   (Continued)

4. Student Affairs Management System
4.1   Student Personal Records
4.1.1 Add Student Personal Records
4.1.2 Modify Student Personal Records
4.1.3 Delete Student Personal Records
4.1.4 Inquire/Report on Student Personal Records
4.2   Student Educational History
4.2.1 Add Student Educational History
4.2.2 Modify Student Educational History
4.2.3 Delete Student Educational History
4.2.4 Inquire/Report on Student Educational History
4.3   Student Academic Qualification
4.3.1 Add Student Academic Qualification
4.3.2 Modify Student Academic Qualification
4.3.3 Delete Student Academic Qualification
4.3.4 Inquire/Report on Student Academic Qualification
4.4   Student Next of Kin Contacts
4.4.1 Add Student Next of Kin Contacts
4.4.2 Modify Student Next of Kin Contacts
4.4.3 Delete Student Next of Kin Contacts
4.4.4 Inquire/Report on Student Next of Kin Contacts
4.5   Student Extracurricular Activities
4.5.1 Add Student Extracurricular Activities
4.5.2 Modify Student Extracurricular Activities
4.5.3 Delete Student Extracurricular Activities
4.5.4 Inquire/Report on Student Extracurricular Activities
4.6   Student Registration Logs
4.6.1 Add Student Registration Logs
4.6.2 Modify Student Registration Logs
4.6.3 Delete Student Registration Logs
4.6.4 Inquire/Report on Student Registration Logs
4.7   Student Academic Performance Logs
4.7.1 Add Student Academic Performance Logs
4.7.2 Modify Student Academic Performance Logs
4.7.3 Delete Student Academic Performance Logs
4.7.4 Inquire/Report on Student Academic Performance
4.8   Student Financial Logs
4.8.1 Add Student Financial Logs
4.8.2 Modify Student Financial Logs
4.8.3 Delete Student Financial Logs
4.8.4 Inquire/Report on Student Financial Logs
. . . .

5. Public Relations & Alumni Management System
5.1   Alumni Information
5.1.1 Add Alumni Information
5.1.2 Modify Alumni Information
5.1.3 Delete Alumni Information
5.1.4 Inquire/Report on Alumni Information
5.2   Alumni Chapters
5.2.1 Add Alumni Chapter Definitions
5.2.2 Modify Alumni Chapter Definitions
5.2.3 Delete Alumni Chapter Definitions
5.2.4 Inquire/Report on Alumni Chapters
5.3   Institutional Sponsors
5.3.1 Add Institutional Sponsors
5.3.2 Modify Institutional Sponsors
5.3.3 Delete Institutional Sponsors
5.3.4 Inquire/Report on Institutional Sponsors
5.4   Alumni Funds Raised
5.4.1 Add Alumni Funds Raised
5.4.2 Modify Alumni Funds Raised
5.4.3 Delete Alumni Funds Raised
5.4.4 Inquire/Report on Alumni Funds Raised
5.5   Other Funds Raised
5.5.1 Add Other Funds Raised
5.5.2 Modify Other Funds Raised
5.5.3 Delete Other Funds Raised
5.5.4 Inquire/Report on Other Funds Raised
5.6   Special Project Definitions
5.6.1 Add Special Project Definitions
5.6.2 Modify Special Project Definitions
5.6.3 Delete Special Project Definitions
5.6.4 Inquire/Report on Special Project Definitions
5.7   Special Project Details
5.7.1 Add Special Project Details
5.7.2 Modify Special Project Details
5.7.3 Delete Special Project Details
5.7.4 Inquire/Report on Special Project Details
5.8   Promotional Initiatives/Activities
5.8.1 Add Promotional Initiatives
5.8.2 Modify Promotional Initiatives
5.8.3 Delete Promotional Initiatives
5.8.4 Inquire/Report on Promotional Initiatives
. . . .

Figure 5.4   (Continued)

6. Human Resource Management System
6.1 Employee Personal Records
6.1.1 Add Employee Personal Records
6.1.2 Modify Employee Personal Records
6.1.3 Delete Employee Personal Records
6.1.4 Inquire/Report on Employee Personal Records
6.2 Employee Academic History
6.2.1 Add Employee Academic History
6.2.2 Modify Employee Academic History
6.2.3 Delete Employee Academic History
6.2.4 Inquire/Report on Employee Academic History
6.3 Employee Work History
6.3.1 Add Employee Work History
6.3.2 Modify Employee Work History
6.3.3 Delete Employee Work History
6.3.4 Inquire/Report on Work History
6.4 Employee Beneficiaries
6.4.1 Add Employee Beneficiary Information
6.4.2 Modify Employee Beneficiary Information
6.4.3 Delete Employee Beneficiary Information
6.4.4 Inquire/Report on Employee Beneficiaries
6.5 Employee Extracurricular Activities
6.5.1 Add Student Extracurricular Activities
6.5.2 Modify Student Extracurricular Activities
6.5.3 Delete Student Extracurricular Activities
6.5.4 Inquire/Report Employee Extracurricular Activities
6.6 Employee Job Definitions
6.6.1 Add Employee Job Definitions
6.6.2 Modify Employee Job Definitions
6.6.3 Delete Employee Job Definitions
6.6.4 Inquire/Report on Employee Job Definitions
6.7 Employee Compensation Packages
6.7.1 Add Employee Compensation Packages
6.7.2 Modify Employee Compensation Packages
6.7.3 Delete Employee Compensation Packages
6.7.4 Inquire/Report Employee Compensation Packages
6.8 Employee Payroll Logs
6.8.1 Add Employee Payroll Logs
6.8.2 Modify Employee Payroll Logs
6.8.3 Delete Employee Payroll Logs
6.8.4 Inquire/Report on Employee Payroll Logs
. . . .

7. Cafeteria Services Management System
7.1 Cafeteria Inventory Items
7.1.1 Add Cafeteria Inventory Items
7.1.2 Modify Cafeteria Inventory Items
7.1.3 Delete Cafeteria Inventory Items
7.1.4 Inquire/Report on Cafeteria Inventory Items
7.2 Cafeteria Meal Plans
7.2.1 Add Cafeteria Meal Plans
7.2.2 Modify Cafeteria Meal Plans
7.2.3 Delete Cafeteria Meal Plans
7.2.4 Inquire/Report on Cafeteria Meal Plans
7.3 Cafeteria Recipes
7.3.1 Add Cafeteria Recipes
7.3.2 Modify Cafeteria Recipes
7.3.3 Delete Cafeteria Recipes
7.3.4 Inquire/Report on Cafeteria Recipes
7.4 Cafeteria Purchase Order Summaries
7.4.1 Add Purchase Order Summaries
7.4.2 Modify Purchase Order Summaries
7.4.3 Delete Purchase Order Summaries
7.4.4 Inquire/Report on Purchase Order Summaries
7.5 Cafeteria Purchase Order Details
7.5.1 Add Purchase Order Details
7.5.2 Modify Purchase Order Details
7.5.3 Delete Purchase Order Details
7.5.4 Inquire/Report on Purchase Order Details
7.6 Cafeteria Sale Order Summaries
7.6.1 Add Sale Order Summaries
7.6.2 Modify Sale Order Summaries
7.6.3 Delete Sale Order Summaries
7.6.4 Inquire/Report on Sale Order Summaries
7.7 Cafeteria Sale Order Details
7.7.1 Add Sale Order Details
7.7.2 Modify Sale Order Details
7.7.3 Delete Sale Order Details
7.7.4 Inquire/Report on Sale Order Details
7.8 Special Catering Projects
7.8.1 Add Special Projects
7.8.2 Modify Special Projects
7.8.3 Delete Special Projects
7.8.4 Inquire/Report on Special Projects
. . . .

Figure 5.4   (Continued)

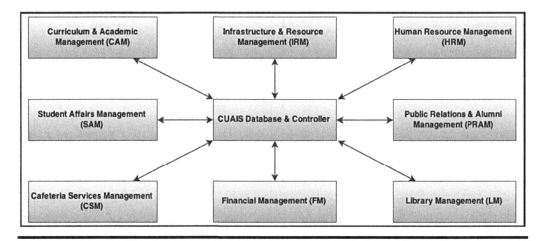

Figure 5.5 OFD for the CUAIS Project, Assuming Repository Model

5.5 Business Area Analysis

In *business area analysis* (BAA), we are interested in the nature of participation that each business area of the organization will have (or already has) on the component systems comprising the information infrastructure. In particular, the information system executive wants to ensure the following:

- The underlying object (database) structure must meet the needs of various business areas.
- The operations provided must meet the requirements of the various business areas.
- Appropriate security mechanism must be put in place to ensure the resources (objects and operations) of the integrated system are accessible to the relevant functional areas of the organization.

Typically, diagrams and/or matrices are constructed to reflect critical information:

- Object structure diagrams indicate the structure and operational services of information object types (in case of object oriented (OO) approach). Alternately, you may use *object/entity specification grids* (O/ESGs). The O/ESG is an innovation from the current author (see [Foster 2016]) that is used to provide detailed specification for objects and/or entities in comprising a software system; this will be further clarified as the course proceeds. Figure 5.6 provides an illustration of O/ESG for four of the object types (or entities) that would comprise a typical manufacturing firm's database. There would be one for each object type (or entity) comprising the system.
- Object-operation matrices show how various object types will be used in different business areas.

- System-component-business-area matrices show what functional areas will use certain system components of the software system, and how.

In constructing the matrices for the BAA, always cluster related components (object types or sub-systems) together. Figure 5.7 illustrates an object-operation matrix while Figure 5.8 illustrates a component-business-area matrix.

```
O1 – Department [RM_Department_BR]
Attributes:
01. Department Number [Dept#] [N4]
02. Department Name [DeptName] [A35]
03. Department Head Employee Number [DeptHeadEmp#] [N7] {Refers to O2.Emp#}
...

Comments:
This table stores definitions of all departments in the organization.
Indexes:
1.   Primary Key Index: RM_Department_NX1 on [01]; constraint RM_Department_PK
2.   RM_Department_NX2 on [02]
Valid Operations:
1.   Maintain Departments [RM_Department_MO]
1.1 Add Departments [RM_Department_AO]
1.2 Update Departments [RM_Department_UO]
1.3 Delete Department [RM_Department_ZO]
2.   Inquire on Departments [RM_Department_IQ]
```

```
O2 – Employee [RM_Employee_BR]
Attributes:
01. Employee Identification Number [Emp#] [N7]
02. Employee Last Name [EmpLName] [A20]
03. Employee First Name [EmpFName] [A20]
04. Employee Middle Initials [EmpMInit] [A4]
05. Employee Date of Birth [EmpDOB] [N8]
06. Employee's Department [EmpDept#] [N4] {Refers to O1.Dept#}
07. Employee Gender [EmpGender] [A1]
08. Employee Marital Status [EmpMStatus] [A1]
09. Employee Social Security Number [EmpSSN] [N10]
10. Employee Home Telephone Number [EmpHomeTel] [A14]
11. Employee Work Telephone Number [EmpWorkTel] [A10]
...

Comments:
This table stores standard information about all employees in the organization.
Indexes:
1.   Primary Key Index: RM_Employee_NX1 on [01]; constraint RM_Employee_NX.
2.   RM_Employee_NX2 on [02, 03, 04]
3.   RM_Employee_NX3 on [09]
Valid Operations:
1.   Manage Employees [RM_Employee_MO]
1.1 Add Employees [RM_Employee_AO]
1.2 Update Employees [RM_Employee_UO]
1.3 Delete Employees [RM_Employee_ZO]
2.   Inquire on Employees [RM_Employee_IO]
3.   Report on Employees [RM_Employee_RO]
```

Figure 5.6   Partial O/ESG for Manufacturing Environment   *(Continued)*

```
┌─────────────────────────────────────────────────────────────────────────┐
│ O3 – Supplier [RM_Supplier_BR]                                            │
├───────────────────────────────────────────────────────────────────────── │
│ Attributes:                                                               │
│ 01. Supplier Number [Suppl#] [N4]                                         │
│ 02. Supplier Name [SupplName] [A35]                                       │
│ 03. Supplier Contact Name [SupplContact] [A35]                            │
│ 04. Supplier Telephone Numbers [SupplPhone] [A30]                         │
│ 05. Supplier E-mail Address [SuppEmail] [A30]                             │
│ ...                                                                       │
├───────────────────────────────────────────────────────────────────────── │
│ Comments:                                                                 │
│ This table stores definitions of all employee classifications.           │
├───────────────────────────────────────────────────────────────────────── │
│ Indexes:                                                                  │
│ 1.   Primary Key Index: RM_Supplier_NX1 on [01]; constraint RM_Supplier_PK. │
│ 2.   RM_Supplier_NX2 on [02]                                              │
│ 3.   RM_Supplier_NX3 on [04]                                              │
├───────────────────────────────────────────────────────────────────────── │
│ Valid Operations:                                                         │
│ 1.  Manage Suppliers[RM_Supplier_MO]                                      │
│ 1.1 Add Suppliers [RM_Supplier_AO]                                        │
│ 1.2 Update Suppliers [RM_Supplier_UO]                                     │
│ 1.3 Delete Suppliers [RM_Supplier_ZO]                                     │
│ 2.  Inquire on Suppliers [RM_Supplier_IO]                                 │
│ 3.  Report on Suppliers [RM_Supplier_RO]                                  │
└───────────────────────────────────────────────────────────────────────── ┘
```

```
┌─────────────────────────────────────────────────────────────────────────┐
│ O4 – Project [RM_Project_BR]                                              │
├───────────────────────────────────────────────────────────────────────── │
│ Attributes:                                                               │
│ 01. Project Number [Proj#] [N4]                                           │
│ 02. Project Name [ProjName] [A15]                                         │
│ 03. Project Summary [ProjSumm] [M]                                        │
│ 04. Project's Manager [ProjManagerEmp#] [N7] {References O2.Emp#}         │
│ ...                                                                       │
├───────────────────────────────────────────────────────────────────────── │
│ Comments:                                                                 │
│ This table stores definitions of all company projects.                    │
├───────────────────────────────────────────────────────────────────────── │
│ Indexes:                                                                  │
│ 1.   Primary Key Index: RM_Project_ NX1 on [01]; constraint RM_Project_PK. │
│ 2.   RM_Project _NX2 on [02]                                              │
├───────────────────────────────────────────────────────────────────────── │
│ Valid Operations:                                                         │
│ 1.  Manage Projects [RM_Project_MO]                                       │
│ 1.1 Add Projects [RM_Project_AO]                                          │
│ 1.2 Update Projects [RM_Project_UO]                                       │
│ 1.3 Delete Projects [RM_Project_ ZO]                                      │
│ 2.  Inquire on Projects [RM_Project_IO]                                    │
│ 3.  Report on Projects [RM_Project_RO]                                    │
└───────────────────────────────────────────────────────────────────────── ┘
```

Figure 5.6   (Continued)

Object Type	Business Areas						
	Finance	Accounting	HRD	Marketing	Production	Engineering	Info. Tech
Chart of Accounts	UQ	AUDQ	Q	Q	Q	Q	Q
Transactions	UQ	AUDQ	Q	Q	Q	Q	Q
Payments Received	Q	AUDQ	Q	AUD0	Q	Q	Q
Payments Made	Q	AUDQ	Q	AUD0	Q	Q	Q
Financial Institution	Q	AUDQ	Q	Q	Q	Q	Q
Budget	AUDQ	AUDQ	AUDQ	AUDQ	AUDQ	AUDQ	AUDQ
Customers	Q	AUDQ	Q	Q	Q	Q	Q
Investments	AUDQ	AUDQ	Q	Q	Q	Q	Q

Key: A = Add   U = Update   D = Delete   Q = Query/Report

Note:
1. The matrix tells what operations of each object type are applicable for different business areas, when this Financial Management System is implemented.
2. The information provided could also be used to help define the security arrangements for the system.
3. A matrix of this sort may be constructed for each component system making up the information infrastructure of the organization.

Figure 5.7   Object-Operation Matrix for a Financial Management System

Component Systems	Cafeteria	Finance & Planning	A academic Admin.	PR & Marketing	Student Services	Academic Depts	Human Resource	Information Technology
Financial Management	Q	AUDQ	Q	Q	Q	Q	Q	Q
Human Resource Management		Q	Q	Q	Q	Q	AUDQ	Q
Library Management		Q	AUDQ	Q	Q	Q		Q
Cafeteria Services Management	AUDQ	Q	Q	Q	Q	Q	Q	Q
Curriculum & Acad. Management		Q	AUDQ	Q	Q	Q	Q	Q
PR & Alumni Management		Q	Q	AUDQ	Q	Q	Q	Q
Student Affairs Management		Q	Q	Q	AUDQ	Q	Q	Q

Key: A = Add   U = Update   D = Delete   Q = Query/Report
Also Note: The matrix tells how each business area will be able to access each component system comprising the information infrastructure.

Figure 5.8   Component-Business-Area Matrix for a University7

5.6 Software System Design

System design relates to design of the component software systems that comprise the information infrastructure of the organization. To be excellent at designing information infrastructures, you need to have mastery of software engineering, database systems, and an understanding of how a typical business works. You also need to have a working knowledge of programming, and outstanding communication and leadership skills. You also need to have mastery of various software diagramming techniques, many of which have been covered in the course. Figure 5.9 provides a checklist of design issues with which you should be familiar; these will be covered as we progress through the course.

Following are some basic guidelines that should guide the design process:

- The design must be accurate and comprehensive, covering both structure and behavior of system components, and with due consideration to the availability of technology.
- The design must adequately take care of current and future requirements of the (component systems in the) organization.
- The design must be pragmatic.
- The design must be informed by the software quality factors emphasized throughout the course (review Chapters 1 and 3).
- The design must uphold established software engineering standards.

Design Activity	Resultant Design Spec Component
Architectural Design	System Architecture Specification
Interface Design	System Interface Specification
Database (Object Structure) Design	Database (Object Structure) Specification
Operations Design	Operations Specification
User Interface Design	User Interface Specification
Documentation Design	System Documentation Specification
Message Design	Message Specification
Security Design	Security Specification

Figure 5.9   Summary of System Design Considerations

5.7 Software System Construction

System construction relates to the development of the component software systems that comprise the information infrastructure of the organization. If enterprise planning, business area analysis, and system design were accurate and comprehensive, system construction will simply be the implementation of a carefully developed plan. You will recall from Chapter 1 (Section 1.6), that system acquisition may take any of several alternatives — from traditional waterfall development to outsourcing.

Part D of the text addresses software construction in more detail. In preparation for this, the following principles are worth remembering:

1. Software system design and construction must embody the software quality factors (review section 1.8), meet user expectations, and conform to established software development standards.

2. Software construction must follow a clearly developed plan with specific targets (deadlines).

3. Prudent project management will be required if targets are to be met.

5.8 Summary and Concluding Remarks

Let us summarize what we have covered in this chapter:

- Information engineering (IE) is the planning, construction, and management of the entire information infrastructure of the organization. Object-oriented information engineering (OOIE) is IE conducted in an object-oriented manner.
- OOIE may be conducted by pursuing four steps: enterprise planning, business area analysis, system design, and system construction.
- Enterprise planning relates to planning the essentials of the organization. In OOIE, we concentrate on issues that relate (directly or indirectly) to the information infrastructure of the organization.
- In business area analysis (BAA), we are interested in the nature of participation that each business area (of the organization) will have (or already has) on the component systems (and system components).
- System design relates to design of the component systems that comprise the information infrastructure of the organization.
- System construction relates to the development of the component systems that comprise the information infrastructure of the organization.
- In conducting OOIE, there must be mastery and appropriate use of various diagramming techniques that are applicable to OOSE.

Since its introduction, information engineering has been overtaken by (or more precisely, morphed into) methodologies such as *data administration*, and *business intelligence* (see [Turban 2007] and [Turban 2008]). However, the principles of IE are still relevant. When OOIE is combined with life cycle models such as phased prototyping, iterative development, agile development, or CBSE, you have formula for successful software engineering.

Once you have obtained a good overview of the information infrastructure of the enterprise, your next step is to zoom in on each component system and provide more detail in terms of the structure and behavior of constituent objects comprising the system. The next chapter will address the relevant issues relating to this.

5.9 Review Questions

1. What is information engineering? I Identify the main objectives or the technique.

2. Provide the main rationale for object-oriented information engineering.

3. Identify the three essentials to be clearly specified during information engineering.

4. What are the main diagramming techniques to be employed during OOIE?

5. What is business area analysis (BAA)? Identify the critical information that should be available after BAA has been conducted.

6. Identify a small business enterprise that you are familiar with. Conduct OOIE on the enterprise, and propose a preliminary set of diagrams about the enterprise.

References and/or Recommended Reading

[Foster 2016] Foster, Elvis C. with Shripad Godbole. 2016. *Database Systems: A Pragmatic Approach*, 2nd ed. New York: Apress. See chapter 5.

[Martin 1989a] Martin, James and James Odell. 1989. *Information Engineering Book I: Introduction*. Eaglewood Cliffs, NJ: Prentice Hall.

[Martin 1989b] Martin, James and Joe Leben. 1989. *Strategic Information Planning Methodologies*, 2nd ed. Eaglewood Cliffs, NJ: Prentice Hall. See chapters 2 and 13.

[Martin 1990a] Martin, James and James Odell. 1990. *Information Engineering Book II: Planning and Analysis*. Eaglewood Cliffs, NJ: Prentice Hall.

[Martin 1990b] Martin, James and James Odell. 1990. *Information Engineering Book III: Design and Construction*. Eaglewood Cliffs, NJ: Prentice Hall.

[Martin 1993] Martin, James and James Odell. 1993. *Principles of Object-Oriented Analysis and Design*. Eaglewood Cliffs, NJ: Prentice Hall. See chapter 5.

[Sprague 1993] Sprague, Ralph H. and C. Barbara. 1993. *Information Systems Management in Practice*, 3rd ed. Eaglewood Cliffs, NJ: Prentice Hall. See chapters 4.

[Turban 2007] Turban, Efraim, Jay Aaronson, Ting-Peng Liang, and Ramesh Sharda. 2007. *Decision Support and Business Intelligence*. Eaglewood Cliffs, NJ: Prentice Hall.

[Turban 2008] Turban, Efraim, Ramesh Sharda, Jay Aaronson, and David King. 2008. *Business Intelligence*. Eaglewood Cliffs, NJ: Prentice Hall.

Chapter 6

The Requirements Specification

This chapter introduces the second major deliverable in a software engineering project—the requirements specification. It provides an overview of the deliverable and prepares you for details that will be covered in the next four chapters. The chapter proceeds under the following subheadings:

- Introduction
- Contents of the Requirements Specification
- Documenting the Requirements
- Requirements Validation
- How to Proceed
- Presenting the Requirements Specification
- The Agile Approach
- Summary and Concluding Remarks

6.1 Introduction

The software *requirements specification* (RS) is the second major deliverable in a software engineering project. It is a document that signals the end of the investigation and analysis phase, and the beginning of the design phase. Its contents will serve as a blueprint during design and development.

Note: Remember, we are not assuming that life cycle model being employed is necessarily the waterfall model, which is irreversible. Rather, please assume that any of the reversible life cycle models is also applicable.

The requirements specification must be accurate as well as comprehensive, as it affects the rest of the system's life in a significant way. Any flaw in the requirements specification will put pressure on the subsequent phases (particularly design and development) to attempt to address it. If the problem is not identified and addressed

DOI: 10.1201/9780367746025-8

by the end of system development, then it must be addressed during the maintenance phase—a particularly undesirable situation. In some instances, this might be too late to attempt saving a reputation of a software product.

The requirements specification serves as a very effective method of system documentation, long after system development. Many texts on software engineering recommend one comprehensive requirements specification document that involves detailed design issues. From experience, this is not a good idea, especially if the software being developed is very large and complex. The approach recommended by this course is to have a requirements specification, which provides comprehensive and accurate coverage of the software system but avoids fine design intricacies (such as characteristics of data elements, operation specifications, use of technical language for business rules, system names for objects, etc.). This blueprint is then used as input to the design phase, where all intricacies are treated in a separate *design specification*.

As you go through this chapter, bear in mind that you will need to gain mastery of a number of techniques in order to develop a requirements specification for a software engineering project. The intent here is to provide you with the big picture, so you will have an appreciation of where you are heading. Chapters 5 and 8 will then work you through the required details.

6.2 Contents of the Requirements Specification

The requirements specification is comprised of a number of important ingredients. These will be briefly described in this section.

Acknowledgments: This is recognition of individuals and/or organizations or offices that significantly contributed to the achievement of the deliverable. Typically, it is written last but appears as the first item in the requirements specification.

Problem Synopsis: This involves the problem statement, proposed system solution, and system objectives: having been drawn from the ISR (see Chapter 3), these would be refined by now. The problem synopsis should also involve a brief review of feasibility findings (the details you can include as appendix).

System Overview: This should provide a broad perspective of the software system. An *information topology chart* (ITC) and/or an *object flow diagram* (OFD) are/is particularly useful here. These techniques will be discussed in Chapter 8.

System Components: Since your prime objective is to provide clarity about the requirements of the software being engineered, it is a good habit to provide a list of the subsystems, modules, entities (object types), and operations.

Detailed Requirements: This involves specification of the software requirements in a concise, unambiguous manner. This involves the use of narrative, flow-charts, diagrams, and other techniques where applicable. These techniques will be discussed in the upcoming chapters. Depending on the size and complexity of the project, the requirements may be specified for the system as a whole (if the system is small), for subsystems comprising the system (in the case of medium sized to large systems), or for modules comprising subsystems (if the subsystems are large). An approach that has been successfully employed by the author is to include the following in the detailed requirements:

- **Storage Requirements:** The information entities (object types) and how they are comprised.
- **Operational Requirements:** The operations to be defined on the information entities (object types).
- **System Rules:** The rules of operation. These include derivation rules, relationship integrity rules, data integrity rules, control conditions, operational rules, and trigger rules. Rules are more thoroughly discussed Chapter 10.

Interface Specification: This involves the guidelines relating to how various subsystems and/or modules will be interconnected. Depending on the project, this may or may not be critical at this point. In many cases, it can be deferred until the design phase (in fact, it will be elucidated later in the course).

System Constraints: This involves the operating constraints of the system. Like the interface specification, this requirement may be trivial and self-evident, or critical to the success of the software. An example of triviality would be in the case of an information system (for instance, and inventory management system) for an organization. An example of situation where system constraints would be critical is a real-time system (for instance managing atmospheric pressure in an aircraft, or water pressure for a water power station).

System Security Requirements: Depending on the intended purpose and use of the software product, system security may be an issue to be considered. In specifying the security requirements, the following must be made clear:

- What users will have access to the system
- What system resources various users can access and how
- What restrictions will apply to various users

Revised Project Schedule: Here, you provide a refined project schedule, which involves a PERT diagram or Gantt chart, or some other equivalent means. This will be discussed in Chapter 8.

Concluding Remarks: Usually, there are some concluding remarks about the requirements specification. These are placed here.

Appendices: The appendices typically include supporting information that are relevant to the software requirements, but for various reasons might have been omitted from the body of the document. Examples of possible information for the appendices are

- Profile of the software engineering team
- Photocopy of approval(s) from end user and management
- Photocopies of source documents used in investigation and analysis
- Detailed feasibility report

6.3 Documenting the Requirements

Documenting the requirements will depend to a large extent on the intended readership and the software tools available. If the intended readership is strictly technical people such as software engineers, then unnecessary details can be avoided, technical jargons are allowed and formal methods may be used (more on this later). If the intended readership includes nontechnical individuals (lay persons), then it is advisable to develop an easy to read document.

The requirements specification may be developed with inexpensive applications such as MS Office, Corel Suite, etc. If a CASE tool is available, it should be used since this will enhance the system development process. In fact, in many cases, the diagrams used in the CASE tool are executable—code is generated from them— so that the distinction of requirements specification from design and design from development are absent (hence the term modeling). However, in many instances, thorough documentation is achieved by employing a mix of software development tools and desktop applications.

Preparation of the requirements specifications is perfected only as the software engineer practices on the job. Whether a CASE tool or desktop publisher is used, you will need to have mastery of fundamental software engineering principles, if the end product is to be credible.

6.4 Requirements Validation

Once the requirements have been specified, they must be validated. This is crucial, since it is these requirements that will be used as the frame of reference for the next deliverable—the design specification. If the requirement specification is lacking in content or accuracy, this will have adverse repercussions on the design and final product. There are two approaches to requirements validation: manual and automatic.

Manual Techniques: In manual review, a resource team is used to check the following:

- Each specification must be traceable to a requirement definition
- Each requirement definition is traceable to a specification
- All required specifications are in place
- All system objectives are met

The techniques available for this (these will be discussed in the next chapter) are

- Interviews
- Questionnaires
- Document review
- Sampling and Experimenting
- Observation
- Prototyping
- Brainstorming
- Manual cross-referencing
- Mathematical proofs

Automatic Techniques: In automatic techniques, an attempt is made to automate the validation process. This involves techniques such as cross-referencing, prototyping, automated simulation models, and, mathematical proofs. Automatic validation is not always feasible, but when applied, is quite helpful.

Discussion:
Which life cycle models would most favor automatic validation techniques?

6.5 How to Proceed

How you proceed to develop the RS is a matter of personal preference. Some people like a top-down approach; others prefer a bottom-up approach. Traditional systems employed functional-oriented methodologies; contemporary systems tend to favor object-oriented methodologies. In this course, you will be exposed to both approaches, even though there is a bias to the object-oriented approach.

In preparing the RS, it is recommended that you be methodical and incremental, revisiting formerly addressed issues whenever you need to. Figure 6.1 describes a recommended approach that you will find useful in various scenarios. You will notice a few newly introduced terms; these will be clarified in the upcoming chapters.

---

1. Refine your problem definition and proposed software solution
2. Conduct your information gathering and identify:
   - Subsystem(s) and/or modules
   - *Object types* or *information entities* (clarified in chapters 7 and 13)
3. Identify or define operations for the various object types.
4. Provide a system overview that includes any convenient combination of the following:
   - System narrative
   - *Information topology chart* (clarified in chapter 8)
   - *Object flow diagram* (clarified in chapter 8)
5. For each subsystem, provide the following:
   - System overview (as in item 4)
   - *Entity-Relationship* Diagram (ERD) or *Object-Relationship* Diagram (ORD); this will be clarified in chapter 13
   - Storage requirements
   - Operational requirements
   - System rules
   - Interface requirements (if applicable)
   - System constraints (if applicable)

The approach for real-time systems would be slightly different. This will be discussed later in the course.

Figure 6.1 How to Prepare the Requirements Specification

6.6 Presentation of the Requirements Specification

It is often required that the requirements specification be presented to a group of experts and possible end-users for scrutiny. Should you have the responsibility to do this, below are some guidelines:

- Have a summarized version for the purpose of presentation.
- Make the presentation interesting by using a balanced mix of the various resources at your disposal.

- Have photocopies for the critical audience.
- Make allocation and be prepared for questions.
- Here are some positive tips:
  - o Be thoroughly prepared
  - o Project loudly
  - o Have a positive eye contact
  - o Make visuals large enough for audience to see
  - o Speak in your natural language, tone style
  - o Be confident and calm

Your presentation must be aesthetically attractive and compelling. As such, you should make use of an appropriate presentation software product (examples include Microsoft PowerPoint, Microsoft Publisher, Corel Draw, Page Maker, etc.).

6.7 The Agile Approach

As mentioned in subsection 1.4.7 of Chapter 1, agile development has emerged as the preferred approach among many software engineering enterprises. While the steps outlined in Figure 6.1 are applicable to the agile approach, the approach has deep philosophical underpinnings and methodological implications to deserve additional attention.

The agile methodology calls for the following important role players: users, product owner, and the development team.

Users: The user role represents the end user needs and requirements of the software system. Identification of user needs and requirements ties back to problem definition of Section 3.2.

Product Owner: Agile methodology thrives on the idea of someone fulfilling the role of championing the needs and requirements of the end user, crafting and enunciating a clear vision for the project, and formulating strategies for realizing that vision. The product owner fulfills this role.

Development Team: The agile development team typically consists of multitalented individuals working with heightened collaboration and cohesion to deliver established software system outcomes with effective and efficient dispatch. The team may consist of role-players such as designers, developers, analysts, and quality assurance engineers. The members meet frequently and work purposefully to meet their established targets.

Development Framework: To reinforce the agile methodologies, a software framework is typically used. Three common candidates are Scrum, Extreme Programming, and Kanban (for more on frameworks, see Chapter 25).

The deliverables of the project proposal (Chapter 3) and the requirements specification could be used to trigger agile development team then taking the candidate project through a series agile design and development activities to ensure incremental development of the software system.

6.8 Summary and Concluding Remarks

Let us summarize what we have covered in this chapter:

- The RS is the second major deliverable of a software engineering project. It must be accurate and comprehensive.
- The RS typically consists of the following components: acknowledgments, problem synopsis, system overview, detailed requirements, interface specification, system constraints, revised project schedule, security requirements, conclusion, and appendices.
- The requirements may be documented using a CASE tool or a desktop processing application.
- Requirements validation ensures that the software requirements are accurate. The process may be manual or automatic.
- In preparing the RS, it is recommended that you be methodical and incremental in your approach.
- You may be required to make a formal presentation of the RS. Preparation is of paramount importance.

Appendix B provides excerpts from the RS of the inventory management system that was mentioned in the previous chapter. Take some time to review it and gain useful insights. But to get to that point where you are able to prepare such a document on your own, you need to learn some more fundamentals, commencing with information gathering techniques. This will be discussed in the next chapter.

6.9 Review Questions

1. Explain the importance of the requirements specification.

2. Identify six important ingredients of the requirements specification. For each, briefly explain what it is and why it is important.

3. How important is requirements validation? Describe two approaches to requirements validation.

4. Outline clearly, how you would proceed to develop a requirements specification for a software engineering project.

References and Recommended Readings

[Kendall 2014] Kendall, Kenneth E., and Julia E. Kendall. 2014. *Systems Analysis and Design*, 9th ed. Boston, MA: Pearson. See chapters 2 & 3.

[Lee 2002] Lee, Richard C. and William M. Tepfenhart. 2002. *Practical Object-Oriented Development with UML and Java.* Upper Saddle River, NJ: Prentice Hall.

[Martin 1993] Martin, James and James Odell. 1993. *Principles of Object-Oriented Analysis and Design.* Eaglewood Cliffs, NJ: Prentice Hall.

[Peters 2000] Peters, James F. and Witold Pedrycz. 2000. *Software Engineering: An Engineering Approach.* New York, NY: John Wiley & Sons. See chapter 3.

[Pfleeger 2006] Pfleeger, Shari Lawrence. 2006. *Software Engineering Theory and Practice*, 3rd ed. Upper Saddle River, NJ: Prentice Hall. Chapter 4.

[Schach 2011] Schach, Stephen R. 2011. *Object-Oriented & Classical Software Engineering*, 8th ed. New York, NY: McGraw-Hill. See chapter 11.

[Sommerville 2016] Sommerville, Ian. 2016. *Software Engineering*, 10th ed. Boston, MA: Pearson. See chapter 4.

Chapter 7

Information Gathering

In order to accurately and comprehensively specify the system, the software engineer gathers and analyzes information via various methodologies. This chapter discusses these methodologies as outlined below:

- Rationale for Information Gathering
- Interviews
- Questionnaires and Surveys
- Sampling and Experimenting
- Observation and Document Review
- Prototyping
- Brainstorming and Mathematical Proof
- Object Identification
- End-User Involvement
- Summary and Concluding Remarks

7.1 Rationale for Information Gathering

What kind of information is the software engineer looking for? The answer is simple, but profound: You are looking for information that will help to accurately and comprehensively define the requirements of the software to be constructed. The process is referred to as *requirements analysis* and involves a range of activities that eventually lead to the deliverable that we call the *requirements specification* (RS). In particular, the software engineer must determine the following:

- **Synergistic interrelationships** of the system components: This relates to the components and how they (should) fit together.
- **System information entities** (*object types*) and their interrelatedness: An entity refers to an object or concept about data is to be stored and managed. Entities and object types will be further discussed in Chapters 12 and 13.

DOI: 10.1201/9780367746025-9

- **System operations** and their interrelatedness: Operations are programmed instructions that enable the requirements of the system to be met. Some operations are system-based and may be oblivious to the end user; others facilitate user interaction with the software; others are internal and often operate in a manner that is transparent to the end user.
- **System business rules:** Business rules are guidelines that specify how the system should operate. These relate to data access, data flow, relationships among entities, and the behavior of system operations.
- **System security mechanism(s)** that must be in place: It will be necessary to allow authorized users to access the system while denying access to unauthorized users. Additionally, the privileges of authorized users may be further constrained to ensure that they have access only to resources that they need. These measures protect the integrity and reliability of the system.

As the software engineer embarks on the path towards preparation of the requirements specification, these objectives must be constantly borne in mind. In the early stages of the research, the following questions should yield useful pointers:

- WHAT are the (major) categories of information handled? Further probing will be necessary, but you should continue your pursuit until this question is satisfactorily answered.
- WHERE does this information come from? Does it come from an internal department or from an external organization?
- WHERE does this information go after leaving this office? Does it go to an internal department or to an external organization?
- HOW and in WHAT way is this information used? Obtaining answers to these questions will help you to identify business rules (discussed in Chapter 10) and operations (discussed in Chapters 14 and 15).
- WHAT are the main activities of this unit? A unit may be a division or department or section of the organization. Obtaining the answer to this question will also help you to gain further insights into the operations (discussed in Chapters 14 and 15).
- WHAT information is needed to carry out this activity? Again here, you are trying to refine the requirements of each operation by identifying its input(s).
- WHAT does this activity involve? Obtaining the answer to this question will help you to further refine the operation in question.
- WHEN is it normally done? Obtaining the answer to this question will help you to further refine the operation in question by determining whether there is a time constraint on an operation.
- WHY is this important? WHY is this done? WHY...? Obtaining answers to these probes will help you to gain a better understanding of the requirements of the software system.

Of course, your approach to obtaining answers to these probing questions will be influenced by whether the software system being researched is to be used for in-house purposes or marketed to the public. The next few sections will examine the commonly used information-gathering strategies.

7.2 Interviewing

Interviewing is the most frequent method of information gathering. It can be very effective if carefully planned and well conducted. It is useful when the information needed must be elaborate, or clarification on various issues is required. The interview also provides an opportunity for the software engineer to win the confidence and trust of clients. It should therefore not be squandered.

7.2.1 Steps in Planning the Interview

In panning to conduct an interview, please observe the following steps:

1. Read background information.
2. Establish objectives.
3. Decide whom to interview.
4. Prepare the interviewee(s).
5. Decide on structure and questions.

7.2.2 Basic Guidelines for Interviews

Figure 7.1 provides some guidelines for successfully planning and conducting an interview. These guidelines are listed in the form of a do-list, and a don't-list.

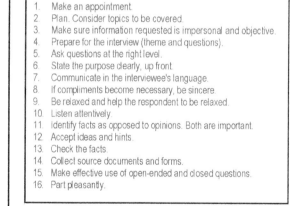

Do-List for Interviews:
1. Make an appointment.
2. Plan. Consider topics to be covered.
3. Make sure information requested is impersonal and objective.
4. Prepare for the interview (theme and questions).
5. Ask questions at the right level.
6. State the purpose clearly, up front.
7. Communicate in the interviewee's language.
8. If compliments become necessary, be sincere.
9. Be relaxed and help the respondent to be relaxed.
10. Listen attentively.
11. Identify facts as opposed to opinions. Both are important.
12. Accept ideas and hints.
13. Check the facts.
14. Collect source documents and forms.
15. Make effective use of open-ended and closed questions.
16. Part pleasantly.

Don't-List for Interviews:
1. Don't be late.
2. " be too formal of too casual.
3. " interrupt the speaker.
4. " use technical jargons.
5. " jump to conclusions.
6. " argue or criticize (constructive or destructive).
7. " make suggestions (not as yet; you will get your opportunity to do so later).

Figure 7.1   Basic Guidelines for Interview

7.3 Questionnaires and Surveys

A questionnaire is applicable when any of the following situations hold:

• A small amount of information is required of a large population.

- The time frame is short but a vast area (and/or dispersed population) must be covered.
- Simple answers are required to a number of standard questions.

7.3.1 Guidelines for Questionnaires

Figure 7.2 provides a set of basic guidelines for preparing a questionnaire. The questionnaire must begin with an appropriate heading and a clearly stated purpose.

```
1.  State purpose clearly          ⎫  Usually in the form of a cover note or letter
2.  Thank the participants         ⎭
3.  Must have a topic or heading which reflects an apt summary of the information sought.
4.  Should adhere to the principles of forms design.
5.  Avoid ambiguity.
6.  Decide when to use open-ended questions, closed questions or scalar questions.
7.  Order questions appropriately.
8.  State questions in a language the respondent will readily understand.
9.  Be consistent in style.
10. Ask questions of importance to the respondents first.
11. Bring up less controversial questions first.
12. Cluster related questions.
```

Figure 7.2   Guidelines for Questionnaires

7.3.2 Using Scales in Questionnaires

Scales may be used to measure the attitudes and characteristics of respondents or have respondents judge the subject matter in question. There are four forms of measurement as outlined below:

1. **Nominal Scale:** Used to classify things. A number represents a choice. One can obtain a total for each classification.
2. **Ordinal Scale:** Similar to nominal, but here the number implies ordering or ranking.
3. **Interval Scale:** Ordinal with equal intervals.
4. **Ratio Scale:** Interval scale with absolute zero.
5. **Likert Scale:** The Likert scale has emerged as the default standard for measuring, quantifying, and analyzing user responses in surveys and research initiatives that involve the use of questionnaires. Likert scales are typically designed to be symmetrical—involving an equal number of possible positive or negative responses, and a neutral point. Likert scale implementations involving five points are seven points appear to be very common. The five-point implementation often appears with responses such as:
   a. Strongly disagree
   b. Disagree
   c. Neutral
   d. Agree
   e. Strongly agree

Example 7.1   The financial status of an individual may be represented as follows:

1 = Extremely Rich; 2 = Very Rich; 3 = Rich; 4 = Not Rich; 5 = poor; 6 = very poor; 7 = Pauper

Example 7.2   Usage of a particular software product by number of modules used (10 means high):

1   2   3   10

Example 7.3   Distance traveled to obtain a system report:

0   1   2   3   ....   10 [Meters]

Example 7.4   Average response time of the system:

0   1   2   3   ....   10 [Minutes]

7.3.3 Administering the Questionnaire

Options for administering the questionnaire include the following:

- Convening respondents together at one time
- Handing out blank questionnaires and collecting completed ones
- Allowing respondents to self-administer the questionnaire at work and leave it at a centrally located place
- Mailing questionnaires with instructions, deadlines, and return postage
- Using the facilities of the World Wide Web (WWW), for example, e-mail, user forums, and/or applications such as Survey Monkey

7.4 Sampling and Experimenting

Sampling is useful when the information required is of a quantitative nature or can be quantified, no precise detail is available, and it is not likely that such details will be obtained via other methods. The findings are then used to influence decisions about software systems of interest. Figure 7.3 provides an example of a situation in which sampling is applicable.

Shipping	Number of Orders	% of Total
As Promised	186	37.2
1 day late	71	14.2
2 days late	49	9.8
3 days late	35	7.0
4 days late	38	7.6
5 days late	28	5.6
6 days late	93	18.6
	500	100.0

Figure 7.3 Examining the Delivery of Orders after Customer Complaints

Sampling is not confined to software engineering alone. Rather, the technique is practiced in most (if not all) professional disciplines. Whenever one desires to draw credible inferences about a large population, analysis is conducted on a representative sample of that population. Sampling theory describes two broad categories of samples:

- *Probability sampling* involving random selection of elements
- *Non-probability sampling* where judgment is applied in selection of elements

7.4.1 Probability Sampling Techniques

There are four types of probability sampling techniques:

- **Simple Random Sampling** uses a random method of selection of elements.
- **Systematic Random Sampling** involves selection of elements at constant intervals. Interval = **N/n** where **N** is the population size and **n** is the sample size.
- **Stratified Sampling** involves grouping of the data in strata. Random sampling is employed within each stratum.
- **Cluster Sampling:** The population is divided into (geographic) clusters. A random sample is taken from each cluster.

The latter three techniques constitute *quasi-random sampling.* The reason for this is that they are not regarded as perfectly random sampling.

7.4.2 Non-Probability Sampling Techniques

There are four types of non-probability sampling techniques:

- **Convenience Sampling:** Items are selected in the most convenient manner available.

- **Judgment Sampling:** An experienced individual selects a sample (e.g. a market research).
- **Quota Sampling:** A subgroup is selected until a limit is reached (e.g. every other employee up to 500).
- **Snowball Sampling:** An initial set of respondents is selected. They in turn select other respondents; this continues until an acceptable sample size is reached.

7.4.3 Sample Calculations

Figure 7.4 provides a summary of the formulas that are often used in performing calculations about samples: mean, standard deviation, variance, standard unit, unit error, and sample size. It is assumed that you are familiar (even if minimally) with these basic concepts.

These formulas are best explained by examining the normal distribution curve (Figure 7.5). From the curve, observe that:

- Prob $(-1 <= Z <= 1) = 68.27\%$
- Prob $(-2 <= Z <= 2) = 95.25\%$
- Prob $(-3 <= Z <= 3) = 99.73\%$

The confidence limit of a population mean is normally given by $X' +/- ZS_E$ where Z is determined by the normal distribution of the curve, considering the percentage confidence limit required. The following Z values should be memorized:

- 68% confidence => Z = 1.64
- 95% confidence => Z = 1.96
- 99% confidence => Z = 2.58

The confidence limit defines where an occurrence X may lie in the range $(X' - ZS_E)$ $\leq X \leq (X' + ZS_E)$, given a certain confidence. As you practice solving sampling problems, your confidence in using the formulas will improve.

Item	Clarification
Mean	$X' = \sum(X_i)/n$ OR $\sum(F_iX_i) / \sum (F_i)$ where n is the number of items (elements); F_i is the frequency of X_i and X_i represents the data values.
Standard Deviation	$S = \sqrt{((\sum F_i (X_i - X')^2) / n)}$ where n is the sample size
Variance	Variance $= S^2$
Standard Unit	$Z = (X_i - X') / S$
Standard Error	$S_E = S / \sqrt{(n)}$
Unit Error	$r = ZS_E = ZS / \sqrt{(n)}$
Sample Size	From the equation for unit error above, $n = (ZS / r)^2$

Figure 7.4   Formulas for Sample Calculations

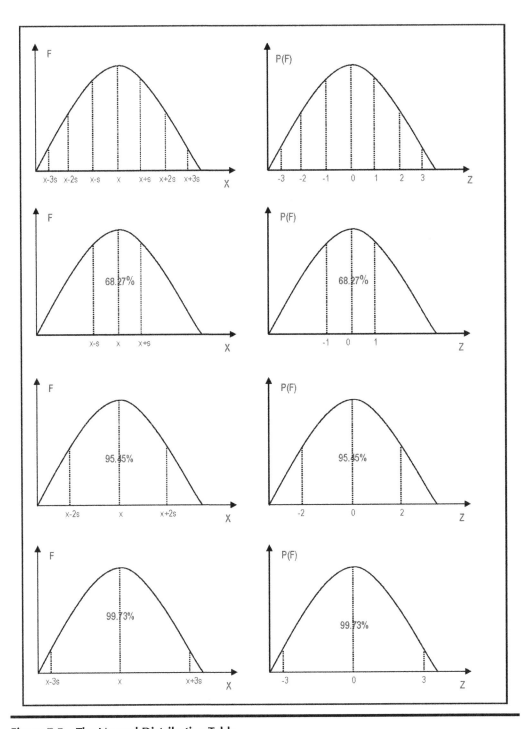

Figure 7.5   The Normal Distribution Table

7.5 Observation and Document Review

Review of source documents will provide useful information about the input, data storage, and output requirements of the software system. This activity could also provide useful information on the processing requirements of the software system. To illustrate, get a hold of an application form at your organization, and attempt to identify information entities (object types) represented on the form. With a little thought, you should be able to identify some or all of the following entities:

- Personal Information
- Family/Kin Contact Information
- Education History
- Employment History
- Professional References
- Extra-Curricular Activities

Internal documents include forms, reports, and other internal publications; external documents are mainly in the form of journals and other professional publications. These source documents are the raw materials for gaining insights into the requirements of the system, so you want to pay keen attention to them.

With respect to observation, it is always useful to check what you have been told against what you observe and obtain clarification where deviations exist. Through keen observation, the software engineer could gather useful information not obtained by other means.

7.6 Prototyping

In prototyping, the user gets a "trial model" of the software and is allowed to critique it. User responses are used as feedback information to revise the system. This process continues until a software system meeting user satisfaction is obtained (review Section 1.4).

The following are some basic guidelines for developing a prototype:

- Work in manageable modules.
- Build prototype rapidly.
- Modify prototype in successive iterations.
- Emphasize the user interface—it should be friendly and meeting user requirements.

There are various types of prototypes that you will find in software engineering. Following is a brief description of five common categories of prototypes.

Patched-up Prototype or Production Model: This prototype is functional the first time, albeit inefficiently constructed. Further enhancements can be made with time.

Nonoperational or Interactive Prototype: This is a prototype that is intended to be tested, in order to obtain user feedback about the requirements of the system represented. A good example is where screens of the proposed system are designed; the user is allowed to pass through these screens, but no actual processing is done. This approach is particularly useful in user interface design (see Chapter 14).

First of Series Prototype: This is an operational prototype. Subsequent releases are intended to have identical features, but without glitches that the users may identify. This prototype is typically used as a marketing experiment: it is distributed free of charge, or for a nominal fee; users are encouraged to use it and submit their comments about the product. These comments are then used to refine the product before subsequent release.

Selected Features Prototype or Working Model: In this prototype, not all intended features of the (represented) software are included. Subsequent releases are intended to be enhancements with additional features. For this reason, it is sometimes referred to as *evolutionary prototype*. The initial prototype is progressively refined until an acceptable system is obtained.

Throw-away Prototype: An initial model is proposed for the sole purpose of eliciting criticism. The criticisms are then used to develop a more acceptable model, and the initial prototype is abandoned.

7.7 Brainstorming and Mathematical Proof

The methodologies discussed so far all assume that there is readily available information which when analyzed, will lead to accurate capture of the requirements of the desired software. However, this is not always the case. There are many situations in which software systems are required, but there is no readily available information that would lead to the specification of the requirements of such systems. Examples include (but are not confined to) the following:

• Writing a new compiler
• Writing a new operating system
• Writing a new CASE tool, RAD tool, or DBMS
• Developing certain expert systems
• Developing a business in a problem domain for which there is no perfect frame of reference

For these kinds of scenarios, a nonstandard approach to information gathering is required. Brainstorming is particularly useful here. A close to accurate coverage of the requirements of an original software product may be obtained through brainstorming: a group of software engineering experts and prospective users come together, and through several stages of discussion, hammer out the essential requirements of the proposed software. The requirements are then documented, and through various review processes, are further refined. A prototype of the system can then be developed and subjected to further scrutiny.

Even where more conventional approaches have been employed, brainstorming is still relevant, as it forces the software engineering team to really think about the requirements identified so far, and ask tough questions to ascertain whether the requirements have been comprehensively and accurately defined.

Mathematical proofs can also be used to provide useful revelations about the required computer software. This method is particularly useful in an environment where formal methods are used for software requirements specification. This approach is often used in the synthesis of integrated circuits and chips, where there is a high demand for precision, and negligible room for error. As mentioned in Chapter 1, formal methods are not applicable to every problem domain. We will revisit the approach later in the course (in Chapter 15).

7.8 Object Identification

We have discussed six different information-gathering strategies. As mentioned in Section 7.1, these strategies are to be used to identify the core requirements of the software system. As mentioned then, one aspect that we are seeking to define is the set of information entities (object types). Notice that the term information entity is used as an alternative to object type. The two terms are not identical, but for most practical purposes, they are similar. An *information entity* is a concept, object, or thing about which data is to be stored. An *object type* is a concept, object, or thing about which data is to be stored, and upon which a set of operations is to be defined.

In object-oriented environments, the term *object type* is preferred to information entity. However, as you will more fully appreciate later, in most situations, the software system is likely to be implemented in a *hybrid environment* as an object-oriented (OO) *user interface* superimposed on a *relational database*. This is a loaded statement that will make more sense after learning more about databases (Chapter 13). For now, just accept that we can use the terms *information entity* and *object type* interchangeably in the early stages of software planning.

Early identification of information entities is critical to successful software engineering in the OO paradigm. This is so because your software will be defined in terms of objects and their interactions. For each object type, you want to be able to describe the data that it will host, and related operations that will act on that data. Approaching the software planning in this way yields a number of significant advantages; moreover, even if it turns out that the software development environment is not object-oriented, the effort is not lost (in light of the previous paragraph).

Several approaches to object identification (more precisely object type identification) have been proposed. The truth, however, is that these approaches do not offer any guarantee of successful identification of all object types required by a given system. Among the approaches that have been proposed are the following:

- Using the Descriptive Narrative Approach
- Using the Rule-of-Thumb Method
- Using Things to be Modeled
- Using Definitions of Objects, Categories, and Types
- Using Decomposition
- Using Generalizations and Subclasses
- Using OO Domain Analysis or Application Framework
- Reusing Individual Hierarchies, Objects, and Classes
- Using Personal Experience
- Using Class-Responsibility-Collaboration Card

7.8.1 The Descriptive Narrative Approach

To use the *descriptive narrative approach*, start with a descriptive overview of the software system. For larger systems consisting of multiple subsystems, prepare a descriptive narrative of each component subsystem. From each descriptive overview, identify nouns (objects) and verbs (operations). Repeatedly refine the process until all nouns and verbs are identified. Represent nouns as object types and verbs as operations, avoiding duplication of effort.

To illustrate, the *Purchase Order and Receipt Subsystem* (of an Inventory System) might have the following overview (Figure 7.6):

> *Purchase orders* are *sent* to *suppliers*, *requesting* inventory items in specific quantities. If a PO is incorrectly *generated*, it is immediately *removed* and a new PO generated. The *purchase invoice* is the official document used to recognize receipt of goods from *suppliers*. All goods received are accompanied by invoices. Once received, the invoice is *recorded*. *Items* received are also *recorded*, and appropriate *inventory adjustments made* to the *inventory item master file*. Receipt quantities can be *adjusted*, but if wrong items are recorded on receipt, or omissions are made, the whole invoice must be *removed* and re-recorded. When a receipt is correctly recorded, the associated *PO status is adjusted*.

Figure 7.6   Descriptive Narrative of Purchase Order and Invoice Receipt Subsystem

From this narrative, an initial list of object types and associated operations can be constructed, as shown below (Figure 7.7). Further refinement would be required; for instance, additional operations may be defined for each object type (left as an exercise); also, the data description can be further refined (discussed in Chapter 13).

Object Type	Data Description	Operations
Purchase Order	Stores Order Number, Order Date, Supplier, Items Ordered and related Quantity Ordered, etc.	Generate, Remove, Adjust-Status
Supplier	Stores Supplier Code, Supplier Name, Supplier Address, Contact Person, Telephone, E-mail, etc.	Sent-Invoice
Purchase Invoice	Stores Invoice Number, Invoice Date, Related Supplier, Items Shipped and related Quantity Shipped, Invoice Amount, Discount, Tax, etc.	Record, Remove, Adjust-Quantity
Inventory Item	Stores Item Code, Item Name, Item Category, Quantity on Hand, Last Purchase Price, etc.	Adjust-Inventory

Figure 7.7   Object Types and Operations for Purchase Order and Invoice Receipt Subsystem

7.8.2 The Rule-of-Thumb Approach

As an alternative to the descriptive narrative strategy, you may adopt an intuitive approach as follows: Using principles discussed earlier, identify the main information entities (object types) that will make up the system. Most information entities that make up a system will be subject to some combination of the following basic operations:

- **ADD:** Addition of data items
- **MODIFY:** Update of existing data items
- **DELETE:** Deletion of existing data items
- **INQUIRE/ANALYZE:** Inquiry and/or analysis on existing information

- **REPORT/ANALYZE:** Reporting and/or analysis of existing information
- **RETRIEVE:** Retrieval of existing data
- **FORECAST:** Predict future data based on analysis of existing data

Obviously, not all operations will apply for all object types (data entities); also, some object types (entities) may require additional operations. The software engineer makes intelligent decisions about these exceptions, depending on the situation. Additionally, the level of complexity of each operation will depend to some extent on the object type (data entity).

In a truly OO environment, the operations may be included as part of the object's services. In a hybrid environment, the information entities may be implemented as part of a relational database, and the operations would be implemented as user interface objects (windows, forms, etc.).

7.8.3 Using Things to be Modeled

This is the preferred approach of experienced software engineers. It is summarized in two steps:

1. Identify tangible objects (more precisely, object types) in the application domain that are to be modeled.
2. Put these objects into logical categories. These categories will become super-classes.

In order for this method to be successful, the software engineer must make a paradigm shift to an object-oriented mindset. For this reason, software engineers who have not made that transition, usually have difficulties employing it.

The main drawback of this approach is that on the surface, it does not help in the identification of abstract (intangible) object types. Of course, the counter argument here is that with experience, identifying abstract object types will not be a problem.

7.8.4 Using the Definitions of Objects, Categories, and Interfaces

This technique assumes that the most effective way to identify object types is to do so directly, based on the software engineer's knowledge of the application domain, as well as his/her knowledge of and experience in object abstraction and object categorization. Objects, categories (classes), and interfaces are identified intuitively.

This approach is very effective for the experienced software engineer who has made the paradigm shift of OO methodologies.

7.8.5 Using Decomposition

This approach assumes that the system will consist of several component objects, and works very well in situations where this assumption is true. The steps involved are:

1. Identify the aggregate objects or categories.
2. Repeatedly decompose aggregate objects into their components until a stable state is reached.

There are two drawbacks with this approach:

- Not all systems have an abundance of aggregate object types.
- If the approach is strictly followed, one might end up proposing component relationships where one-to-many (or many-to-many) relationships would better serve the situation.

7.8.6 Using Generalizations and Subclasses

This technique contends that objects are identified before their categories (object types). The steps involved are:

1. Identify all objects.
2. Identify objects that share the same attributes and services (operations). Generalize these objects into categories (object types).
3. Identify categories (classes) that share common resources (attributes, relationships, operations, etc.).
4. Factor out the common resources to form super-classes, then use generalization or specialization for all categories that share these common resources to form subcategories (subclasses).
5. If the only common factors are service prototypes, use the *interface* to factor out these common factors.

The main advantage of this technique is reuse—it is likely to produce a design that is very compact, due to a high degree of code reuse.

The main disadvantage of the technique is that it could be easily misused to produce a design of countless splinter classes that reuse logically unrelated resources. This could result in a system that is difficult to maintain.

7.8.7 Using OO Domain Analysis or Application Framework

This approach assumes that an OO domain analysis (OODA) and/or application framework of an application in the same problem domain was previously done, and can therefore be used. The steps involved are:

1. Analyze the existing OODA or application framework for the problem.
2. Reuse (with modifications where necessary) objects and/or categories as required.

The main advantage of this approach is that if such reusable components can be identified, system development time can be significantly reduced.

The main drawback of the approach is that it is not always applicable. Most existing systems either have incomplete OODA or no OODA at all. This should not be surprising since software engineering is a fairly youthful discipline. With time, this approach should be quite useful.

7.8.8 Reusing Hierarchies, Individual Objects, and Classes

This technique is relevant in a situation where there is a repository with reusable class hierarchies, which is available. The steps involved are:

1. Use the repository to identify classes and class hierarchies that can be reused.
2. Where necessary, adopt, and modify existing classes to be reused.
3. Introduce new classes (categories) where necessary.
4. If the classes are parameterized, supply the generic formal parameters (this is not currently supported in Java).

The advantages and disadvantages of this approach are identical to those specified in the previous subsection.

7.8.9 Using Personal Experience

This approach is likely to become more popular as software engineers become more experienced in designing software via object technology. The software engineer draws on his/her experience in previously designed systems, to design a new one. The required steps are:

1. Identify objects, categories (classes), and interfaces that correspond to ones used in previous models that are in the same application domain.
2. Utilize these resources (with modification where applicable) in the new system.
3. Introduce new categories (object types) and interfaces where necessary.

This approach could significantly reduce the development time of a new system while building the experience repertoire of the software engineer(s) involved.

Where the relevant prior experience is lacking, this approach breaks down and becomes potentially dangerous. Also, the approach could facilitate the proliferation of shoddy design with poor documentation, or no documentation at all. This could result in a system that is very difficult to maintain.

7.8.10 Using the Class-Responsibility-Collaboration Method

A popular concept in object identification is *responsibility-driven design,* a term used to mean, we identify classes and determine their responsibilities. This must be done well before the internals of the class can be tackled.

The *CRC (class-responsibility-collaborator)* methodology, proposed by Kent Beck and Ward Cunningham (see [Beck & Cunningham 1989]), defines for each class, its responsibilities, and collaborators that the class may use to achieve its objectives (hence the acronym). The responsibilities of a class are the requests it must correctly respond to; the collaborators of a class are other classes it must invoke in order to carry out its responsibilities. In carrying out its responsibilities, a class may use its own internal methods, or it may solicit help via a collaborator's method(s).

The CRC methodology is summarized in the following steps:

1. In a brainstorming session, identify object types (categories) that may be required for the system.
2. For each object type (category), construct a list of (possible) services to be provided (or operations to be performed on an instance of that object type).
3. For each object type (category), identify possible collaborators.
4. Identify missing object types (categories) and interfaces that should be added. To identify new object types, look for attributes and services that cannot be allocated to the current set of object types. To identify new interfaces, look for common service protocols.
5. Develop a CRC card for each object type and place them on a whiteboard (or desk surface). Draw association arcs from object types (classes) to collaborators.
6. Test and refine the model by utilizing *use cases*.

Figure 7.8 provides an illustration of a CRC card. The CRC card must be stored electronically as well as physically, to aid the design process. For instance, CRC cards can be easily classified, printed, and strung out on a table, during a brainstorming session, to assist designers with gaining a comprehensive overview of the system. Class names and responsibilities must be carefully worded to avoid ambiguities.

Figure 7.8 *The CRC Card*

The main advantages of the technique are:

• It is easy to learn and follow.
• When used properly, it helps designers gain a comprehensive overview of the system.

The main drawback of the technique is that it requires experienced software engineer(s) if success is to be achieved. Moreover, as you will soon see (or probably have already observed), CRC cards are no longer used because the principles have been subsumed into the UML (Unified Modeling Language) standards.

7.9 End-User Involvement

OOSE must thoroughly involve end users to ensure their satisfaction and acceptance. The software engineer should not operate in a manner that is oblivious of the end user. Rather, his/her role is to extract information needs from the end users and model it in a way that the users understand and are satisfied with. Workshops are usually effective means. Three types of workshops are:

* JEM—Joint Enterprise Modeling
* JRP—Joint Requirements Planning
* JAD—Joint Application Design

Each session is guided by a facilitator (usually a professional, skilled in OOM). Key end users do most of the talking. The facilitator does the modeling for end users to see and accept or revise. The facilitator should be well experienced in system and software design, an excellent motivator and communicator; must have a good reputation that inspires confidence; a skilled negotiator.

Martin makes a number of specific recommendations about workshops: Workshops typically last for no longer than a week at-a-time, and should not be allowed to spread out over an extended period of time. Nothing should be allowed to stall or slow down the progress of the workshop. Issues that cannot be resolved in a reasonable timeframe must be declared open (by the facilitator), and subject to subsequent analysis and/or discussion leading to satisfactory resolution (see [Martin 1993]).

In order for a user workshop to be successful, thorough preparation must take place. The software engineers on the project must do their homework in obtaining relevant information, analyzing it, and preparing working models and/or proposal that will be examined and/or used in the workshop. Additionally, training of participants may be required prior to the workshop, in order to achieve optimum benefits. The training sessions should be carefully planned and administered. Here, the participants should be briefed on the expectations and activities of the workshops.

7.10 Summary and Concluding Remarks

Here is a summary of what we have covered in this chapter:

* It is important to conduct a research on the requirements of a software system to be developed. By so doing, we determine the synergistic interrelationships, information entities, operations, business rules, and security mechanisms.
* In conducting the software requirements research, obtaining answers to questions commencing with the words WHAT, WHERE, HOW, WHEN, and WHY is very important.
* Information gathering strategies include interviews, questionnaires and surveys, sampling and experimenting, observation and document review, prototyping, brainstorming, and mathematical proofs.
* The interview is useful when the information needed must be elaborate, or clarification on various issues is required. The interview also provides an opportunity for the software engineer to win the confidence and trust of clients. In preparing to conduct an interview, the software engineer must be thoroughly prepared and must follow well-known interviewing norms.
* A questionnaire is viable when any of the following situations hold: A small amount of information is required of a large population; the time frame is short

but a vast area (and/or dispersed population) must be covered; simple answers are required to a number of standard questions. The software engineer must follow established norms in preparing and administering a questionnaire.

- Sampling is useful when the information required is of a quantitative nature or can be quantified, no precise detail is available, and it is not likely that such details will be obtained via other methods. The software engineer must be familiar with various sampling techniques and know when to use a particular technique.
- Review of source documents will provide useful information about the input, data storage, and output requirements of the software system. This activity could also provide useful information on the processing requirements of the software system.
- Prototyping involves providing a trial model of the software for user critique. User responses are used as feedback information to revise the system. This process continues until a software system meeting user satisfaction is obtained. The software engineer should be familiar with the different types of prototypes.
- Brainstorming is useful in situations in which software systems are required, but there is no readily available information that would lead to the specification of the requirements of such systems. Brainstorming involves a number of software engineers coming together to discuss and hammer out the requirements of a software system.
- Mathematical proof is particularly useful in an environment where formal methods are used for software requirements specification.
- One primary objective of these techniques is the identification and accurate specification of the information entities (or object types) comprising the software system. There are various techniques for identifying and refining object types. Among the various object identification techniques that have been proposed are the following: the descriptive narrative strategy; the rule-of-thumb strategy; using things to be modeled; using definitions of objects, categories, and types; using decomposition; using generalizations and subclasses; using OO domain analysis or application framework; reusing individual hierarchies, objects, and classes; using personal experience; using class-responsibility-collaboration cards.
- OOSE must thoroughly involve end users to ensure their satisfaction and acceptance. Three types of workshops that are common are JEM (Joint Enterprise Modeling), JRP (Joint Requirements Planning), and JAD (Joint Application Design).

Accurate and comprehensive information gathering is critical to the success of a software engineering venture. In fact, the success of the venture depends to a large extent on this. Your information gathering skills will improve with practice and experience.

In applying these techniques, the software engineer will no doubt gather much information concerning the requirements of the software system to be constructed. How will you record all this information? If you start writing narratives, you will soon wind up with huge books that not many people will bother to read. In software engineering, rather than writing voluminous narratives to document the requirements, we use unambiguous notations and diagrams (of course, you still need to write but not as much as you would without the notations and diagrams). The next chapter will discuss some of these methodologies.

7.11 Review Questions

1. Why is information gathering important in software engineering?

2. Identify seven methods of information gathering that are available to the software engineer. For each method, describe a scenario that would warrant the use of this approach, and provide some basic guidelines for its application.

3. Suppose that you were asked to develop an inventory management system (with point-of-sale facility) for a supermarket. Your system is required to track both purchase and sale of goods in the supermarket. Do the following:
   • Prepare a set of questions you would have for the purchasing manager.
   • Prepare a set of questions you would have for the sales manager.
   • Apart from interviews, what other information gathering method(s) would you use in order to accurately and comprehensively capture the requirements of your system? Explain.

4. Suppose that you are working for a software engineering firm that is interested in developing software to detect certain types of cancer, based on information fed to it. Your software will also suggest possible treatment for the cancer diagnosed. You are given a list of twelve physicians who are cancer experts; they will form part of your resource team. Answer the following questions:
   • What type of software would you seek to develop and why?
   • What methodology would you use for obtaining critical information from the cadre of physicians?
   • Construct an information-gathering instrument that you would use in this project.

5. What type of prototype would you construct for the following?
   • The cancer diagnosis system of question 4
   • A new compiler that you hope to obtain feedback on
   • A user workgroup designed to elicit useful information for the requirements of a financial management system

References and Recommended Readings

[Beck & Cunningham 1989] Beck, Kent and Ward Cunningham. 1989. "*A Laboratory for Teaching Object-Oriented Thinking.*" *OOPSLA'89 Conference Proceedings*, October 1-6, 1989, New Orleans, Louisiana. http://c2.com/doc/oopsla89/paper.html.

[Bruegge 2010] Bruegge, Bernd and Allen H. Dutoit. 2010. *Object-Oriented Software Engineering*, 3rd ed. Boston, MA: Pearson. See chapters 4 & 5.

[DeGroot 2014] DeGroot, Morris and Mark Schervish. 2014. *Probability and Statistics*, 4th ed. Boston, MA: Pearson.

[Due 2002] Due, Richard T. 2002. *Mentoring Object Technology Projects*. Saddle River, NJ: Prentice Hall. See chapters 3-6.

[Kendall 2014] Kendall, Kenneth E., and Julia E. Kendall. 2014. *Systems Analysis and Design*, 9th ed. Boston, MA: Pearson. See chapters 4 – 6.

[Lee 2002] Lee, Richard C. and William M. Tepfenhart. 2002. *Practical Object-Oriented Development with UML and Java*. Upper Saddle River, NJ: Prentice Hall. See chapter 4.

[Martin 1993] Martin, James and James Odell. 1993. *Principles of Object-Oriented Analysis and Design*. Eaglewood Cliffs, NJ: Prentice Hall. See chapters 4 and 5.

[Pfleeger 2006] Pfleeger, Shari Lawrence. 2006. *Software Engineering Theory and Practice*, 3rd ed. Upper Saddle River, NJ: Prentice Hall. See chapter 4.

[Sommerville 2016] Sommerville, Ian. 2016. *Software Engineering*, 10th ed. Boston, MA: Pearson. See chapter 4.

[Van Vliet 2008] Van Vliet, Hans. *Software Engineering*, 3rd ed. 2008. New York, NY: John Wiley & Sons. See chapter 9.

Chapter 8

Communicating Via Diagrams

In the previous chapter, we discussed information gathering. Remember, as a software engineer, you do not gather information (or anything for that matter) just for the sake of doing so. Rather, there must be a purpose, and as pointed out earlier, your objective is the preparation of the project's second major deliverable—the requirements specification. As you will soon see, this deliverable can be quite bulky, particularly if the software system is quite complex and/or large. In software engineering, we do not like unnecessary fluff; we promote comprehensive but succinct coverage.

Comprehensive coverage and brevity are often not easy to achieve since they are to some extent mutually contradictory. However, to assist in the pursuit of these sometimes-conflicting ideals, we use diagrams. This chapter provides you with a broad overview of various diagramming techniques that are used in documenting the requirements of computer software. It proceeds under the following captions:

- Introduction
- Traditional System Flow Charts
- Procedure Analysis Chart
- Innovation: Topology Charts
- Data Flow Diagrams
- Object Flow Diagram
- Other Contemporary Diagramming Techniques
- Program Flow Chart
- Summary and Concluding Remarks

8.1 Introduction

In the preparation of the requirements specification, the software engineer invariably employs various diagramming techniques in order to convey information about the software. Diagrams provide graphic representation of the flow of information, as well as the inter-relationships among system resources. In developing diagrams, the software engineer uses predefined symbols that have established meanings.

DOI: 10.1201/9780367746025-10

Among the many advantages of system diagrams are the following:

- They aid in the illustration of logical inter-relationships of various components of a system.
- They help in the development of system logic since they show from start to finish, various conditions, actions, and data storage in the system.
- They are traceable.
- They can help to identify bottlenecks and weaknesses in the system.
- In the object-oriented paradigm (particularly with OO-CASE tools), diagrams may be executable.

The diagramming techniques discussed in this chapter are drawn from the traditional function-oriented (FO) software engineering paradigm, as well as the more contemporary object-oriented (OO) paradigm. While contemporary products tend to be designed based on the OO paradigm, there are many legacy systems that still abound. An understanding of both approaches is therefore imperative for the keen software engineer. Figure 8.1 provides a list of some commonly used diagrams and the software engineering paradigm(s) (OO or FO) to which they apply. Some of these diagrams will be discussed in this chapter; others will be discussed in more appropriate sections later in the course.

Diagramming Technique	SE Paradigm
Information Oriented Flowchart (IOF)	FO
Process Oriented Flowchart (POF)	FO
HIPO Chart	FO
Information Topology Chart (ITC)	OO and FO
User Interface Topology Chart (UITC)	OO and FO
Object Flow Diagram (OFD)	OO
Fern Diagram	OO
Object Structure Diagram (OSD)	OO
Object Relationship Diagram (ORD)	OO
Entity Relationship Diagram (ERD)	FO and OO
State Transition Diagram (STD)	FO and OO
Finite State Machine (FSM)	FO and OO
Procedure Analysis Chart (PAC)	FO and OO
Data Flow Diagram (DFD)	FO
Program Flow Chart	FO
Warnier-Orr Diagram	FO and OO
Event Diagram or Activity Diagram	OO
Collaboration Diagram	OO
Unified Modeling Language (UML) Diagram	OO
Decision Table & Decision Tree	FO and OO
Gantt Chart	FO and OO
PERT Diagram	FO and OO

Figure 8.1 Commonly Used Diagramming Techniques

8.2 Traditional System Flowcharts

Traditional system flowcharts include information-oriented flow charts, HIPO charts, and process-oriented flow charts. Figure 8.2 shows symbols used for these traditional system flow charts. This figure provides a good opportunity to clarify a few terms that are commonly used in software requirements specification:

Disk or Data Storage: This represents the storage of a data entity (items) to an appropriate storage device (typically magnetic disk or optical disk). As you will learn later in the course, these entities are comprised of data elements (also called attributes). However, at this stage, we do not concern ourselves with data elements; only the data storage.

Process or Function or Operation: Some texts will distinguish between process and function. Such distinctions are frivolous at best and confusing at worst; this course makes no such distinction; rather, the terms will be used as synonyms (in the context of the FO paradigm). A process (or function) describes a set of related activities that can be conveniently summarized in the descriptive name given to the

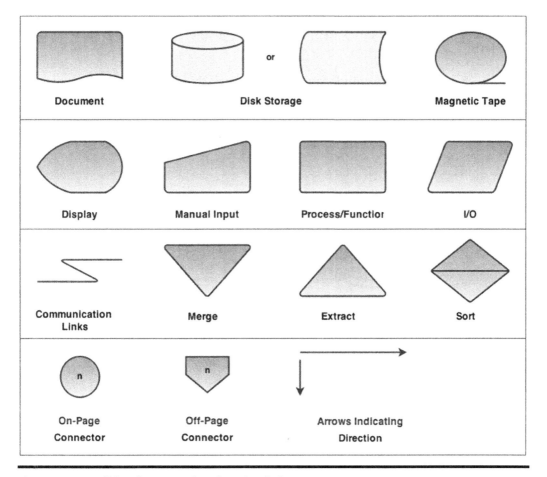

Figure 8.2   Traditional System Flowchart Symbols

process. Processes eventually translate to actual programs that will be written as part of the software system. In the OO paradigm, the term operation is preferred to function or process. Throughout this text, this is the preferred term, except when dealing with traditional (function-oriented) techniques. Finally, it must be borne in mind that operations translate to actual methods of classes, or classes with constituent methods that will be written at development time.

Subsystem: A subsystem is an independent (or almost independent) component of a software system. The subsystem may operate as part of a larger system, or on its own as an independent system.

Module: A module is a subservient component of a software system or subsystem. The module does not operate on its own; rather, it is part of a larger whole.

8.2.1 Information-Oriented Flowchart

The *information-oriented flowchart* (IOF) has the following characteristics:

- It traces the flow of information through the organization.
- It is usually grid structure.
- It uses mainly the document symbol.
- It is normally accompanied by a narrative that describes the information flow steps.

Figure 8.3 shows an example on an IOF. Notice that in this example, only the document symbol is used to show the flow of information across different departments in the organization. This is typical of IOFs.

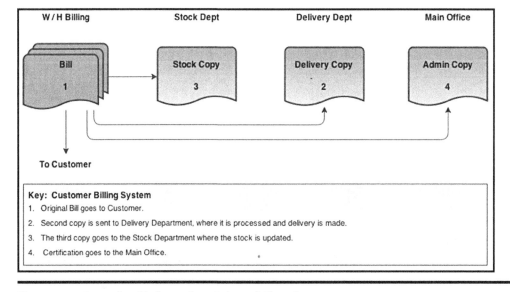

Figure 8.3   Example of Information-Oriented Flowchart

8.2.2 Process-Oriented Flow Chart

The *process-oriented flowchart* (POF) has the following characteristics:

- It traces the processing of information throughout the organization.
- It may be highly summarized or fairly detailed.
- It uses mainly the I/O symbol, the process (or operation) symbol, the document symbol, and the storage symbol.
- It is normally accompanied by a descriptive narrative.
- It is useful at the analysis, system specification, and design stages.

Figure 8.4 shows an example of a POF. Notice from the example that the POF conveys a number of important pieces of information about the software system as mentioned below:

- The inputs to the system
- The important processes/operations comprising the system
- The important data storages comprising the system (these translate to information entities in the underlying database)
- The critical outputs from the system

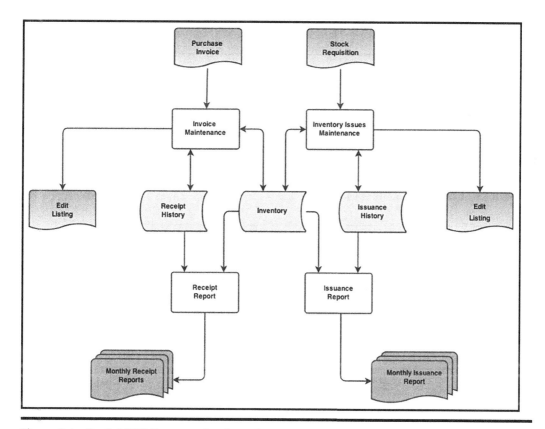

Figure 8.4   Partial POF Representing Part of an Inventory Management System

The POF, though traditional, embodies sound software engineering principles that can be applied to contemporary practice of the discipline. If you choose to use a POF in a contemporary setting, then here are two suggested tweaks:

- Instead of *process*, use the term *operation*
- Instead of *data storage*, use the term *entity*

8.2.3 Hierarchy- Input–Process-Output Chart

The *hierarchy-input-process-output* (HIPO) chart has the following characteristics:

- It presents the system and its main functional components in a hierarchical manner, so that relationships among them can be easily depicted.
- The name of the software system is at level-1; the second and/or intermediate levels contain major functional components of the system (subsystems, modules, and eventually functions/processes); for the final level, each function is broken down into component activities.
- A second IPO chart can show more details about each functional module, outline inputs, processing steps, and outputs. This is useful at the design phase.

Figures 8.5a and 8.5b illustrates a HIPO chart for the earlier mentioned inventory management system. Notice that on the HIPO chart, system functions are usually indicated at the lowest level of the hierarchy.

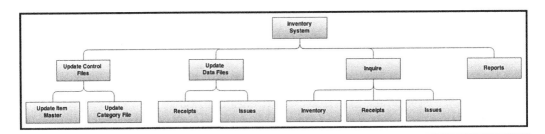

Figure 8.5a Example of a HIPO Chart

| Function: Update Item Master | | |
| Operation Description: Allows maintenance of the information stored in the Inventory Master File. | | |
Input	Processing	Output
Inventory Items, Inventory Categories	1. Accept Item Number	Edit Listing, Inventory Master File
	2. If this is a new Item Number, allow **addition** of a new inventory item	
	3. If this is a preexisting Item Number, find out if the user desires **modification** or **deletion**	
	4. If modification is chosen, allow modification of the inventory record; otherwise, allow deletion of the employee record	

Figure 8.5b Example of an IPO Chart

8.3 Procedure Analysis Chart

The *procedure analysis chart* (PAC) is used to conduct process analysis in the organization. Its main objective is to identify bottlenecks and solve them by any of the following strategies:

- Eliminating delays
- Merging processes
- Eliminating redundant processes
- Introducing new processes if necessary

The procedure analysis chart is particularly useful in process re-engineering in the organization. Figure 8.6 provides an example. From the chart, you will observe

Summary	Present		Proposed		Difference	
	No.	Time	No.	Time	No.	Time
# Activities	7	7.42				
# Transportations	4	0.47				
# Inspections	2	3.05				
# Delays	1	0.02				
# Storages	1	0.3				

Note: 1. Time given in Minutes
2. Distance given for transportation in meters (second figure in transportation column).

Action	Activity	Transportation	Inspection	Delay	Storage	Action Change (combine, delete, improve)
Select Next Sale Order	0.02					
Examine Credit Request			3.00			
Calculate Sale Amount	0.50					
Find Customers File	0.30					
Record Customers Balance	0.10					
Obtain Supervisor's Approval		0.12, 20m				
Note Unprocessed Memos	5.00					
Return to Desk		0.25, 40m				
Calculate New Customer Bal.	0.50					
Compare New Bal. to Limit			0.05			
Approve or disapprove	1.00					
Place 3 Copies of Order in Out-tray				0.02		
Take Copy to Customer File		0.05, 15m				
File Customer Information					0.30	
Return to Desk		0.05, 15m				

Figure 8.6   Procedure Analysis Chart for a Credit Check Procedure

that there are five types of actions that are studied: activity, transportation, inspection, delay, and storage. The summary provides the total time by each type of action, for that particular job function. By examining the chart, the software engineer can make recommendations as to whether improvements are needed in that particular functional area of the system. For this reason, PACs are extremely useful in business process reengineering (BPR).

Observe:
You can easily construct the equivalence of a PAC by simply using a spreadsheet. The symbols may be different, but the important thing is to have the feature that allows easy summation of columns and other calculations.

8.4 Innovation: Topology Charts

The following two topology charts have been proposed by the current writer and may be employed in both (OO and FO) paradigms of software engineering.

8.4.1 Information Topology Chart

The *information topology chart* (ITC) shows information levels of the software system in a top-down manner—the system is at the highest level; the subsystems and entities are at the intermediate level(s); and data elements, if present, appear at the lowest level. It presents information to be managed in the system in a logical and modular way and therefore allows for easy analysis and identification of omissions or redundancies.

The ITC is particularly useful in providing a global view of the system, including all significant components. It is useful for analysis and specification, as well as the design phases of the SDLC. The technique differs from the HIPO chart in the following way: The HIPO chart is a functional representation of processes in a system. On the other hand, the ITC is a conceptual representation of component information entities (object types) of the system.

The ITC also differs from other techniques such as fern diagram, object-relationship diagram, and object flow diagram. A comparison of the ITC with these methodologies is available in reference [Foster 1999]. This information has been excluded because it is not considered necessary for this course. However, it is useful to mention some benefits of the approach [Foster, O'Dea, & Dumas 2015]:

* The ITC is a useful design and documentation aid.
* The technique is easy to learn, involving minimal use of symbols.
* The technique is useful in conceptualizing (the entire) system scope.
* The technique is useful in illustrating how information (object types) will be managed.
* The technique is usefully applicable in the FO paradigm as well as OO paradigm.

The ITC is more useful in OOSE, but can also be used in the traditional approach. Figure 8.7 illustrates a portion of an ITC for a generic *College/University Administrative*

Information System (CUAIS). From this diagram, it can be seen that the CUAIS is an integrated software system consisting of nine independent but interrelated subsystems. Software systems of this sort belong to a family called *enterprise resource planning* (ERP) systems, which in turn belongs to a larger family called *management support systems* (MSSs). Each subsystem is responsible for the management of various information entities, some of which have been included in the diagram. Appendices B and C provide additional ITC illustrations.

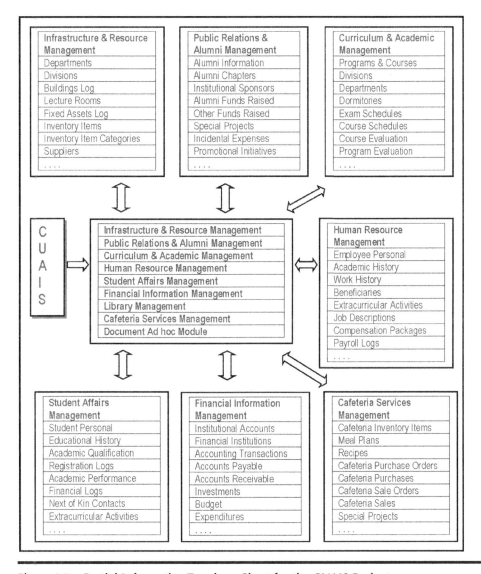

Figure 8.7   Partial Information Topology Chart for the CUAIS Project

8.4.2 User Interface Topology Chart

The *user interface topology chart* (UITC) is logically constructed from the ITC and is comparable to Schneideman's *Object-Action Interface* (OAI) model for user interfaces [Schneiderman 2017]. It shows the operational levels of the system in a top-down manner: the system is represented at the highest level; subsystems (may) appear at the intermediate levels; actual operations are represented at the lowest level. It is similar to a HIPO chart, except that it favors an OO approach to software design.

The UITC presents operations of the system in a logical manner, showing inter-relationships, and how they fit in the overall system architecture. It also presents the end user with a panoramic perspective of the entire system. Hence, as the name suggests, it is useful is portraying a blown-out static picture of the (menu driven or graphical) user interface of the system. Figure 8.8 illustrates a partial UITC for the CUAIS project. Additional illustrations of UITCs are available in Appendix C.

8.5 Data Flow Diagrams

A data flow diagram (DFD) reflects the main information storage, information flow, and processes of a software system all in one. It has relevance in the FO paradigm, as a useful analysis technique for representing the logic underlying a system. Figure 8.9 shows the symbols used. Notice the similarity between the process and the data storage symbols. In the original proposals for DFD, the process symbol is a rectangle with rounded corners, and a data storage symbol is a rectangle with one side opened. In the interest of simplicity, these intricacies have been relaxed.

The DFD provides the following advantages:

- It enhances understanding of the interrelatedness of systems and subsystems.
- It is an effective tool in communicating to users.

In drawing a DFD, the following conventions are employed:

- Indicate process number at the top of the process rectangle. Use decimals to indicate sub-processes.
- Indicate data store number at the left of the data store rectangle.
- Label entities, processes, and data stores on the inside of respective rectangles.
- Develop DFD in a top-down manner; level-0 to level-n. Explode from one level to the next in order to reveal more detail about the processes. At each level, identify external entities, processes, data flows, and data storages.

By way of illustration, let us revisit the inventory management system (IMS) introduced earlier in the chapter, and take a closer look. Consider the case where the manager of a supermarket, variety store, or auto-store desires such a system that will allow him/her to effectively keep track of products needed to keep the store in operation. In each of these cases, the system implementation will have several points of similarities with the other two scenarios, and a few points of differences. Let us for the moment concentrate on some of the similarities. This illustration does not attempt to cover all aspects of the IMS; rather it covers some of the salient features:

CUAIS Main Menu
1 Infrastructure & Resource Management System
2 Curriculum & Academic Management System
3 Financial Information Management System
4 Student Affairs Management System
5 Public Relations & Alumni Management System
6 Human Relations Management System
7 Cafeteria Services Management System
8 Library Information Management System
9 Other Ad-Hoc Services

1. Infrastructure & Resource Management System
1.1 Department Definitions
1.1.1 Add Department Definitions
1.1.2 Modify Department Definitions
1.1.3 Delete Department Definitions
1.1.4 Inquire/Report on Department Definitions

1.2 School/Division Definitions
1.2.1 Add School/Division Definitions
1.2.2 Modify School/Division Definitions
1.2.3 Delete School/Division Definitions
1.2.4 Inquire/Report on School/Division Definitions

1.3 Building Definitions
1.3.1 Add Building Definitions
1.3.2 Modify Building Definitions
1.3.3 Delete Building Definitions
1.3.4 Inquire/Report on Building Definitions

1.4 Lecture Room Definitions
1.4.1 Add Lecture Room Definitions
1.4.2 Modify Lecture Room Definitions
1.4.3 Delete Lecture Room Definitions
1.4.4 Inquire/Report on Lecture Room Definitions

1.5 Fixed Asset Logs
1.5.1 Add Fixed Assets
1.5.2 Modify Fixed Assets
1.5.3 Delete Fixed Assets
1.5.4 Inquire/Report on Fixed Assets

1.6 Inventory/Resource Items
1.6.1 Add Inventory/Resource Items
1.6.2 Modify Inventory/Resource Items
1.6.3 Delete Inventory/Resource Items
1.6.4 Inquire/Report on Inventory/Resource Items

1 Infrastructure & Resource Management System
1.7 Supplier Definitions
1.7.1 Add Supplier Definitions
1.7.2 Modify Supplier Definitions
1.7.3 Delete Supplier Definitions
1.7.4 Inquire/Report on Supplier Definitions

1.8 Purchase Order Summaries
1.8.1 Add Purchase Order Summaries
1.8.2 Modify Purchase Order Summaries
1.8.3 Delete Purchase Orders
1.8.4 Inquire/Report on Purchase Order Summaries

1.9 Purchase Order Details
1.9.1 Add Purchase Order Details
1.9.2 Modify Purchase Order Details
1.9.3 Delete Purchase Order Details
1.9.4 Inquire/Report on Purchase Order Details

1.10 Purchase Invoice Summaries
1.10.1 Add Purchase Invoice Summaries
1.10.2 Modify Purchase Invoice Summaries
1.10.3 Delete Purchase Invoices
1.10.4 Inquire/Report on Purchase Invoice Summaries

1.11 Purchase Invoice Details
1.11.1 Add Purchase Invoice Details
1.11.2 Modify Purchase Invoice Details
1.11.3 Delete Purchase Invoice Details
1.11.4 Inquire/Report on Purchase Invoice Details

1.12 Purchase Return Summaries
1.12.1 Add Purchase Return Summaries
1.12.2 Delete Purchase Returns
1.12.3 Inquire/Report or Purchase Return Summaries

1.13 Purchase Return Details
1.13.1 Add Purchase Return Details
1.13.2 Modify Purchase Return Details
1.13.3 Delete Purchase Return Details
1.13.4 Inquire/Report on Purchase Return Details

1.14 Supplier-Resource Mappings
1.14.1 Add Supplier-Resource Mappings
1.14.3 Delete Supplier-Resource Mappings
1.14.4 Inquire/Report on Supplier-Resource Mappings

2 Curriculum & Academic Management System
2.1 Academic Program Definitions
2.1.1 Add Academic Program Definitions
2.1.2 Modify Academic Program Definitions
2.1.3 Delete Academic Program Definitions
2.1.4 Inquire/Report on Academic Program Definitions

2.2 Course Definitions
2.2.1 Add Course Definitions
2.2.2 Modify Course Definitions
2.2.3 Delete Course Definitions
2.2.4 Inquire/Report on Course Definitions

2.3 Academic Department Definitions
2.3.1 Add Academic Department Definitions
2.3.2 Modify Academic Department Definitions
2.3.3 Delete Academic Department Definitions
2.3.4 Inquire/Report on Academic Department Definitions

2.4 Dormitory Definitions
2.4.1 Add Dormitory Definitions
2.4.2 Modify Dormitory Definitions
2.4.3 Delete Dormitory Definitions
2.4.4 Inquire/Report on Dormitory Definitions

2.5 Course Schedules
2.5.1 Add Course Schedules
2.5.2 Modify Course Schedules
2.5.3 Delete Course Schedules
2.5.4 Inquire/Report on Course Schedules

2.6 Examination Schedules
2.6.1 Add Examination Schedules
2.6.2 Modify Examination Schedules
2.6.3 Delete Examination Schedules
2.6.4 Inquire/Report on Examination Schedules

2.7 Course Evaluations
2.7.1 Add Course Evaluations
2.7.2 Modify Course Evaluations
2.7.3 Delete Course Evaluations
2.7.4 Inquire/Report on Course Evaluations

2.8 Academic Program Evaluations
2.8.1 Add Academic Program
2.8.2 Modify Academic Program
2.8.3 Delete Academic Program
2.8.4 Inquire/Report on Academic Program

3 Financial Information Management System
3.1 Chart of Accounts
3.1.1 Add Account Definitions
3.1.2 Modify Account Definitions
3.1.3 Delete Account Definitions
3.1.4 Inquire/Report on Account Definitions

3.2 Financial Institutions
3.2.1 Add Financial Institution Definitions
3.2.2 Modify Financial Institution Definitions
3.2.3 Delete Financial Institution Definitions
3.2.4 Inquire/Report on Financial Institution Definitions

3.3 Financial Transactions
3.3.1 Add Financial Transactions
3.3.2 Modify Financial Transactions
3.3.3 Delete Financial Transactions
3.3.4 Inquire/Report on Financial Transactions

3.4 Purchase Orders — Summaries & Details
3.4.1 Add Purchase Orders
3.4.2 Modify Purchase Orders
3.4.3 Delete Purchase Orders
3.4.4 Inquire/Report on Purchase Orders

3.5 Purchase Invoices — Summaries & Details
3.5.1 Add Purchase Invoices
3.5.2 Modify Purchase Invoices
3.5.3 Delete Purchase Invoices
3.5.4 Inquire/Report on Purchase Invoices

3.6 Sale Orders — Summaries & Details
3.6.1 Add Sale Orders
3.6.2 Modify Sale Orders
3.6.3 Delete Sale Orders
3.6.4 Inquire/Report on Sale Orders

3.7 Sale Invoices — Summaries & Details
3.7.1 Add Sale Invoices
3.7.2 Modify Sale Invoices
3.7.3 Delete Sale Invoices
3.7.4 Inquire/Report on Sale Invoices

3.8 Investments
3.8.1 Add Investment Entries
3.8.2 Modify Investment Entries
3.8.3 Delete Investment Entries
3.8.4 Inquire/Report on Investments

(Continued)

Figure 8.8 Partial User Interface Topology Chart for the CUAIS Project

4. Student Affairs Management System

4.1 Student Personal Records
4.1.1 Add Student Personal Records
4.1.2 Modify Student Personal Records
4.1.3 Delete Student Personal Records
4.1.4 Inquire/Report on Student Personal Records

4.2 Student Educational History
4.2.1 Add Student Educational History
4.2.2 Modify Student Educational History
4.2.3 Delete Student Educational History
4.2.4 Inquire/Report on Student Educational History

4.3 Student Academic Qualification
4.3.1 Add Student Academic Qualification
4.3.2 Modify Student Academic Qualification
4.3.3 Delete Student Academic Qualification
4.3.4 Inquire/Report on Student Academic Qualification

4.4 Student Next of Kin Contacts
4.4.1 Add Student Next of Kin Contacts
4.4.2 Modify Student Next of Kin Contacts
4.4.3 Delete Student Next of Kin Contacts
4.4.4 Inquire/Report on Student Next of Kin Contacts

4.5 Student Extracurricular Activities
4.5.1 Add Student Extracurricular Activities
4.5.2 Modify Student Extracurricular Activities
4.5.3 Delete Student Extracurricular Activities
4.5.4 Inquire/Report on Student Extracurricular Activities

4.6 Student Registration Logs
4.6.1 Add Student Registration Logs
4.6.2 Modify Student Registration Logs
4.6.3 Delete Student Registration Logs
4.6.4 Inquire/Report on Student Registration Logs

4.7 Student Academic Performance Logs
4.7.1 Add Student Academic Performance Logs
4.7.2 Modify Student Academic Performance Logs
4.7.3 Delete Student Academic Performance Logs
4.7.4 Inquire/Report on Student Academic Performance

4.8 Student Financial Logs
4.8.1 Add Student Financial Logs
4.8.2 Modify Student Financial Logs
4.8.3 Delete Student Financial Logs
4.8.4 Inquire/Report on Student Financial Logs

5. Public Relations & Alumni Management System

5.1 Alumni Information
5.1.1 Add Alumni Information
5.1.2 Modify Alumni Information
5.1.3 Delete Alumni Information
5.1.4 Inquire/Report on Alumni Information

5.2 Alumni Chapters
5.2.1 Add Alumni Chapter Definitions
5.2.2 Modify Alumni Chapter Definitions
5.2.3 Delete Alumni Chapter Definitions
5.2.4 Inquire/Report on Alumni Chapters

5.3 Institutional Sponsors
5.3.1 Add Institutional Sponsors
5.3.2 Modify Institutional Sponsors
5.3.3 Delete Institutional Sponsors
5.3.4 Inquire/Report on Institutional Sponsors

5.4 Alumni Funds Raised
5.4.1 Add Alumni Funds Raised
5.4.2 Modify Alumni Funds Raised
5.4.3 Delete Alumni Funds Raised
5.4.4 Inquire/Report on Alumni Funds Raised

5.5 Other Funds Raised
5.5.1 Add Other Funds Raised
5.5.2 Modify Other Funds Raised
5.5.3 Delete Other Funds Raised
5.5.4 Inquire/Report on Other Funds Raised

5.6 Special Project Definitions
5.6.1 Add Special Project Definitions
5.6.2 Modify Special Project Definitions
5.6.3 Delete Special Project Definitions
5.6.4 Inquire/Report on Special Project Definitions

5.7 Special Project Details
5.7.1 Add Special Project Details
5.7.2 Modify Special Project Details
5.7.3 Delete Special Project Details
5.7.4 Inquire/Report on Special Project Details

5.8 Promotional Initiatives/Activities
5.8.1 Add Promotional Initiatives
5.8.2 Modify Promotional Initiatives
5.8.3 Delete Promotional Initiatives
5.8.4 Inquire/Report on Promotional Initiatives

6. Human Resource Management System

6.1 Employee Personal Records
6.1.1 Add Employee Personal Records
6.1.2 Modify Employee Personal Records
6.1.3 Delete Employee Personal Records
6.1.4 Inquire/Report on Employee Personal Records

6.2 Employee Academic History
6.2.1 Add Employee Academic History
6.2.2 Modify Employee Academic History
6.2.3 Delete Employee Academic History
6.2.4 Inquire/Report on Employee Academic History

6.3 Employee Work History
6.3.1 Add Employee Work History
6.3.2 Modify Employee Work History
6.3.3 Delete Employee Work History
6.3.4 Inquire/Report on Work History

6.4 Employee Beneficiaries
6.4.1 Add Employee Beneficiary Information
6.4.2 Modify Employee Beneficiary Information
6.4.3 Delete Employee Beneficiary Information
6.4.4 Inquire/Report on Employee Beneficiaries

6.5 Employee Extracurricular Activities
6.5.1 Add Student Extracurricular Activities
6.5.2 Modify Student Extracurricular Activities
6.5.3 Delete Student Extracurricular Activities
6.5.4 Inquire/Report on Employee Extracurricular Activities

6.6 Employee Job Definitions
6.6.1 Add Employee Job Definitions
6.6.2 Modify Employee Job Definitions
6.6.3 Delete Employee Job Definitions
6.6.4 Inquire/Report on Employee Job Definitions

6.7 Employee Compensation Packages
6.7.1 Add Employee Compensation Packages
6.7.2 Modify Employee Compensation Packages
6.7.3 Delete Employee Compensation Packages
6.7.4 Inquire/Report on Employee Compensation Packages

6.8 Employee Payroll Logs
6.8.1 Add Employee Payroll Logs
6.8.2 Modify Employee Payroll Logs
6.8.3 Delete Employee Payroll Logs
6.8.4 Inquire/Report on Employee Payroll Logs

7. Cafeteria Services Management System

7.1 Cafeteria Inventory Items
7.1.1 Add Cafeteria Inventory Items
7.1.2 Modify Cafeteria Inventory Items
7.1.3 Delete Cafeteria Inventory Items
7.1.4 Inquire/Report on Cafeteria Inventory Items

7.2 Cafeteria Meal Plans
7.2.1 Add Cafeteria Meal Plans
7.2.2 Modify Cafeteria Meal Plans
7.2.3 Delete Cafeteria Meal Plans
7.2.4 Inquire/Report on Cafeteria Meal Plans

7.3 Cafeteria Recipes
7.3.1 Add Cafeteria Recipes
7.3.2 Modify Cafeteria Recipes
7.3.3 Delete Cafeteria Recipes
7.3.4 Inquire/Report on Cafeteria Recipes

7.4 Cafeteria Purchase Order Summaries
7.4.1 Add Purchase Order Summaries
7.4.2 Modify Purchase Order Summaries
7.4.3 Delete Purchase Order Summaries
7.4.4 Inquire/Report on Purchase Order Summaries

7.5 Cafeteria Purchase Order Details
7.5.1 Add Purchase Order Details
7.5.2 Modify Purchase Order Details
7.5.3 Delete Purchase Order Details
7.5.4 Inquire/Report on Purchase Order Details

7.6 Cafeteria Sale Order Summaries
7.6.1 Add Sale Order Summaries
7.6.2 Modify Sale Order Summaries
7.6.3 Delete Sale Order Summaries
7.6.4 Inquire/Report on Sale Order Summaries

7.7 Cafeteria Sale Order Details
7.7.1 Add Sale Order Details
7.7.2 Modify Sale Order Details
7.7.3 Delete Sale Order Details
7.7.4 Inquire/Report on Sale Order Details

7.8 Special Catering Projects
7.8.1 Add Special Projects
7.8.2 Modify Special Projects
7.8.3 Delete Special Projects
7.8.4 Inquire/Report on Special Projects

Figure 8.8 (Continued)

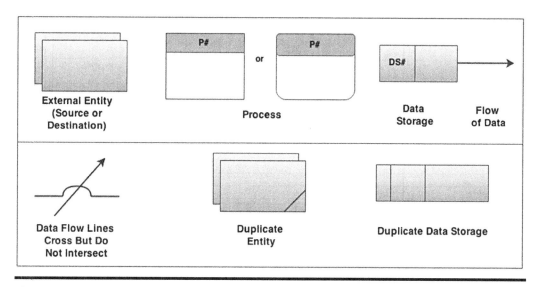

Figure 8.9   Symbols Used in DFD

- Two obvious external entities that are crucial to the implementation of the software system are suppliers of inventory items and the customers who purchase from the business.
- Figure 8.10 provides a list of some important operations/processes and data entities (storage objects) that would comprise the IMS (this is not a comprehensive list). As you view this, remember that the storage objects typically translate to information entities in the underlying database.

Operation/Process	Description
P01: Prepare Purchase Orders	Facilitates the generation of purchase orders that are sent to suppliers.
P02: Modify Purchase Orders	Facilitates modification of purchase orders before sending them off to suppliers.
P03: Remove Purchase Orders	If a purchase order was incorrectly generated, this process facilitates its removal.
P04: Query Purchase Orders	Facilitates the querying of one or more purchases orders.
P05: Receive Purchase Invoices	Facilitates the recording of purchase invoices received from suppliers.
P06: Modify Purchase Invoices	Facilitates modification of purchase invoices where errors might have been made.
P07: Remove Purchase Invoices	Facilitates the removal of incorrectly recorded purchase invoices.
P08: Query Purchase Invoices	Facilitates the querying of one or more purchases invoices.
P09: Pay Creditors	Facilitates recording the payment of funds to suppliers and other creditors (for instance the bank, etc.).
P10: Add New Inventory Items	Facilitates the addition of new inventory items to the system.
P11: Modify Inventory Items	Facilitates modification of inventory item(s).
P12: Remove Inventory Items	Facilitates the removal of incorrectly entered inventory items.
P13: Query Inventory	Facilitates querying one or more inventory items.
P14: Generate Sale Invoices	Facilitates modification of sale invoices before they are sent to customers
P15: Modify Sale Invoices	Facilitates the generation of receipts
P16: Remove Sale Invoices	Facilitates the removal of sale invoices that were incorrectly generated.
P17: Query Sales	Facilitates querying one or more sale activities.
P18: Receive Payments	Facilitates the recording of payments received for goods sold (on credit).
P19: Perform Cash Sales	Facilitates the processing of cash sales.

Figure 8.10a   Important Operations in an Inventory Management System

Data Entity/Storage	Description
D01: Inventory Items	Stores a record of each inventory item; includes category, on hand quantity, purchase price, sale price, and other related data.
D02: Purchase Orders	Stores records of each purchase order generated; includes order number, supplier sent to, order date, items ordered and related quantities, etc.
D03: Purchase Invoices	Stores records of all purchases made; includes invoice date, invoice number, related supplier and purchase order, items received and related quantities and amounts, etc.
D04: Accounts Payable	Stores information on amounts owed to creditors.
D05: Payments Made	Stores information on payments made to creditors.
D06: Sale Invoices	Stores records of all sales made; similar to D03.
D07: Accounts Receivable	Stores information on amounts owed to the store by customers.
D08: Payments Received	Stores information on amounts paid to the store; includes cash sales.
D09: Customers	Stores information on customers of the store
D10: Suppliers	Stores information on suppliers of the store.

Figure 8.10b   Important Data Entities in an Inventory Management System

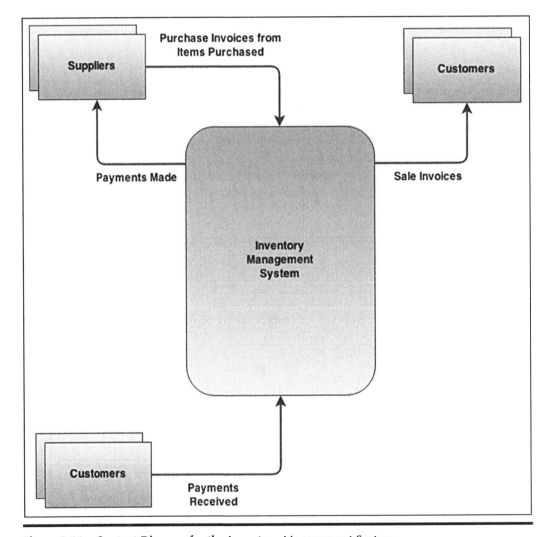

Figure 8.11   Context Diagram for the Inventory Management System

Figure 8.11 shows a level 0 DFD called a *context diagram*. Notice that at this level, the software system is perceived as a large process. Figure 8.12 shows a (partial) level-1 DFD of the system, highlighting processes P01, P05, P09, P11, P14, P18, and P19 (see Figures 8.10a and 8.10b). These were chosen because they represent the most critical processes in the system. You would then have a choice as to how to continue refining your requirements: for each of the processes specified in your level-1 diagram (Figure 8.12), you could specify a level-2 DFD as illustrated in Figure 8.13, a program flowchart (see Section 8.8), an IPO chart (review Section 8.2.3), a Warnier-Orr diagram (discussed in Chapter 15), an activity diagram (discussed in Chapter 12), or an extended operation specification (discussed in Chapter 15).

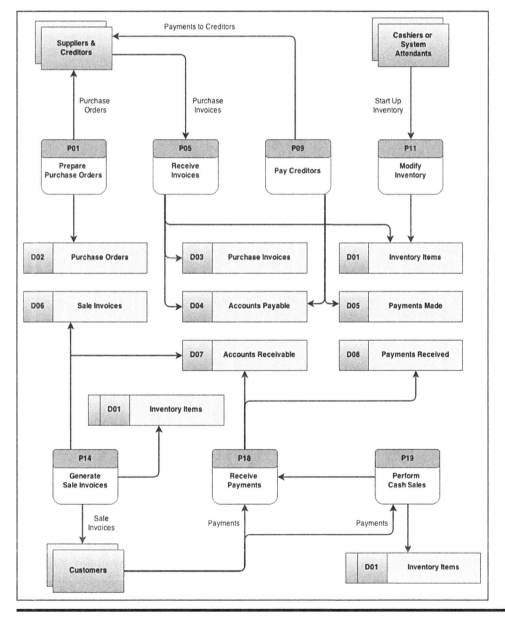

Figure 8.12   Level-1 DFD for the Inventory Management System

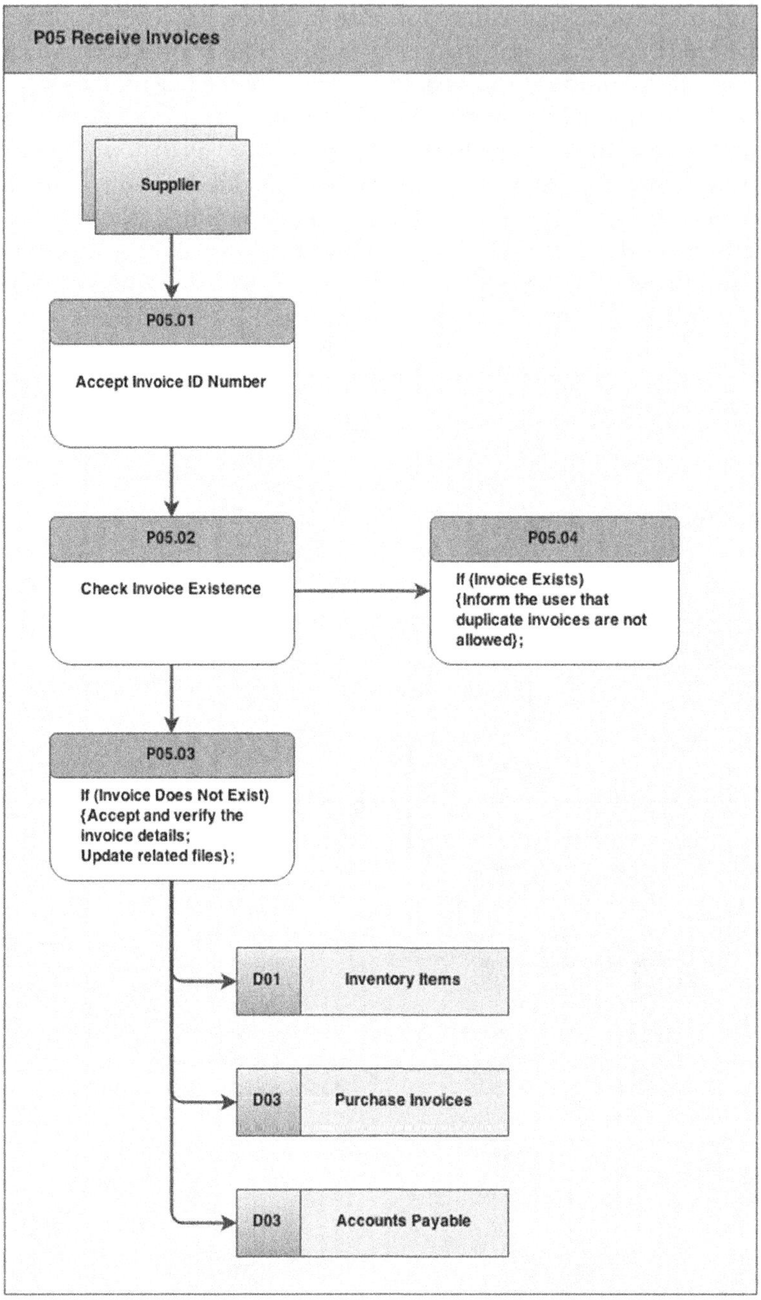

Figure 8.13   Level-2 DFD for Process P05 of the Inventory Management System

8.6 Object Flow Diagram

An *object flow diagram* (OFD) is typically used in OOSE to show how major software system components (superclasses or subsystems) communicate with one another. It is normally used at a high level of system specification. Here, it has similarities to a DFD level-0 and an ITC at the highest level. Figure 8.14 illustrates an OFD for a financial management system. From this diagram, it is apparent that the central subsystem relates to financial transactions, and all other subsystems are related to this. For additional illustrations, see Figures 8.7 and 8.11 also.

While the OFD is not mandatory, it is very useful in providing an overview of the architecture of the software system. The technique is normally used in an OO environment and may be accompanied by an ITC. In a more traditional (FO) environment, a DFD would have been used.

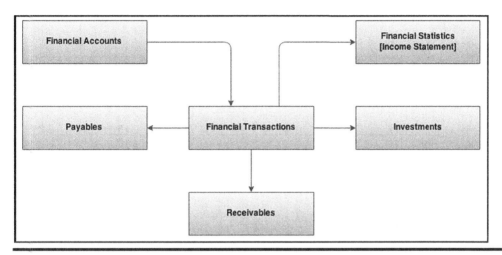

Figure 8.14   OFD for a Financial Management System

8.7 Other Contemporary Diagramming Techniques

Still, other diagramming techniques abound:

- The *entity-relationship diagram* (ERD) will be introduced in Chapter 13, but a full treatment is left for a course in database systems. The corresponding *object-relationship diagram* (ORD) of the OO paradigm will also be introduced in Chapter 13.
- The *fern diagram, event diagram,* and other OO techniques will be discussed in the upcoming chapter.
- The UML (*Unified Modeling Language*) notation includes *object structure diagram* (OSD), *use-case diagram*, ORD, *activity diagram, collaboration diagram*, among other techniques. These will be introduced in the upcoming chapter and reinforced in subsequent chapters.
- The *Warnier-Orr diagram* will be discussed in Chapter 15.

The remainder of this section will briefly introduce the *state transition diagram* and the *finite state machine* while leaving the excessive details for more advanced courses.

8.7.1 State Transition Diagram

A state transition diagram is used, primarily in OOSE, to represent the allowable state transitions associated with an object. Figure 8.15 provides an illustration of a state transition diagram for an **Employee** object type. All instances of this object type will be subject to the transitions represented in the diagram.

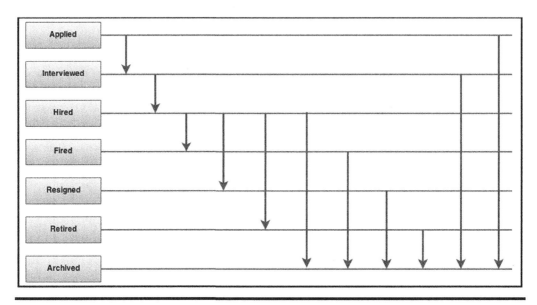

Figure 8.15   State Transition Diagram of Employee Object

8.7.2 Finite State Machine

An alternative to the state transition diagram is the finite state machine (FSM), also called the state diagram in some texts. Though existent before object technology (for instance in compiler design), FSMs find very useful application in here.

In an FSM (also referred to as state diagram) nodes are states; arcs are transitions labeled by event names (the label on a transition arc is the event name causing the transition). The state-name is written inside the node. Figure 8.16 illustrates an FSM for the state transition diagram of Figure 8.15.

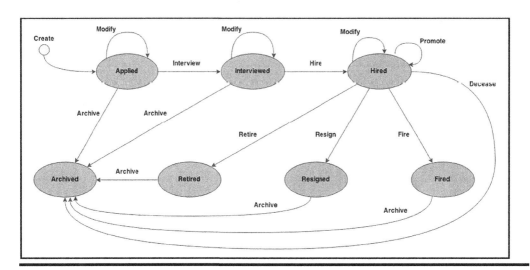

Figure 8.16   FSM for Employee

As you can see, a significant advantage of the FSM over the state transition diagram is that you can show operations that cause change of state as well as those that do not cause change of state.

8.8 Program Flowchart

The program flowchart is a traditional technique of the FO paradigm that represents functional logic. Its use in contemporary software design has been overtaken by other techniques such as algorithms (in pseudo-code), Warnier-Orr diagrams, and activity diagrams. It is assumed that from your earlier courses, you have gained mastery of algorithm development; as for Warnier-Orr diagrams, and activity diagrams, these will be covered later in the course (Chapter 15). The symbols used for program flowcharting are shown in Figure 8.17. The main structures (simple sequence, selection, and iteration) are also illustrated in Figure 8.18.

The main advantages of programming flow chart are easy debugging, economy in writing and easy tractability. The main flaw is its limitation to non-parallel logic.

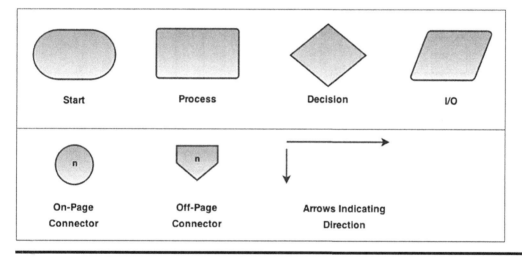

Figure 8.17   Symbols Used in Program Flow Chart

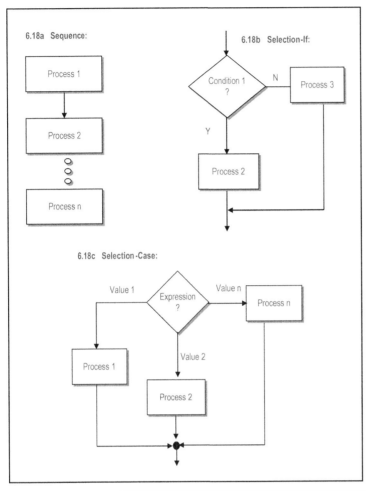

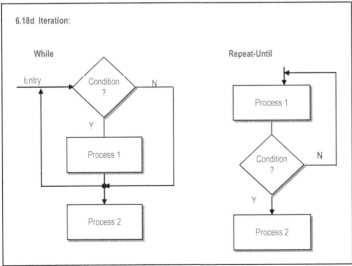

Figure 8.18 Control Structures for Program Flow Chart

8.9 Summary and Concluding Remarks

It is time once again to summarize what we have covered in this chapter:

- The software engineer relies on diagrams to document and communicate information concerning the requirements of a software system.
- Traditional system flow charts include information-oriented flow charts, process-oriented flow charts, HIPO charts, and data flow diagrams (DFDs).
- Traditional logic charts include decision tables, decision trees, and program flow charts.
- The procedure analysis chart (PAC) is useful in both FO and OO paradigms, particularly for business process reengineering (BPR).
- OO system diagrams include object flow diagrams (OFDs), O-R diagrams, state transition diagrams, finite state machine (FSMs), information topology charts (ITCs), and user interface topology charts (UITCs), fern diagrams, object structure diagrams, and UML diagrams.
- OO logic diagrams include event diagrams and activity diagrams.
- Diagrams that apply to both FO and OO paradigms include Warnier-Orr diagrams, ERDs, ORDs, finite state machine (FSMs), information topology charts (ITCs), and user interface topology charts (UITCs), procedure analysis charts (PACs), decision tables, and decision trees.

There is only one way to gain mastery of these techniques—by applying them to problems. Take the time to review Appendix B and Appendix C once more; you will see some applications of some of the techniques. The next chapter will continue the discussion of some techniques not covered in this chapter.

8.10 Review Questions

1. What are the advantages of using diagrams?

2. What information is conveyed by the following diagrams?
   - Information-oriented flowchart
   - Process-oriented flowchart
   - Hierarchy input process output chart
   - Procedure analysis chart
   - Information topology chart
   - User interface topology chart
   - Data flow diagram
   - Object flow diagram
   - State transition diagram
   - Finite state machine
   - Program flowchart

3. Compare the following diagrams:
   - Information topology chart versus hierarchy input process output chart

- Process-oriented flow chart versus data flow diagram
- State transition diagram versus finite state machine

4. A supermarket wishes to computerize its operation. The main issues to address are as follows:
   - Purchases from external vendors are represented by invoices
   - Purchases affect the Purchase Log (file), as well as Accounts Payable (file) and the Inventory (file)
   - Inventory Management is affected by purchases and sales
   - Every inventory item belongs to a category
   - Sale of goods to customers (credit sales) as well as cash sales affects the Sales Log (file), the Accounts Receivable, as well as the Inventory

   From the information given, propose the following:
   a. A POF or DFD of the system
   b. An ITC and UITC

5. Suppose that a Library Management System (LMS) is being constructed. The software engineer is focusing on the information entity called **Book**. It was discovered that a **Book** object could be in any of the following states: **ordered**, **invoiced**, **loaned**, **shelved**, **archived**, **lost**, **stolen**.
   - Propose a finite state machine and state transition diagram for the Book object type.
   - Compare the information conveyed by both diagrams.

References and Recommended Readings

[Foster 1999] Foster, Elvis C. 1999. *Labour Market Information System: Thesis*. Mona, Jamaica: Department of Mathematics and Computer Science, University of the West Indies. See section 4.4.1.

[Foster, O'Dea, & Dumas 2015] Foster, Elvis C., Thomas Dea & Myles Dumas. 2015. "Three Innovative Software Engineering Methodologies." *Global Online Conference on Information and Computer Technology*. Sullivan University. Available at http://www.elcfos.com/papers-in-cs.

[Kendall 2014] Kendall, Kenneth E., and Julia E. Kendall. 2014. *Systems Analysis and Design*, 9th ed. Boston, MA: Pearson. See chapter 7.

[Pfleeger 2006] Pfleeger, Shari Lawrence. 2006. *Software Engineering Theory and Practice*, 3rd ed. Upper Saddle River, NJ: Prentice Hall. Chapter 4.

[Schach 2011] Schach, Stephen R. 2011. *Object-Oriented & Classical Software Engineering*, 8th ed. New York: McGraw-Hill. See chapters 11 & 12.

[Schneiderman 2017] Schneideman, Ben, et al. 2017. *Designing the User Interface*, 6th ed. Boston, MA: Pearson.

Chapter 9

More Diagramming

This chapter continues the discussion of diagramming techniques. The focus here is on OOSE techniques for categorizing objects and specifying object behavior. As we work toward constructing a software system via the object-oriented paradigm, identifying the object types and their interrelationships becomes very crucial.

The chapter includes:

- Introduction
- Fern Diagram
- Unified Modeling Language—a Cursory Introduction
- Object Relationship Diagrams—a Cursory Introduction
- Representing Details about Object Types
- Avoiding Multiple Inheritance Relationships
- Top-Down versus Bottom-Up
- Use-cases
- Event Diagrams
- Triggers
- Activity Diagrams
- Sequence Diagrams and Collaboration Diagrams
- Summary and Concluding Remarks

9.1 Introduction

OO modeling has two aspects, which may be examined separately—object structure and object behavior.

Martin's OO modeling pyramid (see [Martin 1993]) identifies four levels of *information engineering* activities in the organization: enterprise modeling, business area analysis, system design, and system construction; it also shows how *object structure analysis* (OSA) and *object behavior analysis* (OBA) are related to these activities. Figure 9.1 illustrates an OO modeling hierarchy based on Martin's OO modeling pyramid:

DOI: 10.1201/9780367746025-11

- OSA relates to the structure of object types and classes comprising the software system; it utilizes diagrams that convey object types, object associates, compositions, and generalization. Correspondingly, class structure design includes class identification, inheritance, and data structure.
- OBA concerns itself with the behavior of interacting objects in the software system; it utilizes object flow diagrams, state diagrams, event diagrams, and operation specifications. Correspondingly, method design includes operation identification and method creation.

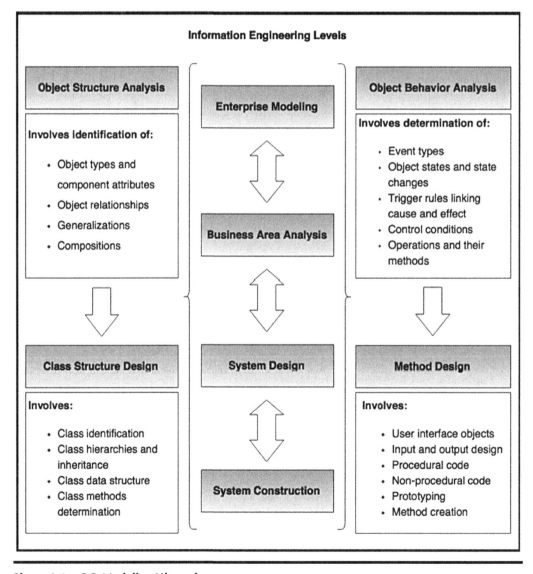

Figure 9.1   OO Modeling Hierarchy

Against this background, Figure 9.2 provides an overview of selected OO diagramming techniques that may be employed. Some of the techniques included in this list have already been discussed in the previous two chapters. This chapter continues the discussion of diagramming techniques.

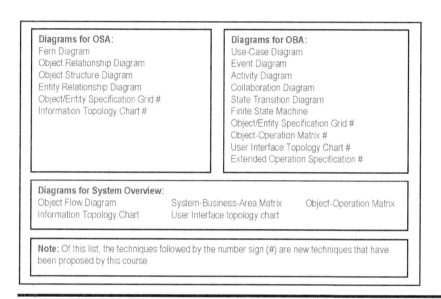

Diagrams for OSA:
Fern Diagram
Object Relationship Diagram
Object Structure Diagram
Entity Relationship Diagram
Object/Entity Specification Grid #
Information Topology Chart #

Diagrams for OBA:
Use-Case Diagram
Event Diagram
Activity Diagram
Collaboration Diagram
State Transition Diagram
Finite State Machine
Object/Entity Specification Grid #
Object-Operation Matrix #
User Interface Topology Chart #
Extended Operation Specification #

Diagrams for System Overview:
Object Flow Diagram   System-Business-Area Matrix   Object-Operation Matrix
Information Topology Chart   User Interface topology chart

Note: Of this list, the techniques followed by the number sign (#) are new techniques that have been proposed by this course.

Figure 9.2   Recommended OOM Diagramming Techniques

9.2 The Unified Modeling Language—A Cursory Introduction

The Unified Modeling Language (UML) was developed by three leading prodigies in the area of object technology—James Rumbaugh, Grady Booch, and Ivar Jacobson of Rational Software (now a division of IBM). This language was developed as an OO modeling language and is widely used in actual software development as well as research.

UML defines standards for object structure as well as object behavior. As such, we use UML standards for constructing object structure diagram (OSD), object-relationship diagram (ORD), use-case diagram (USD), activity diagram, state transition diagram (STD), and collaboration diagram. Moreover, UML standards are included in many of the software modeling/design tools that are available in the marketplace (review Section 2.4 of Chapter 2). You will be further exposed to this language in Chapters 13 and 15 (see the recommended readings also).

9.3 Object-Relationship Diagrams—A Cursory Introduction

An object-relationship diagram (ORD) depicts the relationship among object types in a software system. The possible types of relationships that are covered are:

- One-to-one (1:1) relationship
- One-to-many (1:M) relationship

- Many-to-one (M:1) relationship
- Many-to-many (M:M) relationship
- Component relationship (if OO database)
- Aggregation relationship (if OO database)
- Super-type-sub-type relationship (if OO database)

An object-relationship diagram (ORD) is similar to an entity-relationship diagram (ERD). The conventions used in both techniques are also similar for the most part. However, there are a few subtle differences relating to how relationships are represented versus the representation of object types. For these reasons, a full discussion of ORDs is deferred for Chapter 13 (database design).

When diagramming ORDs, the UML conventions are normally used. Figure 9.3 shows the main symbols used in an ORD. To entice your appetite, Figure 9.4 shows the ORD for an Inventory Management System (IMS) which will be further clarified in Appendices B and C. We will revisit this example in Chapter 13.

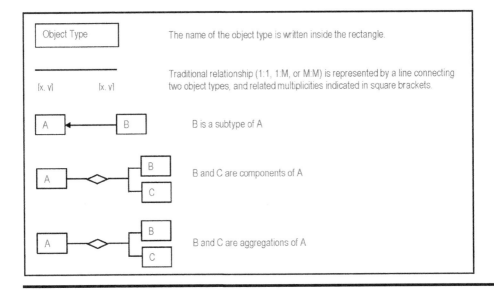

Figure 9.3   Symbols used in Constructing an ORD

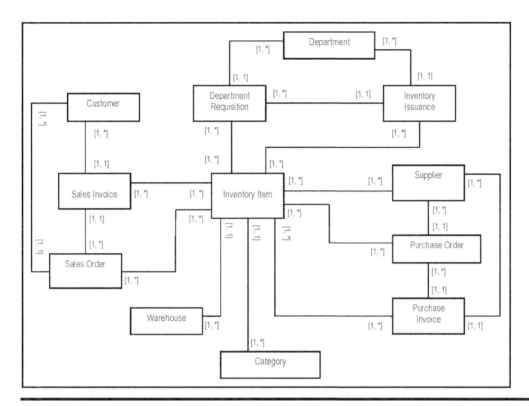

Figure 9.4   ORD for an Inventory Management System

9.4 Fern Diagram

Fern diagrams are useful in depicting the object types that make up the system. A fern diagram may be tree structured (where there is no multiple inheritance) or network structured (where there is multiple inheritance). It typically includes aggregation, component, and inheritance relationships but makes no distinction among them.

The fern diagram is usually read from left to right or top to bottom (no arrowheads required). It is a useful technique, particularly where system is large and complex. Figures 9.5 and 9.6 provide two illustrations.

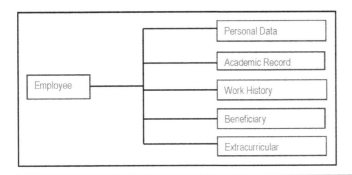

Figure 9.5   A Tree-Structured Fern Diagram

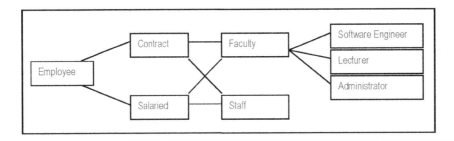

Figure 9.6   A Network Structured Fern Diagram

Advantages of fern diagrams:

- They are easy to draw and maintain.
- They are useful in assisting in the categorization of objects.

The main drawbacks of the fern diagram are:

- No distinction is made between a subtype relationship and an aggregation relationship.
- Neither does it show other types of relationships.
- The diagram may become cluttered and unwieldy as the system's size and complexity increases.

At a cursory glance, you may be tempted to compare the ITC of the previous chapter (Section 8.4) with the fern diagram. Here is the difference: The fern diagram is used to illustrate object categorization (including component and inheritance relationships). In contrast, the ITC's primary purpose is to illustrate how information will be classified and managed in the software system. As such, its focus is comprehensive coverage of all object types (or information entities) for the software system being modeled. Component relationships are typically covered but that is not the primary focus of the technique. The ITC also differs from the HIPO chart (of Chapter 8) in that whereas the HIPO chart is a functional representation of processes in traditional software systems, the ITC focuses on entities comprising the software system.

Sometimes it is useful to show instances on a fern diagram. This is done with the aid of broken lines. In the main, showing instances is impractical, but there are times when instances may have particular meaning in the design as in Figure 9.7, where **Bruce** is a software engineer and **Karen** is a staff member.

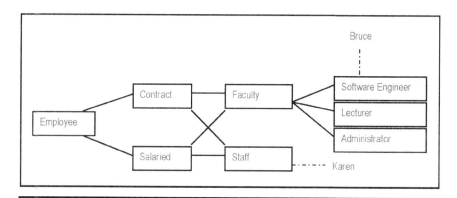

Figure 9.7   Fern Diagram with Instances

9.5 Representing Details about Object Types

Two methodologies for representing details about object types are the object structure diagram (OSD) and the class-responsibility-collaboration (CRC) card. *This course introduces a third approach — the object/entity specification grid (O/ESG)*, which will be discussed in Chapter 13 (Section 13.2.7) and illustrated in Appendix C. We shall briefly discuss the OSD and CRC card here.

9.5.1 Object Structure Diagram

The object structure diagram (OSD) is just an alternate term for the class diagram, so you have already been introduced to it from your object-oriented programming. The recommended standard for OSDs is the UML notation. Typically, you will not find stand-alone OSDs for each object type comprising a software system; rather, OSDs are incorporated in ORDs in order to convey useful information about the structure and interrelatedness of object types comprising a software system. Note, however, that from time to time, it might be necessary to highlight the OSD for a set of object types. One case in point would be where a software engineer is desirous of writing or modifying code for a specific set of object types. In situations where you are modeling a database, alternate methodologies such as ERDs and/or O/ESGs may be considered (as discussed in Chapter 13).

OSDs and ORDs (via the UML notation) are widely supported in contemporary software planning and development tools (review Section 2.4.5 of Chapter 2). The technique itself is quite simple and easy to follow. Figure 9.8 shows an excerpt of the ORD for the CUAIS project of earlier mention, depicting an inheritance relationship between object type **CollegeMember** (the super-type) on the one hand, and object types **Employee** and **Student** (the subtypes) on the other.

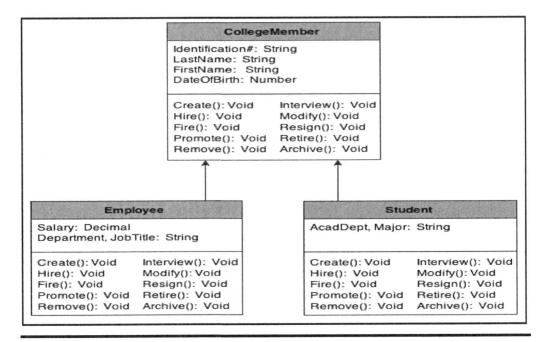

Figure 9.8   ORD Depicting Inheritance in a College Community

Moving to another example, Figure 9.9 illustrates a configuration of five object types in what is called a *star schema*: A central object type (**SalesSumary**) is surrounded by a set of object types (in this case **Location**, **TimePeriod**, **Product**, and **ProductLine**). Each object type forms a 1:M relationship with the central object type. The star schema represented in the figure relates to tracking sales by a marketing firm based on dimensions such as time, location, product line, and product. Star schemas are widely used in data modeling. However, a full discussion of this topic is not necessary for this course. For more information on the matter, see the recommended readings ([Foster 2016] and [Hoffer 2013]).

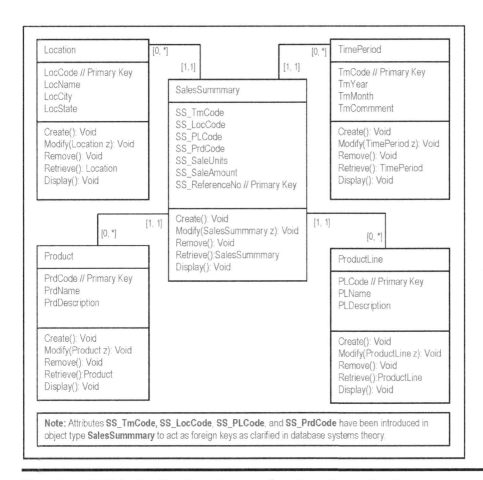

Figure 9.9   ORD for Tracking Sales Summary for a Large Marketing Company

Note: Since ORD can grow bulky rather quickly, it is common practice to deemphasize (or even sometimes omit) the details relating to attributes and operations on the ORD. Some tools provide a plus sign (+) to expand a related section of the diagram, or a minus sign (−) to contract a related section.

9.5.2 CRC Card

Recall from Section 7.8.10 that the class-responsibility-collaboration (CRC) card can also be very useful in providing details about a class (which is the implementation of an object type). For the purpose of comparison, Figure 9.10 summarizes the information contained in an OSD as well as a CRC card.

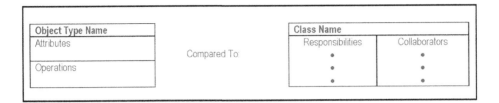

Figure 9.10   Object Structure Diagram versus the CRC Card

Traditionally, CRC cards were used manually to assist in the analysis of the software system. Designers would literally prepare a deck of CRC cards for the object types comprising the system (one CRC card per object type), and use them during brainstorming sessions to assist in refining and finalizing the structure and role of each object type of the system (review Section 7.8.10). To bring this technique to a contemporary scenario, the CRC card can be easily stored electronically, and used in not only refining but also modeling the software system.

9.6 Avoiding Multiple Inheritance Relationships

Dealing with multiple inheritances can be a challenge. You will recall from your OO programming, that they can cause confusion. Because of this, some OO programming languages (OOPLs) do not support them. James Rumbaugh in [Rumbaugh 1991] describes three techniques for circumventing multiple inheritances; they are paraphrased here. The techniques (called *workarounds*) allow for avoidance of multiple inheritances in one of three ways:

* Delegation using aggregation
* Delegation and inheritance
* Nested generalization

A fourth approach for circumventing multiple inheritances is the use of interfaces as described in Chapter 4. This approach is supported quite nicely in the Java programming language.

Figure 9.11 illustrates a multiple inheritances problem to be addressed. Let us examine how this can be resolved using the above-mentioned approaches, and as described in [Rumbaugh 1991]. The first thing to note is that based on the figure, any alternate configuration should facilitate at least five categories of employees: tenured faculty, tenure track faculty, tenure track contractor, salaried contractor (no tenure track), and hourly paid staff. Now let us examine Rumbaugh's workarounds.

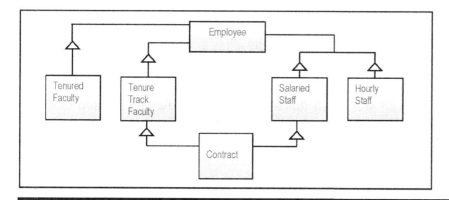

Figure 9.11   A Multiple Inheritance Problem

9.6.1 Delegation Using Aggregation

The delegation via aggregation technique involves the introduction of abstract object types that are composed of other types. Figure 9.12 illustrates a solution to the multiple inheritance problem of Figure 9.11, using aggregation. Notice the splitting of **Salaried Staff** into two separate object types, namely **Staff** and **Salaried**. This is necessary since there could be salaried or hourly-paid staff members. The abstract object types introduced are **Faculty** and **Staff**. A quick visual examination will also reveal that the minimum five categories of employees are facilitated in the figure.

Note: The introduced abstract object type needn't have actual data attributes, and may merely consist of abstract operations (methods), which are overridden in the respective subtypes. This is particularly advantageous if you are using a purely object-oriented implementation language such as Java. This is the preferred approach for dealing with multiple inheritances.

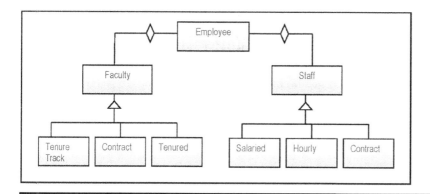

Figure 9.12   Multiple Inheritances using Delegation

9.6.2 Delegation and Inheritance

In the delegation and inheritance technique, we inherit the most important class and delegate the rest. Figure 9.13 illustrates a solution to the multiple inheritance problem of Figure 9.11, using this approach. The original problem did not indicate which class is the most important, a judgment call was made to inherit on the faculty side, and delegate on the staff side. With this approach, you must be prepared to make such judgments. The role of the abstract class **Staff** is identical to the explanation in the previous subsection. Also note that the minimum five categories of employees are again facilitated.

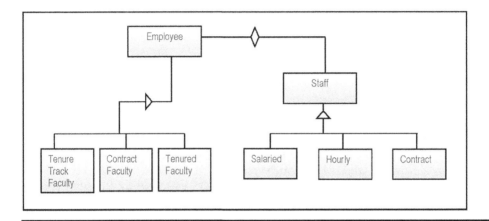

Figure 9.13   Multiple Inheritances via Inheritance and Delegation

9.6.3 Nested Generalization

The nested generalizations technique involves factoring one generalization first, then the other, until all possibilities are covered. It involves the introduction of abstract classes where necessary, in order to facilitate useful generalizations. Figure 9.14 illustrates a solution to the multiple inheritance problem of Figure 9.11, using nested generalization. Notice the introduction of three abstract object types, namely **Faculty**, **Contract**, and **Staff**. Also observe that as in the two previous approaches, the minimum five categories of employees are facilitated.

While this approach is very straightforward, it is not always recommended, since it often violates the principle of Occam's Razor by significantly increasing the number of classes to be managed.

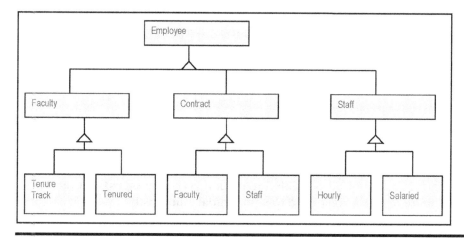

Figure 9.14 Multiple Inheritances via Nested Generalization

9.7 Top-Down versus Bottom-Up

You may conduct your object categorization by using either a top-down approach or a bottom-up approach. In practice, it is a good habit to use both approaches—one as a check-and-balance mechanism to the other. Invariably, your implementation will be bottom-up (you have to create classes before you can use them).

9.7.1 Top-Down Approach

For the top-down approach, use the following guidelines:

1. Start by looking at a summarized picture: determine what are the main facets of information are to be managed.
2. Break down the facets into constituents and sub-constituents as necessary (avoiding unnecessary indentation levels).
3. Consider the facets as system modules or sub-systems, depending on the size of your project. Then consider the constituents and sub-constituents as object types.
4. A final step—not required for your ITC, but required for your database specification—is to identify and define for each object type, a set of properties (data attributes and allowable operations).

9.7.2 Bottom-Up Approach

For the bottom-up approach, use the following guidelines:

1. Start out by identifying object types (tangible as well as intangible ones).
2. Identify and define for each object type, a set of properties (data attributes and allowable operations).
3. For each object type, provide an appropriate descriptive name.

4. Organize related object types into logical groups. These groups will constitute your super-types, system modules, and/or subsystems (depending on the complexity of the project).
5. Integrate all modules and/or subsystems into one integrated system.

9.8 Use-Cases

The technique called *use-case* was first introduced by Ivar Jacobson in [Jacobson 1992], and has since then been embraced by the software engineering industry. A use-case is a representation of all possible interactions (via messages) among a system and one or more *actors* in response to some initial stimulus by one of its actors. The use-case describes the functionality provided by a system, in order to yield a visible result for one of its actors.

In constructing use-cases, a few terms need to be clarified:

- **Actor:** An actor is an external entity that interfaces with the system. Actors could be individuals as well as other external systems that interact with the system in question.
- **Scenario:** A *scenario* is a specific instance of a use-case. The use-case represents a set of scenarios that involve an actor's engagement.
- **Use-case Bundle:** A use-case bundle refers to a group of related use-cases. Bundling may be based on the fact that the use-cases share the same actor and/or state, common entities, or common workflow.

The use-case allows the user to view the system from a high level, focusing on possible interactions between the system and its actors. In a way, use-cases should remind you of data flow diagrams (DFDs) in functional design.

9.8.1 Symbols Used in Use-case

Figure 9.15 includes the symbols used in use-case diagrams. As an example, consider the CUAIS project, and let us focus on the scenario where a student may register for a course, change his/her registration (for instance, auditing a course previously registered for), or drop a course. In the same vein, a department representative could add a course, cancel a course, or modify a course offering (changing its room and/or time period). Figure 9.16 illustrates a use-case diagram for this scenario.

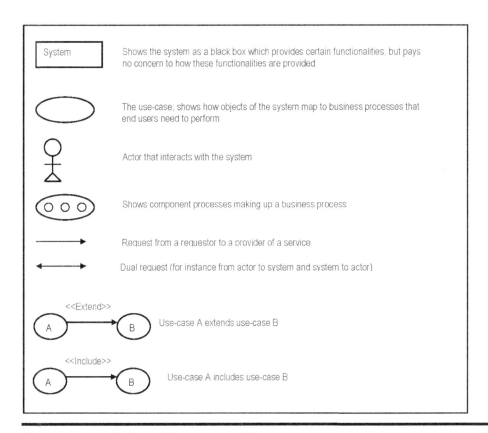

Figure 9.15   Symbols Used in Use-case Diagrams

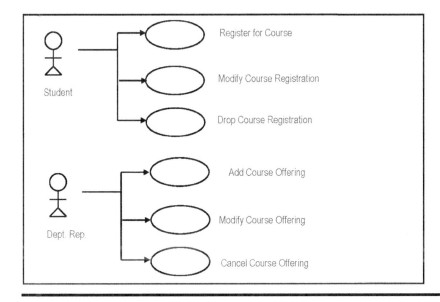

Figure 9.16   Use-case for Registration Process

9.8.2 Types of Use-cases

Use-case diagrams can be constructed for different levels of system functionality (they could be *high-level* or *low-level*):

*   A high-level operation provides the essence of the business values provided.
*   A low-level operation provides more detail about component or concurrent activities and their order.

Use-cases may also be classified in terms of *primary* or *secondary* operations that they represent:

*   Primary operations relate directly to the essential business functionalities of the system. They are the options provided to the end user.
*   Secondary operations are those operations, which, though not directly accessible by the end user, are essential for the delivery of a coherent, robust system. They are also referred to as subservient operations.

A third perspective is to differentiate *essential operations* from *concrete operations:*

*   Essential operations are business solutions that are platform independent.
*   Concrete operations are design dependent.

A use-case may be high-level or low-level; primary or secondary; essential or concrete. It may also involve a mix of these classifications. Additionally, a use-case may *include* a previously defined use-case, or it may *extend* a previously defined use-case.

9.8.3 Information Conveyed by a Use-case

A use-case may be used to convey the following information:

*   **Actors:** These are the participating entities.
*   **Relationships with Other Use-cases:** Two kinds of relationships are possible—the *include* relationship and the *extend* relationship.
*   **Pre-conditions:** A pre-condition is a condition that must be satisfied prior to the invocation of the use-case.
*   **Post-conditions:** A post-condition is a condition (or state) that must hold after the execution of the use-case.
*   **Details:** This entails the step-by-step interactions among participating objects in the use-case.
*   **Constraints:** Any constraint (apart from pre-conditions and post-conditions) that applies to the use-case must be specified. Constraints may be with respect to values and resources available for manipulation.
*   **Exceptions:** This involves the identification of all possible errors that might occur in the use-case.
*   **Variants:** This includes all variations that may apply.

Not all of this information may be relevant for each use-case; Figure 9.17 provides some guidelines. Of course, the judgment of the software engineer will be important in evaluating the scenario at hand and making the appropriate decisions.

Information	HPE Use-Case	LPSE Use-Case	LPC Use-Case	LPSC Use-Case
Actors	✓	✓	✓	✓
Relationships		✓		✓
Pre-conditions	✓	✓	✓	✓
Post-conditions	✓	✓	✓	✓
Details	✓	✓	✓	✓
Constraints	✓	✓	✓	✓
Exceptions		✓		✓
Variants	✓	✓	✓	✓

Key:
HPE = High-level primary essential
LPSE = Low-level primary/secondary essential
LPC = Low-level primary concrete
LPSC = Low-level primary/secondary concrete

Figure 9.17   Information Associated with Different Kinds of Use-cases

9.8.4 Bundling Use-cases and Putting Them to Use

As pointed out earlier, a use-case bundle refers to a group of related use-case. Typically, the component use-cases make sense when the bundle is viewed from a business perspective. Use-cases may be bundled using any of the following three criteria:

Same Actor, Same State: All use-cases involving the same actor and the same state are grouped. The use-cases of Figure 9.16 meet this criterion.

Common Entities: Use-cases are bundled based on the fact that they use and/ or impact the same entities. The rule-of-thumb strategy for object identification, discussed in Chapter 7 (Section 7.8.2), flows naturally into this approach. For example, in designing an inventory management system, one may identify and refine use-cases relating to the purchase of an invoice. These would typically involve adding, modifying, deleting, and inquiring on invoices.

Specific Workflow: Use-cases that relate to a particular actor (user) carrying out a functional responsibility are bundled together. More often than not, same-actor-same-state use-vases qualify as specific workflow use-cases. Figure 9.16 therefore applies here also. The idea here is that if we can identify, define, and implement all the operations which each actor of the system requires, we would achieve a comprehensive system meeting all related user requirements.

Once the use-cases have been identified and represented (via use-case diagrams), the next step is to use the information to refine your system model. Remember, your focus is object behavior. The rest of the chapter will concentrate on some methodologies for analyzing and modeling system behavior.

9.9 Event Diagrams

[Martin 1993] describes event diagrams as a means of modeling the interrelationship between operations and events, as summarized in this section. An *event* is a noteworthy change in the state of an object. Events trigger operations. An operation may change the state of an object thereby causing another event. The event diagram shows these events and their associated operations.

9.9.1 Basic Event Diagrams

An event is represented by a filled in triangle; an operation is represented by a round-cornered box (shadowed if operation is external). When an event results from an operation, the triangle is attached to the operation box. An example is shown in Figure 9.18, where an event diagram is shown for the processing of a purchase order.

The event diagram relates to the state transition diagram in the following way: the latter depicts state changes; the former depicts causes for those state changes. On the state transition diagram, each transition may be elaborated by an event diagram. State transition diagram and event diagram are complementary. The one assists in ensuring the accuracy of the other. With an OO-CASE tool, the designer chooses to work with either. The operation is isolated from cause and effect. Each object type (class) has an associated state transition diagram and/or finite state machine (review Section 8.7), as well as related event diagram(s).

The event diagram is to OOSE what the program flowchart is to the more traditional (function-oriented) approach to software engineering. In more complex environments, operation specifications may be required to compliment the event diagrams.

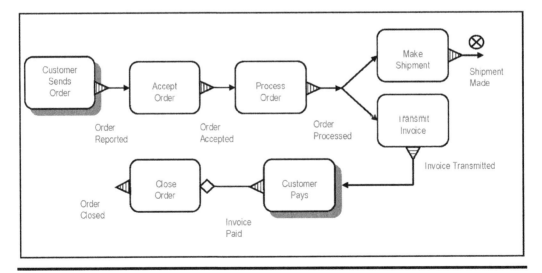

Figure 9.18   An Event Diagram for Processing a Purchase Order

An operation may have pre-condition(s) and post-condition(s). Pre-conditions must be true before the operation can take place; post-conditions must hold after the operation is completed. Pre-conditions and post-conditions are referred to as control conditions. The diamond symbol of Figure 9.18 indicates that pre-condition(s) exist for the operation **Close Order.** With an OO-CASE, pre-conditions can be clicked for details.

9.9.2 Event Types

Just as there are object types, there are event types. The analyst is more interested in event types than the number of occurrences of each event. Typically, event types describe the following kinds of changes:

- Creation of new object
- Termination of object
- An object changes classification
- An object's attribute(s) is (are) changed
- Object is classified as an instance of another type; for example: **Lecturer** becomes **Department Chair**
- Object is declassified; for example: **Lecturer** ceases to be **Department Chair**

Control conditions are particularly useful when multiple trigger rules lead to a specific operation, as illustrated in Figure 9.19. In the figure, **Operation A**, **Operation B**, and **Operation C** must occur before **Operation D** can occur.

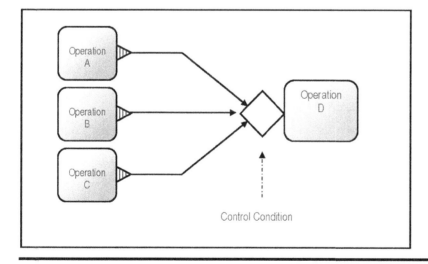

Figure 9.19   Illustrating Control Condition

Events may also have subtypes and super-types as illustrated in Figure 9.20. In this figure, **Operation A** has two sub-events.

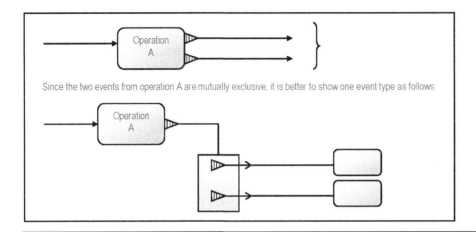

Figure 9.20 Illustrating Event Sub-types

9.10 Triggers

[Martin 1993] also describes and trigger rules—two related concepts that have become commonplace in contemporary software engineering environments. The line going from an event to the operation(s) it triggers, represents a trigger rule. The trigger rule defines the causal relationship between event and operation(s). An event may trigger one or more operations; also, an operation may be caused by multiple triggers; Figure 9.21 illustrates.

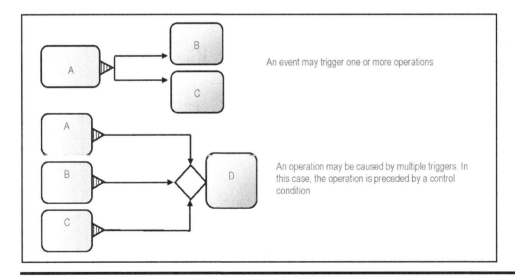

Figure 9.21 Illustrating the Relationship between Triggers and Operations

Like objects types, operations may have subtypes and super-types as illustrated in Figure 9.22. In the figure, trigger A has sub-triggers A1, A2, and A3.

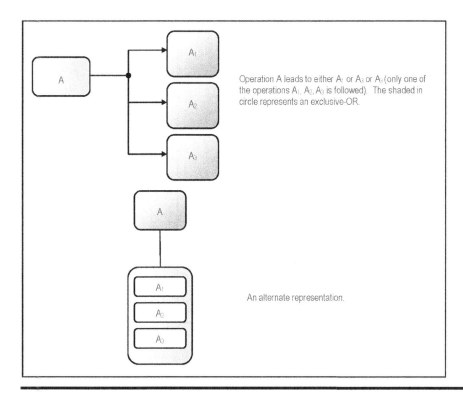

Operation A leads to either A_1 or A_3 or A_2 (only one of the operations A_1, A_2, A_3 is followed). The shaded in circle represents an exclusive-OR.

An alternate representation.

Figure 9.22   Illustrating Operation Sub-types

9.11 Activity Diagrams

As an alternative to event diagrams, UML supports *activity diagrams*. An activity diagram describes how related activities (computations and workflow) are coordinated. Bear in mind that activities are not existence independent, but are typically subservient to operations and other activities.

The activity diagram (graph) depicts activity states and the transition among activities in a workflow. An activity state represents the execution of a statement in a procedure, or the performance of an activity in a workflow. The activity diagram conveys information about sequential activities as well as concurrent activities (threads) related to an operation. Figure 9.23 shows the UML notations for activity diagrams, while Figure 9.24 provides an example.

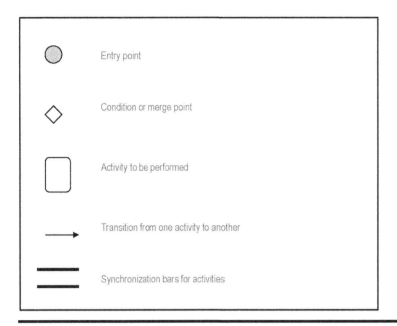

Figure 9.23   UML Notations for Activity Diagram

The activity diagram may contain *swim-lanes*: Whenever the situation arises where it is desirable to group related activities of a process into distinct partitions (these partitions may represent functional areas in a business), vertical or horizontal lines are used to define the partitions. The partitions are called swim-lanes. Swim-lanes could have been introduced for the three concurrent threads in Figure 9.24 (they were omitted since their introduction would have cluttered the diagram).

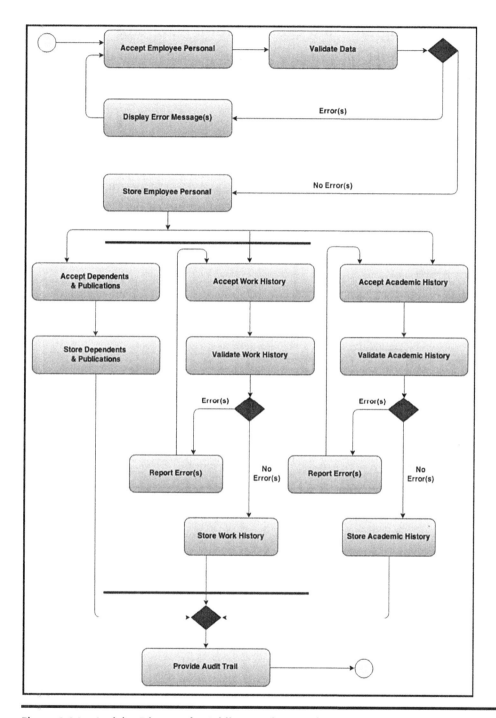

Figure 9.24   Activity Diagram for Adding Employee Information

9.12 Sequence Diagrams and Collaboration Diagrams

Another UML notation is the *collaboration diagram*. A collaboration diagram (also described in older literature on object technology as an *object interaction diagram*) is a special class diagram that is used to depict various interactions among objects that participate in an operation.

The collaboration diagram is useful in illustrating the implementation of a class operation, or the set of operations of a class. It includes permanent objects as well as objects created and/or destroyed during the process. Figure 9.25 illustrates a collaboration diagram for an inquiry on student information in the CUAIS project of earlier discussions. A message from one object to another is represented as a directed arrow, labeled with the name of the service requested and an argument list.

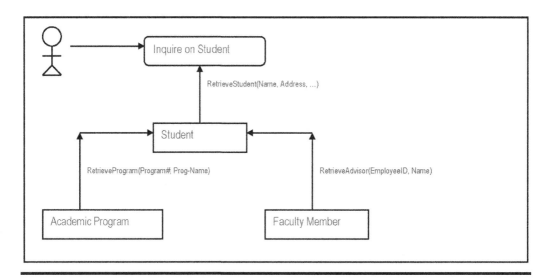

Figure 9.25   Collaboration Diagram for Inquiry on Student Information

A *sequence diagram*—another UML notation—shows how objects interact with each other (via messages) with respect to time, in a two-dimensional chart. The vertical dimension indicates time, which increases down the axis. The horizontal dimension indicates the objects in the interaction. Except for real-time systems, sequence diagrams are not as widely used as the other techniques discussed; they are therefore not discussed any further. For additional information, see the recommended readings [Lee 2002] and [Rumbaugh 1999].

9.13 Summary and Concluding Remarks

Here is a summary of what has been discussed in this chapter:

• Object structure analysis (OSA) relates to the structure of object types and classes comprising the software system.

- Object behavior analysis (OBA) concerns itself with the behavior of interacting objects in the software system.
- The UML defines standards for object structure as well as object behavior, and represents the de facto standard for specifying software requirements.
- A fern diagram is an older technique for aggregation, component, and inheritance relationships among object types, but makes no distinction among them.
- An ORD is a UML technique used to depict how object types relate to each other in a software system.
- An OSD is a UML technique used to represent the structure of an object type or class.
- A CRC card is an older technique for representing an object type and/or class.
- When designing a software system, multiple inheritances should be avoided. Three techniques for resolving them are delegation using aggregation, delegation and inheritance, and nested generalization.
- Object-oriented design may proceed in a top-down or bottom-up manner, each strategy complimenting the other.
- A use-case is a UML representation of all possible interactions (via messages) among a system and one or more *actors* in response to some initial stimulus by one of its actors.
- An event diagram shows these events and their associated operations. The event diagram typically includes events, triggers, and operations.
- An activity diagram describes how related activities (computations and workflow) are coordinated; this is the UML alternative to event diagrams.
- A collaboration diagram is a special class diagram that is used to depict various interactions among objects that participate in an operation.
- A sequence diagram is another UML technique that shows how objects interact with each other (via messages) with respect to time, in a two-dimensional chart.

Still, there are additional diagramming techniques that will be covered in upcoming chapters. However, this and the previous two chapters have covered quite a bit. The next chapter focuses on decision models for system logic.

9.14 Review Questions

1. Identify a medium-sized business enterprise that you are familiar with and answer the following related questions:
   - Identify the main business areas for your organization/institution of focus.
   - Identify at least twelve (12) object types that would be included in an integrated information system for the institution. Preferably, the object types should relate to a specific business area.
   - Construct an object-operation matrix for these object types.1-4 Propose an ORD for the object types that you have identified.
   - Construct a detailed OSD for each object type that you have identified.
   - Propose a set of use-cases and activity diagrams for selected object types in your model.
   - Propose a collaboration for selected major operations in your system.

2. Why might it be undesirable to have several cases of multiple inheritance in a database design?

3. What are the three strategies that can be employed to avoid this situation?

4. Figure 9.26 contains a case of multiple inheritance. Propose a modified diagram that resolves the multiple inheritance.

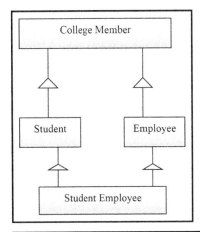

Figure 9.26   A Multiple Inheritance Problem

References and Recommended Reading

[Bruegge 2010] Bruegge, Bernd and Allen H. Dutoit. 2010. *Object-Oriented Software Engineering*, 3rd ed. Boston, MA: Pearson. See chapter 2.

[Foster 2016] Foster, Elvis C. with Shripad V. Godbole. 2016. *Database Systems: A Pragmatic Approach*, 2nd ed. New York, NY: Apress. See chapters 3–5.

[Hoffer 2013] Hoffer, Jeffrey A., Ramesh Venkataraman, & Heikki Topi. 2013. *Modern Database Management*, 11th ed. Boston, MA: Pearson. See chapters 2–4.

[Jacobson 1992] Jacobson, Ivar. 1992. *Object Oriented Software Engineering—A Use Case Approach*. Reading, MA: Addison-Wesley.

[Lee 2002] Lee, Richard C. and William M. Tepfenhart. 2002. *Practical Object-Oriented Development With UML and Java*. Upper Saddle River, NJ: Prentice Hall. See chapters 3, 5, 6-8.

[Martin 1993] Martin, James and James Odell. 1993. *Principles of Object-Oriented Analysis and Design*. Eaglewood Cliffs, NJ: Pretence Hall. See chapters 6–9.

[Rumbaugh 1991] Rumbaugh, James, et al. 1991. *Object Oriented Modeling And Design*. Eaglewood Cliffs, NJ: Pretence Hall. See chapters 4 – 6.

[Rumbaugh 1999] Rumbaugh, James, Ivar Jacobson and Grady Booch. 1999. *The Unified Modeling Language Reference Manual*. Reading, MA: Addison-Wesley. See chapter 8.

Chapter 10

Decision Models for System Logic

The previous chapter examined various system flow charts, which form part of the inventory of diagramming techniques employed by the software engineer. This chapter continues the discussion of diagrams by focusing on techniques used to represent and manage system logic. In the traditional FO paradigm, decisions are represented and analyzed by use of structured language (pseudo-code), decision tables, or decision trees. In the more contemporary OO paradigm, the traditional methods are still applicable, but in addition to system rules. The main challenge is the identification of these decision issues; they are not always obvious. Here, skill and experience in information gathering are precious virtues.

The chapter will proceed with discussions in the following areas:

- Structured Language
- Decision Tables
- Decision Trees
- Which Technique to Use
- Decision Techniques versus Flowcharts
- System Rules
- Summary and Concluding Remarks

10.1 Structured Language

Decisions can be represented by use of structured English in pseudo-code-like manner. Use of the standard control structures for selection, iteration, and recursion is common (you should be familiar with these you're your earlier programming courses). As usual, indentation improves readability. It is assumed that you know how to write a pseudo-code; therefore, nothing further will be said on this matter. Decisions may also be presented in a tabular manner as illustrated in Figure 10.1.

DOI: 10.1201/9780367746025-12

If	Condition	Action
Student Balance:	Semester Tuition or more	Disallow registration
	Less than Semester Tuition	Allow registration for the difference
Student GPA:	Less than 2.0	Disallow registration
	2.0 or more	Allow registration
Course Prerequisite:	Done successfully	Allow registration
	Not done successfully	Disallow registration

Figure 10.1   Decisions to be Taken on Student Registration at a College or University

10.2 Decision Tables

A decision table is a tabular technique for describing logical rules. It serves as an aid to creative analysis and expresses a business situation in a cause–effect relationship. The decision table provides an excellent means of communication between users and software developers. It forces the software engineer to be objective in the decision-making process, and to consider all the decision alternatives. Figure 10.2 shows the basic components of a decision table.

Condition Stub	Condition Entries (Y/N)
Action Stub	Action Entries (Xs)

Figure 10.2   Structure of a Decision Table

10.2.1 Constructing the Decision Table

The following guidelines are useful in the construction of a decision table:

- Make small tables (maximum four conditions).
- All possible rules must be represented. A rule is simply a condition entry combined with an action entry (i.e., a column of the decision table).
- Every rule must have an action.
- Rules must be unique and independent.
- Define all alternate outcomes. In particular, note that n conditions imply 2^n outcomes (rules).
- Develop a set of conditions that yields each outcome.
- Assign a decision for each outcome.

Example 10.1   The Airline Flight Reservation Problem

Consider as an example, an airline ticket seller is given the following guidelines:

1. There are two classes of tickets—first class and economy.
2. If a request is for first class and the space is available then reserve first-class seats. If request is for economy and space available, reserve economy seat.

Figure 10.3 shows a faulty decision table for the situation described. In the figure, the guideline that states that every rule must have an action is not met; the table is therefore incorrect. Figure 10.4 shows a correct decision table. Notice that two additional conditions have been introduced.

Airline Booking				
If 1st Class	Y	Y	N	N
If Space Avail.	Y	N	Y	N
Reserve First Class	X			
Reserve Economy			X	

There are two problems with this decision table:
1. All possible conditions are not shown.
2. Every rule must have an action.

Figure 10.3   Incorrect Decision Table for Airline Problem

Conditions/Actions	Rules															
	1	2	3	4	5	6	7	8	9	10	11	12	13	14	15	16
Request 1st class	Y	Y	Y	Y	Y	Y	Y	Y	N	N	N	N	N	N	N	N
Req. Space Available	Y	Y	Y	Y	N	N	N	N	Y	Y	Y	Y	N	N	N	N
Accept Alternative Class	Y	Y	N	N	Y	Y	N	N	Y	Y	N	N	Y	Y	N	N
Alt. Space Available	Y	N	Y	N	Y	N	Y	N	Y	N	Y	N	Y	N	Y	N
Reserve 1st class	X	X	X	X									X			
Reserve Economy					X				X	X	X	X				
Standby 1st class							X	X								
Standby Economy															X	X
Standby Either					X									X		

Figure 10.4   Correct Decision Table for Airline Problem

10.2.2 Analyzing and Refining the Decision Table

Two principles are used to analyze and refine the decision table: elimination of redundancies, and avoidance of ambiguities. Let us examine each.

Elimination of Redundancy: *Redundancy* occurs when two rules result in the same action and the condition entry responses are the same except for the last condition. The two rules can be combined (i.e. one is discarded). In Figure 10.4, rules 1&2, 3&4, 7&8, 9&10, 11&12, 15&16 constitute redundancies. For each pair, the second rule is eliminated; the revised table is shown in Figure 10.5.

Avoidance of Ambiguity: If two equivalent sets of conditions require different actions, the table is said to be *ambiguous*. Ambiguity occurs when *don't care* situations exist (indicated by dash (–)). Figure 10.6 illustrates. Ambiguities are also referred to as contradictions. When they occur, you must revisit the analysis that led to them, and make appropriate adjustments.

Conditions/Actions	1	5	6	7	9	13	14	15
Request 1st class	Y	Y	Y	Y	N	N	N	N
Requested Class Available	Y	N	N	N	Y	N	N	N
Accept Alternative Class	-	Y	Y	N	-	Y	Y	N
Alternative Class Available	-	Y	N	-	-	Y	N	-
Reserve 1st class	X					X		
Reserve Economy		X			X			
Standby 1st class				X				
Standby Economy								X
Standby Either			X				X	

Note: Don't Care condition entries are indicated by the dash (-).

Figure 10.5   The Enhanced Decision Table for Airline Problem

Condition/Actions	1	2	3	4	5	6	7	8
Condition 1	Y	Y	Y	Y	N	N	N	N
Condition 2	Y	Y	N	N	Y	Y	N	N
Condition 3	-	N	-	N	Y	N	Y	N
Action 1	X			X	X			
Action 2			X			X		
Action 3		X					X	

Ambiguity ⬆ ⬆   ⬆ ⬆ Ambiguity

Figure 10.6   Illustration of Ambiguity

10.2.3 Extended-Entry Decision Table

In situations where the number of conditions exceeds four and/or each condition can have more than two outcomes, the decision table becomes large and unwieldy. In these cases, an extended-entry decision table can be useful.

Extended-entry decision tables are normally used to save space. Figure 10.7 provides an example. Note that statement(s) made in the condition stub are incomplete so that both stub and entry must be combined in order to obtain the intended message.

Conditions/Actions	R1	R2
If Assets > =	10,000	8,000
Grant 5,000 loan	X	
Grant 4,000 loan		X

Figure 10.7   Example of Extended Entry Decision Table

Extended-entry tables also reduce the possibility of redundancy and contradiction as Figure 10.8 illustrates. In this table, seven scenarios for action are represented, along with a default scenario, if none of the other seven scenarios is true. As an exercise, try to convert this to the most optimized system logic possible (you may represent the logic using pseudo-code, Pascal, C++ or Java). The exercise should also help you to see that the table can be further refined.

Conditions/Actions	R1	R2	R3	R4	R5	R6	R7	ELSE
Cost of The Item A: Cost < $ 15 B: $ 15 <= Cost <= $ 60 C: Cost > $ 60	-	A	A	B	B	C	C	
Order Quantity D: Qty. < 100 E: 100 <= Qty. <= 175 F: Qty. > 175	D	E	F	E	F	E	F	
Order Immediately		X		X		X		
Check Supervisor			X		X		X	X
Wait for Regular Order	X							

Figure 10.8   Extended Entry Decision Table for Inventory Ordering

10.3 Decision Trees

The decision tree is useful when complex branching occurs in a structured decision process, and it is essential to keep a string of decisions in a particular sequence. It therefore shows the relationship among decision factors. This is not shown by a basic decision table and is poorly handled by a flowchart.

The decision tree used in software engineering and systems analysis is not as complicated as that used in management science, where probabilities and monetary expectations are shown. Essentially, the tree shows conditions and actions in a completely structured decision process.

Two symbols used are in the construction of a decision tree: an oval shape is used to symbolize **if condition**; a square shape symbolizes **then action**. Figure 10.9 illustrates a decision tree for a sale operation.

Decision trees are preferred to decision tables when any of the following situations hold:

- The process(es) concerned is (are) accomplished in stages.
- The logic is asymmetrical.
- The decision conditions and actions are related.

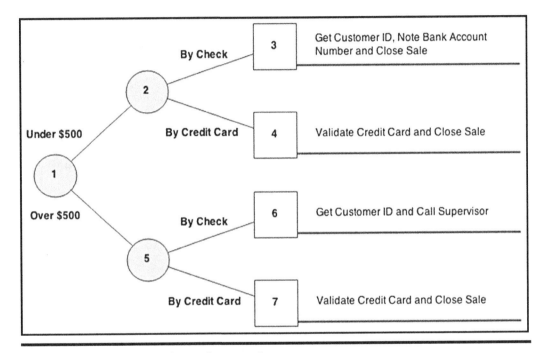

Figure 10.9   Decision Tree for a Sale Operation

How would you compare decision tables with decision trees? To get you started, here are some pointers:

- The decision tree emphasizes the sequential relation of decision conditions. This is impossible with the decision table.
- The decision tree handles quite effectively, situations where certain conditions and actions apply, given specific circumstances, but do not apply, given other circumstances. This is more difficult to illustrate via the decision table.
- The decision tree is more readily understood by users than the decision table.
- Unlike the decision table, the decision tree effectively handles scenarios involving more than four conditions.
- The decision table is more effective than the decision tree at avoiding redundancies and ambiguities. It is also more concise than the decision tree.
- The decision table is more easily incorporated in a CASE tool than a decision tree.

10.4 Which Technique to Use

Use structured language when any of the following holds:

- There are many repetitious conditions.
- Communication to end users is paramount.

Use decision tables when any of the following holds:

- Complex combinations of conditions, actions, and rules exist.
- A method that effectively avoids impossible situations, redundancies, and contradictions is required.
- There are no more than four condition entries.

Use extended decision tables when any of the following holds:

- There are more than four condition entries.
- It is desirous to save space by representing the decision scenarios in a concise manner.
- There are (or may be) conditions that have more than two possible outcomes.
- It is desirous to reduce/eliminate redundancies and ambiguities.

Use decision trees when any of the following holds:

- The sequence of conditions and actions is critical.
- Not every condition is relevant to every action (branches are different).
- There are more than four condition entries.

10.5 Decision Techniques versus Flowcharts

Following is a comparison between decision techniques as discussed in this chapter and flow charts as discussed in the previous chapter. The comparison is based on advantages and disadvantages of both sets of methodologies.

Flowcharts provide the following advantages:

- Flowcharts are useful in identifying bottlenecks, delay factors, redundancies, and other system flaws.
- They are effective in depicting the movement of data.
- They are also effective in depicting transitions between system states.
- They are useful in the representation of system logic.
- They are superb in emphasizing the interrelatedness of system components—subsystems, operations, data storage, user interfaces, etc.
- They are excellent for easy communication with end users.

The following disadvantages are associated with flowcharts (if a CASE tool that supports the diagrams is used, these disadvantages are minimized):

- Flowcharts can be difficult to draw, especially for large, complex systems. Diagrams may be quite complex, particularly when there are several paths to consider.
- It may be difficult to determine whether the total problem is covered or whether superior alternate methods exist.
- Revision may be as difficult as drawing a complicated chart.

Decision techniques provide the following advantages:

- Structured format makes for easy drawing and clarity of presentation.
- Applications involving complex interactions of input variables are well documented and represented in a simple manner.
- Diagrams used are easy to revise and maintain with or without CASE.
- There is easy communication with end users.

The following disadvantages are associated with decision techniques:

- With decision techniques, there is no indication of the movement of data.
- They do not assist in identifying system bottlenecks, redundancies, or delays as flowcharts.
- They do not represent system state transitions.
- They do not show the interrelatedness of the system components, only decision conditions.

Decision techniques and flow chart techniques are most effective when they are used to complement each other. The software engineer decides which techniques will be used to represent different aspects of the system. In an integrated CASE environment, these tools are all integrated as the system is modeled.

10.6 System Rules

Figure 10.10 proposes a ten-step approach for the development of software using the OO paradigm. The upper portion of the diagram relates to the requirements specification (RS); this was introduced in Chapter 6 (Figure 6.1). The lower portion relates to the *design specification* (DS) and the actual development; discussion of these issues commences in Chapter 12. As can be seen, the specification of system rules begins in the first portion of the schedule (item 5). These rules are subsequently refined and applied appropriately during the second portion of the schedule (item 8). Rules form an integral part of the system logic. The rest of this section focuses on essential details you should know about system rules.

The OO modeling hierarchy based on Martin's OO modeling pyramid (first introduced in Chapter 9) is repeated in Figure 10.11 for ease of reference. The conceptualization depicted in the figure should help you to have a better appreciation of the schedule presented in Figure 10.10.

Requirements Specification Activities
1. Refine your problem definition & proposed software solution.
2. Conduct your information gathering and identify: 　■ Subsystem(s) and/or modules 　■ Object types (information entities)
3. Identify or define operations for the various object types.
4. Provide a system overview which includes any convenient combination of the following: 　■ System narrative 　■ Information topology chart (ITC) 　■ Object flow diagram (OFD)
5. For each subsystem, provide the following: 　■ System overview (item 4) 　■ Entity-Relationship (E-R) or Object-Relationship (O-R) diagram 　■ Storage requirements 　■ Operational requirements 　■ System rules 　■ Interface requirements (if applicable) 　■ System constraints (if applicable)

Design Specification to Actual Development Activities
6. Refine object types (in terms of structure and operation). Note: If an OO-ICASE tool is used, each object type is defined in terms of name, attributes, and operations. They are automatically linked by the software system as defined in the O-R diagram. In the absence of an OO-ICASE tool, the linkages are not automatic. Alternately, prepare an O/ESG for each object type.
7. For each object type, define the following: 　■ a state transition diagram, or an FSM (preferably FSM); 　■ a collaboration diagram; 　■ a set of activity diagrams. **Note:** 　■ Your use of diagrams may be constrained by the CASE tool being employed. 　■ In the context of an OO-CASE tool, some of the diagrams may be automatically linked, so that update of one ripples to the others. 　■ The state transition diagram (or FSM) should precede the activity diagram; the former aids the development of the latter.
8. Refine rules for each object type (class). Rules may also be refined and applied to operations, reflecting the business policy of the organization. Note: If you are using an OO-GUI superimposed on a RDBMS, it may be prudent to include as many of the business rules as possible in the relational database.
9. Develop operation specification for each operation comprising the software system.

Figure 10.10　The OO Software Construction Process

10.6.1 Rule Definition

A *rule* is a guideline for the successful and acceptable implementation of a process or set of operations. Failure to observe the rule could result in unacceptable system results. With traditional software engineering methodologies, rules are usually hidden in the code of system functions. In the OO paradigm (thanks to non-procedural languages), we want to define rules as an integral (encapsulated) part of the definition of (the operations of) a class. Code can then be automatically generated from this.

Example 10.2　Examples of Non-procedural Languages

Examples of non-procedural languages are LISP, Haskell, Prolog, Structured Query Language (SQL), and Knowledge Query Language (KQL).

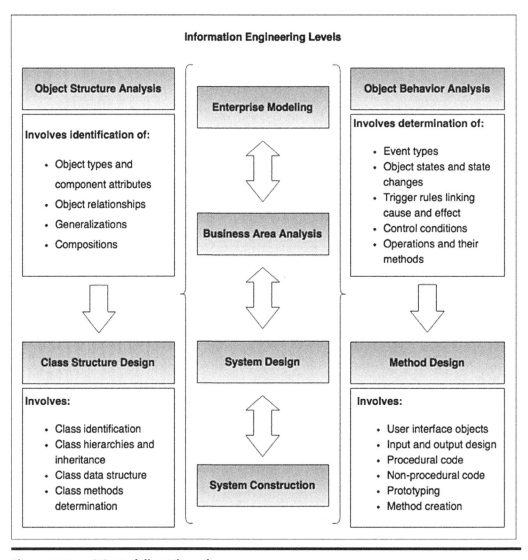

Figure 10.11   OO Modeling Hierarchy

Rules are particularly helpful in designing the desired behavior of the system objects. They represent the laws about how the business is run. Rules may be in any of the following broad categories:

- Relating to object structure analysis (OSA) and inter-object relationships (business rules).
- Relating to object behavior analysis (OBA) and business policies of the organizations.

Rules must be rigorous, precise, concise, clear, and easily understood by end users. Otherwise, they are not very useful and could lead to more complications in the software system.

Example 10.3   A Simple OBA Rule

The following is an example of an OBA rule:
    When a student's GPA is 3.6 or greater, he/she goes on the Dean's list.

10.6.2 *Declarative versus Procedural Statements*

Conventional programming languages are procedural. Declarative languages declare facts and rules, without specifying processing procedure. Declarative statements are more concise and easier to understand and validate than procedural statements.

Example 10.4   Illustrating a Declarative Statement

Suppose that we are storing information about students in a database table called **Student.** Assume further that the property **StudGPA** stores the GPA of the student. The following SQL statement produces a logical view (called **DeanList**) of all students with a GPA of at least 3.6:
    Create View DeanList AS SELECT * FROM Student WHERE StudGPA >= 3.6;

Because declarative statements are more flexible, it is preferable to build systems from declarative languages linked to OO-CASE tools, rather than procedural languages. With an OO-CASE tool, rules are constructed with the use of a rule editor. Code can be automatically generated so the rule can be executed immediately.

10.6.3 *Types of Rules*

Lee's [Lee 2002] refinement of Martin's [Martin 1993] categories of system rules include eight (8) categories as described below:

Data Integrity Rule: A rule of this type states that stipulated condition(s) about an attribute must be true. For example, and attribute for marital status must have values corresponding to one of the following: **married, single, widowed, separated,** or **divorced.** As an alternate description, data integrity rules are often called data validation rules.

Relationship Integrity Rule: Such a rule states that something about a relationship between two object- types must be true. For example, a department may be restricted to no more than 25 employees. Or, there may be database integrity rules (typically, entity integrity rule and/or referential integrity rule) that apply to certain data fields (for more on this, see [Foster 2016]).

Derivation Rule: A derivation rule prescribes a relationship between a dependent variable and a set of independent variables; it typically relates to calculated values (for example **InvoiceAmount** = Summation of **Price * Quantity** for each line item on the invoice).

Pre-condition Rule: This type of rule states that a condition must be true before an operation is executed (for example, an employee object cannot be fired unless it is in a state of **Hired**).

Post-condition Rule: This type of rule states the condition(s) which must exist after an operation is performed (for example, after the **FireEmployee** operation is executed, the stat of the employee object is **Fired**).

Action Trigger Rule: A rule of this type defines the causal relationship between an event and the Operation that triggers it (for example, when the result of an interview is recorded, the appropriate action, commensurate with the recommendation, must be taken).

Data Trigger Rule: This type of rule defines the causal relationship between an attribute's status and an action (for example, when a student's GPA falls below 2.0, send an alert letter).

Control Condition Rule: Such a rule handles the situation in which an operation is caused by multiple triggers. The control condition rule determines what combination of circumstances will cause the operation to execute (e.g., if the student's GPA is below 2.0 AND the student has registered for more than 12 credits, him/her student on probation).

Traditionally, data integrity rules, relationship integrity rules, and derivation rules are described as *business rules*. The term *stimulus/response rules* is sometimes used to refer to action trigger rules, data trigger rules, and control condition rules. Figure 10.12 provides the recommended constructs for the different types of rules. Figure 10.13 provides some examples.

1. For stimulus/response rules:
   IF <Condition> THEN <Action>   // Preferred for data triggers
        Or
   WHEN <Event> IF <Condition> THEN <Action> // Preferred for action triggers & control conditions

2. For business rules:
   IT MUST ALWAYS BE THAT <Statement of Fact>

3. For pre-condition rules:
   BEFORE <Operation> IT MUST BE THAT <Statement of Fact>

4. For post-condition rules:
   AFTER <Operation> IT MUST BE THAT <Statement of Fact>

5. For derivation rules:
   WHEN <Event or Condition> THEN <Statement of Fact>

Figure 10.12   Recommended Constructs for Different Types of Rules

IT MUST ALWAYS HOLD THAT
    An employee belongs to one and only one department.

IT MUST ALWAYS HOLD THAT
    The number of employees who are managers with salary greater than $2.5 M is less than 5.

IT MUST ALWAYS HOLD THAT
    Sex is either 'M'ale or 'F'emale.

IF Student.GPA ≥ 3.90
    THEN classify student as "Summa Cum Laude"

BEFORE Bruce.Interview() IT MUST BE THAT (Bruce.GetStatus() = "Applied")

Figure 10.13   Illustrating How Rules may be Specified

10.7 Summary and Concluding Remarks

Let us summarize what has been discussed in this chapter:

- The traditional methods of representing decisions are via structured language, decision tables and decision trees.
- The structured language may be in the form of pseudo-code or a tabular representation. It is applicable in simple situations where communication to end users is of paramount importance.
- The decision table has four main sections: the condition stub, the condition entries, the action stub, and the action entries. The decision table must be free of redundancies and ambiguities. Decision tables are useful in situations where there are complex combinations of four or less conditions, and it is desirable to efficiently represent all the possibilities while avoiding ambiguities as well as redundancies.
- The extended decision table is a modification of the basic decision table in order to provide more flexibility. It is particularly useful in any combination of the following situations: more than four conditions exist; not all the conditional possibilities are relevant; and it is desirable to be concise but accurate.
- The decision tree is a hierarchical graphical representation of the decision problem. It is useful in any combination of the following situations: the sequence of conditions and actions is critical; not every condition is relevant to every action (i.e. the branches are different); there are more than four condition entries.
- A system rule is a guideline for the successful and acceptable implementation of a process or set of processes.
- There are eight types of rules: data integrity rules, relationship integrity rules, derivation rules, pre-condition rules, post-condition rules, action trigger rules, data trigger rules, and control condition rules.
- Rules must be clearly specified according to predetermined formats.

Traditionally, expert systems were based on an important component called the *inference engine*.

An inference engine is a collection of facts and rules about a specific area of knowledge and facilitates deductions based on established techniques of logical reasoning. Special software was used for recording of such facts.

Object technology facilitates easier, more pragmatic construction of inference engines for expert systems. It's also more exciting.

Some rules are to be visible to end users, so that they may check them in a workshop. These include business policy rules, derivation rules. There are other rules that are best kept invisible (transparent) to the end user, to avoid confusion. These include rules for technical design, integrity rules, rules internal to the software (OO-CASE tool and/or repository) being used.

Mastery of the techniques discussed since Chapter 3 places the software engineer in an excellent position to prepare an impressive RS. This time, your RS should have a refined project schedule. The next chapter will provide guidelines on how to achieve this. Before you proceed, take a second look at the sample requirements specification included in appendix 2; it should provide you with useful insights.

10.8 Review Questions

1. What are the decision techniques that have been discussed in this chapter? What circumstance(s) would warrant the use of each technique?

2. Consider the registration process at your institution, or a similarly complex process at an organization that you are familiar with:
   • Identify the different circumstances (conditions) that must be considered.
   • Determine an appropriate action for each circumstance.
   • Represent your findings using one of the decision techniques discussed in the chapter.

3. Develop a software application to address the problem, using Delphi, Visual C++, C++ Builder, or any other RAD tool that you are familiar with. What is a system rule? Briefly describe the different types of rules discussed.

4. Still considering the registration process at your institution, answer the following questions:
   4a. Assume that there is an operation called **Student.RegisterCourse**. What precondition might exist for registering for a particular course?
   4b. Assume that there is an operation called **Student.FindGPA** that calculates the GPA of a student from and array field **Student.Grades** which contains the student's grades. Propose a derivation rule for the GPA.

References and/or Recommended Readings

[Foster 2016] Foster, Elvis C. with Shripad Godbole. 2016. *Database Systems: A Pragmatic Approach*, 2nd ed. New York, NY: Apress. See chapters 1–4.

[Jacobson 1992] Jacobson, Ivar. 1992. *Object Oriented Software Engineering: A Use Case Approach*. Boston, MA: Addison-Wesley.

[Kendall 2014] Kendall, Kenneth E., and Julia E. Kendall. 2014. *Systems Analysis and Design*, 9th ed. Boston, MA: Pearson. See chapter 9.

[Lee 2002] Lee, Richard C. and William M. Tepfenhart. 2002. *Practical Object-Oriented Development With UML and Java*. Upper Saddle River, NJ: Prentice Hall. See chapters 9.

[Martin 1993] Martin, James and James Odell. 1993. *Principles of Object-Oriented Analysis and Design*. Eaglewood Cliffs, NJ: Pretence Hall. See chapter10.

[Pfleeger 2006] Pfleeger, Shari Lawrence. 2006. *Software Engineering Theory and Practice*, 3rd ed. Upper Saddle River, NJ: Prentice Hall. See sections 4.4 and 14.3.

Chapter 11

Project Management Aids

In this chapter, we examine three project management aids—PERT/CPM, Gantt charts, and project management software. The techniques are useful for managing resources, as well as monitoring of targets and expenditure during a software engineering project. The techniques are also useful in planning the software engineering project—a process that commences with the initial software requirement (ISR), is refined in the requirements specification, and further refined in the design specification. The chapter proceeds under the following captions:

- PERT and CPM
- The Gantt Chart
- Project Management Software
- Summary and Concluding Remarks

11.1 PERT and CPM

PERT is an acronym for *Program Evaluation and Review Technique*; CPM is an acronym for *Critical Path Method*. Developed in the 1950s by the US Navy, the two techniques are normally used together; they constitute the most popular methodology for managing projects.

Projects managed with PERT/CPM need not be related to information systems or software engineering; the technique is applicable to all disciplines.

Among the advantages of PERT/CPM are the following:

- The project is represented graphically, showing important events in some form of chronology.
- The critical path is identified (no delay is allowed along the critical path).
- The technique (CPM) allows for analysis and management of resource scheduling.

DOI: 10.1201/9780367746025-13

- The technique shows areas where tradeoffs in time or resources might increase the possibility of meeting major schedule targets.

In planning and preparing the PERT/CPM model for a project, the following steps are required:

1. Itemize activities in a tabular form showing for each activity its immediate predecessor, description, estimated time (duration).
2. Draw PERT Diagram using either the *activity-on-arrow* (AOA) approach or the *activity-on-node* (AON) approach.
3. For each activity (event), calculate the earliest start time (ES), the earliest finish time (EF), the latest start time (LS), and the latest finish time (LF). Also indicate the activity's duration (D).
4. Determine the critical path—the path through the network that has activities that cannot be delayed without delaying the entire project.
5. Conduct a sensitivity analysis to aid and inform the resource management process.

Example 11.1   A Simple PERT/CPM Example

Let us suppose that we are involved in a software engineering project to design, construct, and implement a software product. We will construct an activity table and then refine it into a PERT diagram.

11.1.1 Step 1: Tabulate the Project Activities

Figure 11.1 shows the schedule for our software engineering project. Notice that except for the starting activity, each activity has a predecessor. Each activity also has an estimated duration (typically expressed in days, but any unit of time may be used).

Activity	Predecessor	Description	Estimated Time (days)
A	--	Design System Architecture	30
B	A	Design Operation Specifications	12
C	B	Design Control Operations	8
D	B	Design Modification Operations	15
E	B	Design Inquiry/Report Operations	7
F	C	Code Control Operations	2
G	C	Prepare System. User Guide	5
H	F	Test Control Operations	2
I	D	Code Modification Operations	6
J	I	Test Modification Operations	4
K	H, J	Test Control/Modification Operations	2
L	E	Code Report Operations	3
M	L	Test Inquiry/Report Operations	1
N	K, M	Integration Test	4

Figure 11.1   Schedule of Activities for a Project

11.1.2 Step 2: Draw the PERT Diagram

The PERT Diagram is shown in Figure 11.2. Note that the AON convention is followed. Note also the use of *dummy nodes* (activities) to improve the clarity and readability of the diagram. In keeping with convention, a start node and a stop node have been introduced. They are dummy nodes that have no duration.

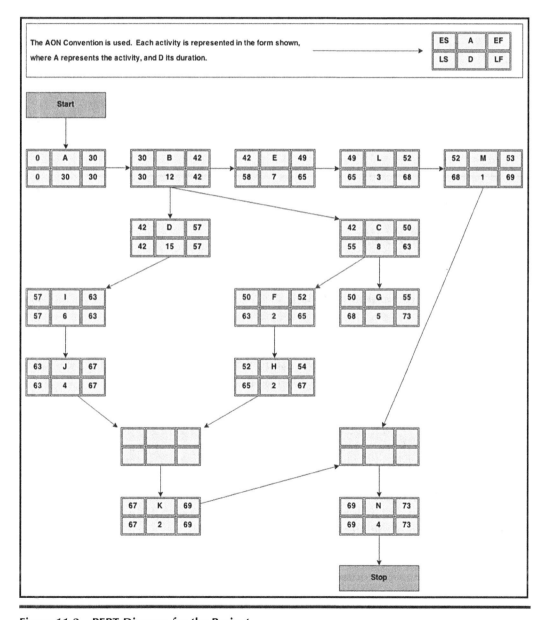

Figure 11.2 PERT Diagram for the Project

11.1.3 Step 3: Determine ES, EF, LS, and LF for Each Activity

The ES, EF, LS, LF values for each activity is shown in Figure 11.2. These values are to be calculated based on the formulae and rules shown in Figure 11.3.

ES (Activity) = EF (Immediately Previous Activity)   // if only one activity immediately precedes
            = Max [EF of all activities immediately preceding]   // if multiple activities precede

EF (Activity) = ES (Activity) + Duration of Activity

LS (Activity) = Min [LS of immediately succeeding activities] - Duration of Activity
            = LF (Activity) - Duration of Activity

LF (Activity) = LS (Activity) + Duration
            = Min [LS of all activities immediately following]

Slack (Activity) = LS (Activity) - ES (Activity)

Note:
1. ES of first activity is 0; LF of last activity, or terminal activities is equal to the earliest finish of the last activity.
2. ES and EF calculation originate at start and proceed to the end. LS and LF calculations originate at the finish and proceed backwards to the start.
3. A terminal activity is an activity that no other activity follows. In our example, G and N are terminal activities. For such activities, LF = EF of the last activity.

Figure 11.3   Basic PERT Diagram Calculations

11.1.4 Step 4: Determine the Critical Path

The critical path is the path with zero slack. No delay can be allowed on the critical path. In the example, the critical path is **A B D I J K N**. The project should not be allowed to overrun the critical path time.

11.1.5 Step 5: Conduct a Sensitivity Analysis

PERT sensitivity analysis relates to two main matters: determination of the estimated time for each activity, and analysis of the prospect of *crashing* the project.

Determination of estimated time is usually done using the formulae shown in Figure 11.4. It must be stated, however, that in many cases, rather than calculating the estimated time on each activity, and experienced software engineer will place an estimated time period on each activity, based on the rigor of that activity, the talents and skills of the software engineering team, and comparative knowledge of the duration of similar activities on other project(s).

Crashing is the process of shortening the project at an increased cost. The cost of crashing is evaluated and weighed against the cost of missing the project deadline. A final decision is then made. Please note:

- Only activities on the critical path must be crashed.
- Crashing may be considered when the cost of crashing is less than the overhead cost for the *crash period* contemplated. The total cost of the project would then be reduced.

$$Total \ Cost = Overhead \ Cost + Crashing \ Cost$$

$$t_e = 1/6 \, (t_o + 4t_m + t_p)$$
$$SD = (t_p - t_o)/6$$

Where
t_e = estimated time;
t_o = optimistic time;
t_p = pessimistic time;
t_m = most likely time.
SD = standard deviation.

Figure 11.4   Calculating Estimated Time for an Activity

The critical path can be also used as a guide in resource planning and management, as well as cost management. Since activities on this path cannot have non-zero slack, the project manager is informed on what activities have to start on time, and what activities may not.

Example 11.2   Illustrating a Basic Sensitivity Analysis on Example 11.1

Suppose that the project cost is $ 3,000 per week for two professionals, and that management can spend an additional $ 1,000 to have the project reduced by one week. Would crashing be feasible?

Projected duration of project is 73 days i.e. 14.6 weeks, to be reduced to 13.6 weeks.

$$\text{Original Project Cost} = 14.6 * \$3,000 = \$43,800$$

$$\text{Limit management can spend} = \$43,800 + \$1,000 = \$44,800$$

Payment rate is $ 1,500 per week, per professional
Crashing by one week would attract one additional professional i.e. $ 1,500, making

$$\text{Total Project Cost} = 13.6 * \$3,000 + \$1,500 = \$42,300.$$

Since this is less than the maximum the company can spend, crashing is feasible.

Note: If (Crash Cost + Overhead Cost > Max. Management Can Spend), then crashing is not feasible.

11.2 The Gantt Chart

The Gantt chart was devised by Henry Gantt (1920's). The project is represented on a bar chart based on the following guidelines:

- Tasks are listed vertically.
- A horizontal time scale is used to indicate time duration.
- A bar is used to depict activities.
- Some indication is given of the percentage completion of each activity (e.g. by color code).

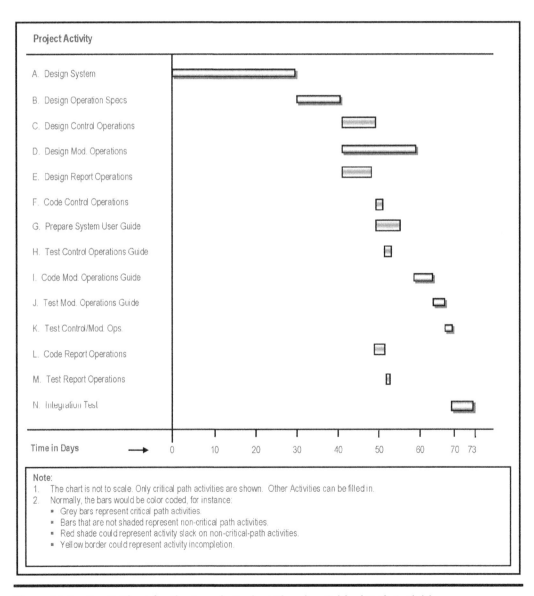

Figure 11.5   Gantt Chart for the Sample Project Showing Critical Path Activities

With the advance of PERT/CPM, the Gantt chart is best used to emphasize the critical path activities. The main advantages are:

• The chart is simple and easy to read.
• The chart can be used to manage the progress of activities on a software engineering project.

Figure 11.5 illustrates the Gantt chart for the project discussed in the previous section. On the diagram, the critical path activities are represented by grey bars, while non-critical-path activities are represented by bars that are not shaded. A third color code is often used to indicate the level of completion of certain activities.

The Gant chart, combined with the PERT diagram constitutes a powerful project management tool. Following are some ways the two techniques may be employed:

• The lead software engineer can make informed decisions about assigning certain responsibilities to members of the software engineering team. Since critical path activities cannot be delayed, such activities are usually assigned to stronger members of the team. Conversely, non-critical activities with longer slack times can be assigned to weaker members of the team.
• The project manager may use the techniques to assist in monitoring the progress of the project on a day-to-day basis.
• Should this become necessary, informed decisions about crashing can be made by identifying the critical path activities.

11.3 Project Management Software

You might be wondering, am I required to draw these PERT diagrams and Gant charts from scratch? The honest answer to this question is yes, you should know how to draw these diagrams from scratch. However, I believe what you want to know is whether there are software systems that can assist you in using these techniques. The answer to this inquiry is also a resounding yes. There is a wide range of project management software systems that are available in the marketplace. Reference [Capterra 2021] provides a comprehensive list of several such software alternatives that you can peruse.

In choosing a project management software system, here are some desirable features that should be facilitated:

• Management of multiple projects
• Specification of project activities for each project
• Choice from techniques such as PERT diagram, Gantt chart, and any other technique available
• Automated generation of the desired diagram
• Ability to print diagrams or save them in different formats
• Management of other resources such as budget, as well as physical resources
• Management of project requirements
• Project estimations

- Scheduling of activities for team members
- Progress tracking for overall project
- Progress tracking for team members
- Cost-to-completion tracking for each project
- Risk assessment and management
- Ability to probe what-if scenarios

In addition to project management software products, some of the more sophisticated software planning and development tools that are available (as discussed in Section 2.4.5) have project management facilities incorporated in them. Thus, the software engineer does not have to do all the hard work from scratch; you can use technology to make an otherwise very challenging job exciting and enjoyable.

11.4 Summary and Concluding Remarks

Here is a summary of what has been discussed in this chapter:

- A PERT diagram is a graphical representation of a project schedule that facilitates easy project management.
- Each node on the PERT diagram must have its name, duration (D), earliest start time (ES), earliest finish time (EF), latest start time (LS), and latest finish time (LF) clearly indicated.
- ES and EF calculations originate at the start point and proceed to the final point of the project. LS and LF calculations originate at the final point and work backward to the starting point.
- The critical path through a PERT diagram is the path that has zero slack on all its activities.
- Crashing is the process of shortening a project at increased cost. If the sum of the cost of crashing and the overhead cost is within the limit that the organization is prepared to spend, then crashing is feasible.
- The Gantt chart is best used to emphasize the critical path activities.
- A project management software is a software system that facilitates easy management of a project. These products are readily available in the marketplace.

Armed with this knowledge, you can now incorporate an impressive project schedule in your requirements specification, and thus complete your second major deliverable. The next section of the course discusses software design issues.

11.5 Review Questions

1. Clearly outline the steps involved in conducting a PERT-CPM analysis.

2. Explain how a PERT diagram and a Gantt chart may be used to assist in the management of a software engineering project.

3. A project involving the installation of a computer system consists of eight activities as shown in the activity table below:

Activity	Predecessor	Time (weeks)
A Equipment acquisition and Preparation	--	3
B Data conversion and Testing	--	6
C User Training	A	2
D Parallel Run	B, C	5
E Troubleshooting and Observation	D	4
F Fine-tuning	E	3
G Advanced User Training	B, C	9
H Signoff	F, G	3

a. Draw the PERT network for this project.
b. Identify the critical path.
c. What is the expected completion time for the project?
d. Identify three activities that can be delayed and for each, determine the slack.

4. LMX Software considering developing a new software product. The project activities identified so far are shown below:

Activity	Predecessor	Time (weeks)
A Prepare Requirements Spec	--	06
B Prepare Design Spec	--	08
C Prepare Development Team	A, B	12
D Develop and Test Module 1	C	04
E Develop and Test Module 2	C	06
F Develop and Test Module 3	D, E	15
G Conduct Comprehensive Sys Test	E	12
H Refine and Hand Over System	F, G	08

a. Develop a PERT network for the project.
b. Identify the critical path.
c. Determine the total project duration.
d. Identify three activities that can be delayed and for each, determine the slack.

References and Recommended Readings

[Anderson 2016] Anderson, David, et al. 2016. *Introduction to Management Science: Quantitative Approaches to Decision Making*, 14th ed. Boston, MA: Cengage Learning. See chapter 9.

[Capterra 2021] Capterra. 2021 Project Management Software. Accessed June 2017. https://www.capterra.com/project-management-software/.

[Peters 2000] Peters, James F. and Witold Pedrycz. 2000. *Software Engineering: An Engineering Approach*. New York, NY: John Wiley & Sons. See chapter 4.

[Pfleeger 2006] Pfleeger, Shari Lawrence. 2006. *Software Engineering Theory and Practice*, 3rd ed. Upper Saddle River, NJ: Prentice Hall. See chapter 3.

[Sommerville 2016] Sommerville, Ian. 2016. *Software Engineering*, 10th ed. Boston, MA: Pearson. See chapters 22 & 23.

SOFTWARE DESIGN   C

The next six chapters will focus on the Design Phase of the SDLC. The objectives of this phase are as follows:

- To use the findings of the requirements specification to construct a model of the software that will serve as the blueprint to the actual development of the software product
- To thoroughly document this model

The activities of this phase culminate into the third major deliverable of a software engineering project—the design specification. Chapters to be covered include the following:

- Chapter 12: Overview of Software Design
- Chapter 13: Database Design
- Chapter 14: User Interface Design
- Chapter 15: Operations Design
- Chapter 16: Other Design Considerations
- Chapter 17: Putting the Pieces Together

DOI: 10.1201/9780367746025-14

Chapter 12

Overview of Software Design

This chapter provides you with an overview of the software design process. It is the first on your experience towards the preparation of the next major deliverable in a software engineering project—the design specification. The chapter proceeds under the following captions:

- The Software Design Process
- Design Strategies
- Architectural Design
- Integration/Interface Design
- Software Design and Development Standards
- The Design Specification
- Summary and Concluding Remarks

12.1 The Software Design Process

Software design is a creative process that is perfected with experience. It cannot be fully learnt from a reading a series of chapters or chapters of a book; however, guidelines can be given. The objective of the software design process is the construction of a model that is representative of the required software system. The model must be accurate and comprehensive; it must conform to established software design and development standards; it must also meet the quality factors discussed in Chapters 1 and 3.

The design process starts with an informal design outline and undergoes several stages of refinement until a final model is obtained. The key deliverable from the design phase is the *design specification* (DS). Much feedback (backtracking) occurs among the stages of the design phase, as well as between the preparation of the requirements specification and the preparation of the design specification. In fact, it is a good idea to work on both deliverables concurrently, since changes in either will trigger adjustments in the other.

Figure 12.1 shows activities that go on during the design phase and what they lead to. Bear in mind that typically, some of these activities go on in parallel, rather than sequentially, so that feedback is constantly obtained and utilized. Below, these activities are clarified, but fuller discussion (on each activity) will follow in this and the next four chapters.

Design Activity	Resultant Design Spec Component
Architectural Design	System Architecture Specification
Integration/Interface Design	System Integration/Interface Specification
Database (Object Structure) Design	Database (Object Structure) Specification
Operations Design	Operations Specification
User Interface Design	User Interface Specification
Documentation Design	System Documentation Specification
Message Design	Message Specification
Security Design	Security Specification

Figure 12.1   Important Software Design Activities

Architectural Design: Architectural design relates to the subsystems and/or modules making up the system, as well as their interrelationships. Information topology charts (ITCs) and object-flow diagrams (OFDs) are useful diagramming tools that are used to represent the software model.

Integration/Interface Design: For each component (subsystem or module), its interface with other related components is designed and documented. The spin-off of interface design is the *system integration specification,* which must be unambiguous. This process is particularly useful in the scenario where the software system or component in being integrated into a larger software/information infrastructure, or where the CBSE life cycle model is being used.

Object Structure Design: Object structure design relates to the data structures used in the system—object types (information entities), relationships, integrity constraints, data dictionary, etc. Depending on the software engineering paradigm employed, it may be referred to as database design (in the case of traditional or hybrid approach), or object structure design (in the case of purely object-oriented approach). Object structure design results in the *object structure specification* of the system. This will be further discussed in Chapter 13.

User Interface Design: User interface design relates to screen design, menu structure(s), input and output design and in some systems, a user interface language. This will be further discussed in Chapter 14.

Operations Design: This involves preparing operation specifications for all the operational components of the software. Operations design also involves categorization of the operations as well as design of algorithms used in the operations. It will be further elaborated in Chapter 15.

Documentation Design: Documentation is an important part of software engineering. It includes all forms of product documentation (help system, users' guide, system manuals, etc.) and will be elucidated in Chapter 16.

Message Design: One important method of software communication with end users is via (error and status) messages. This aspect of software planning and construction is often ignored or belittled, to the horror of those who subsequently have to maintain the software. As you will see later (Chapters 15 and 16), message design is a very important component of operations design.

Security Design: This relates to the various authority constraints that different users of the software product will have. How this is designed and the kind of security mechanisms that may be required will depend to a large extent on the type of software and its intended users. These issues will be addressed in Chapter 16.

The importance of software design as a precursor to software construction cannot be over emphasized. Success in the former often leads to success in the latter. Additionally, flawed design inevitably leads to flawed development. Note, however, that in both cases, the implication is not necessarily reversible: You can have a flawed construction after a good design. Here are three important rules worth remembering (the first was learned from a former professor, E. K. Mugisa; the other two were constructed out of experience).

Rule 1: The sooner you run to the computer, the longer you stay there.

Rule 2: If it does not work on paper, it simply does not work.

Rule 3: Keep your design simple but not simplistic.

12.2 Design Strategies

The design approach may be *top-down* or *bottom-up*. Top-down design is traditionally associated with *function-oriented design* (FOD) and bottom-up with *object-oriented design* (OOD). This is somewhat misleading, however, since it is possible to have top-down design that is object oriented (this is sometimes recommended), and bottom-up design that is function oriented.

Until the early 1990s, FOD was the more widely used strategy. However, since the 1990s, OOD has gained widespread popularity, and is regarded today, as the preferred approach for many scenarios. Whichever strategy is employed, the quality factors mentioned in Chapters 1 and 3 apply here. Quality must be built into the design from the outset. In the interest of clarity, these quality factors are restated here:

Software Quality Factors: Efficiency, Reliability, Flexibility, Security, User-friendliness, Integrity, Growth Potential, Maintainability, Adequacy of Documentation, Functionality, Cohesiveness, Adaptability, Productivity, and Comprehensive Coverage

12.2.1 Function-Oriented Design

In FOD, the system is designed from a functional viewpoint, starting at the highest level, and progressively refining this into a more detailed design. The software engineer explicitly specifies the *what, wherefore and how* of the system; subsequent development must also deal with the *what, wherefore and how* as separate issues to be tied together.

FOD commences with the development of DFDs or POFs and other function-oriented system flow charts mentioned in Chapter 8.

FOD conceals details of an algorithm in a function, but the system state (data) is not hidden. In fact, there is a centralized system state, shared by all functions. Further, changes to a given function can affect the behavior of other functions, and by extension, the entire system.

FOD leads to a system of interacting functions, acting on files and data stores (system state). Here, the principle of *data independence* (immunity of application programs to structural changes of an underlying database) is very important. As you will see later, violation of this principle could be catastrophic.

12.2.2 Object-Oriented Design

Chapter 4 outlines some fundamental tenets of OOSE. In this subsection, let us recall and build on those concepts, by adding some additional insights.

In OOD, the system is designed from an object-oriented viewpoint, and consists of *objects* that have hidden states (data) and methods; each object belongs to an *object type*. The software engineer explicitly specifies the *what* of the system (in the form of object types), but focuses on *encapsulating the wherefore and how* of the system into its objects.

This course recommends that your OOD should commence with the development of an information topology chart (ITC) and/or object flow diagram (OFD), then employ other object-oriented diagramming techniques (such as ORDs, OSDs, transition diagrams, activity diagrams and other OO techniques) mentioned in Chapters 5, 8, and 9.

OOD conceals details of an object by encapsulating these details in an object type, implemented as a class, but the behavior is not hidden. Moreover, changes to the internal structure of an object type (class) are isolated from all other system objects.

OOD leads to a system of interacting objects, each with its own internal structure and operations (interaction is facilitated through the operations). Here, the principle of *encapsulation* (information hiding) is very important. Encapsulation to the OO paradigm is what data independence is to the FO paradigm.

As you are aware, an *object type* is an entity that has a state (data structure) and a defined set of operations that operate on that state. The state is represented by a set of attributes, thus giving the object a structure. Object behavior is represented by operations, which are implemented by methods. The attributes and operations of an object (type) are referred to as *properties*. Object types are implemented as *classes* that encapsulate both data structure (attributes) and behavior (operations). These concepts are illustrated in Figure 12.2; this figure shows the UML (Unified Modeling Language) representation of an object type called **Employee**. Such a diagram is called an *object structure diagram* (OSD), or simply a *class diagram*.

Objects communicate with each other via messages. Messages are usually implemented as procedure (or function or method) calls with appropriate parameters.

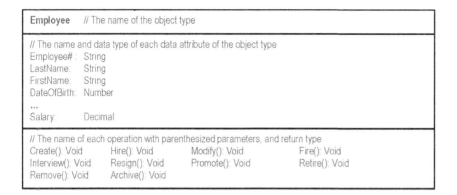

Figure 12.2 UML Diagram for an Object Type

OOD facilitates *inheritance*:

- Every object must belong to a class from which it inherits all its *properties* (attributes and operations). In object-oriented environments, a class is also an object, and may therefore inherit properties from another class called the *super-class* (also called the *parent class* or *base class*). Inheritance hierarchies can therefore be established as illustrated in Figure 12.3. The inheriting class is called the *subclass*, *child class*, or *derived class*.
- A class could inherit properties from more than one super-class. This situation is referred to as *multiple inheritance* and is illustrated in Figure 12.4. Multiple inheritance, while providing flexibility, can be a source of confusion; for this reason, it is avoided in certain environments (e.g., Java).
- A class could be a super-class and a sub-class at the same time; this is also illustrated in Figure 12.4.

OOD facilitates *amalgamation*: An object type may be defined as the amalgamation of several constituent object types. This is illustrated in Figure 12.4, where **Employee** is the composition of **EmpPersonalInfo**, **EmpEmploymentHistory**, **EmpAcademicRecord**, **EmpPublicationLog**, and **EmpExtraCurricular**. If the amalgamation is optional (i.e. the constituents can exist on their own), it is called an *aggregation*, and the diamond shape is not shaded in; if the amalgamation is mandatory (i.e. the constituents cannot exist on their own), it is called a *composition*, and the diamond shape is shaded in (as in the figure). Quite often, the distinction is not very clear, and the software engineer has to make a judgment call on the matter.

Let us now revisit the concept of *polymorphism*—the act of an object or operation taking on a different form, depending on the prevailing circumstances. To illustrate, consider an operation, **Create()**, which creates a college member object (refer to Figure 12.4). This operation could be made to behave differently, depending on whether the object being created is a **CollegeMember**, an **Employee**, or a **Student**. Programmatically, polymorphism may be realized via any of the following strategies: object casting, method overloading, methods with default parameters, method overriding, operator overloading, and generic program components (methods or classes).

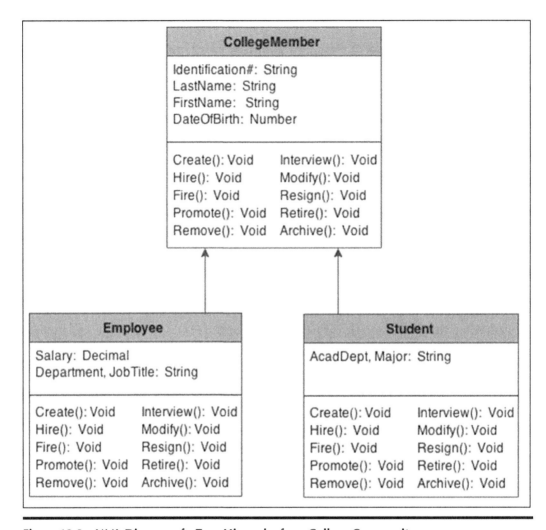

Figure 12.3   UML Diagram of a Type Hierarchy for a College Community

By observation, most OO environments are actually hybrid environments, where an OO *user interface* is superimposed on a *relational database* (examples include MySQL, DB2, and Oracle). This will be further elucidated later (subsection 12.2.4).

12.2.3 Advantages of Object-Oriented Design

Object-oriented design brings a number of advantages to the software engineering arena. Some of the commonly mentioned ones are as follows:

- Reusability of code
- Easier maintenance
- Enhancement of the understandability of the system
- Higher quality design

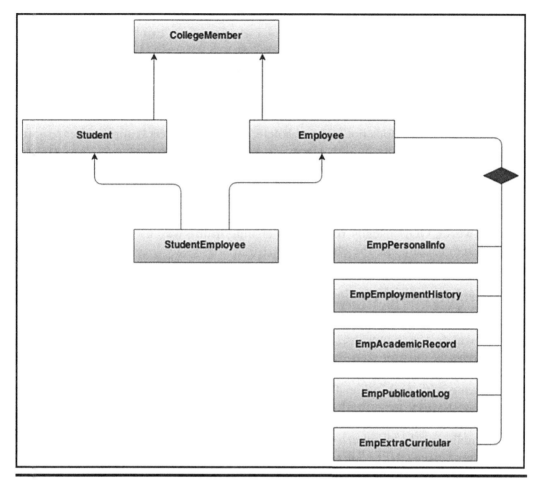

Figure 12.4 Abbreviated UML Diagram of an Inheritance Network for a College Community

- Design independence—classes which are independent of platforms can be developed
- Large, complex systems are simplified to interacting objects
- More powerful OO CASE tools and RAD tools have evolved. For many of these tools, the SDLC is simplified, merging actual design and development into what we call *modeling*, since many of the diagrams are said to be *executable diagrams* (meaning that actual code is generated from the diagram).

12.2.4 Using Both FO and OO Strategies

In large and complex projects, FOD and OOD can be skillfully employed to be complimentary rather than competing. Borrowing principles from both strategies often leads to better software design.

OOD seems most natural and beneficial at the highest and lowest levels of system design. At the highest level, it is more convenient to perceive a system as a set of interrelating component subsystems (which can be implemented as super-classes). At the lowest level, objects and operations can be implemented as programming classes.

FOD seems most natural and beneficial at the intermediate level, where the system can be viewed as a set of interacting operations (implemented as programs). Also, remember that at some stage, methods for objects must be specified; this often involves some amount of procedural programming.

Object technology does not make obsolete, all traditional (FO) approaches to software construction. Rather, it adds clarity, creativity, and convenience to the software engineering discipline. In many hybrid environments, an OO user interface is often superimposed (made to access) a relational database. In fact, many of the available contemporary software development products reflect this approach.

One final note: FOD may or may not result in a change in the way users do their work. However, OOD usually results in a change in the way users do their work… and the users usually like the change.

12.3 Architectural Design

Complex software systems are typically composed of subsystems and/or modules, some of which may themselves be complete software systems. These components must seamlessly integrate into the larger system. Architectural design addresses this challenge by determining what the components are and how they integrate and communicate with each other. The following are some considerations (factors) that influence system architecture decisions:

Performance: If performance is a critical requirement, the architecture should seek to minimize the number of components that have to cross-communicate (i.e. employ more *large-grain* components than *fine-grain* components).

Security: If security is critical, the architecture should provide a stringent security mechanism (preferably layered with the most critical resources at the innermost layer).

Availability: If availability is a critical requirement, the architecture should include controlled redundancies by making components as independent as possible.

Maintainability: If maintainability is a critical requirement, the architecture should employ fine-grain, self-contained components that can be readily changed.

These requirements typically do not synchronize with each other, so trade-offs will be necessary. Two issues of paramount importance in architectural design are *resource sharing* and *system control*. We will discuss these issues in the next two subsections, then close the section with a brief discussion on system components.

12.3.1 Approaches to Resource Sharing

There are four approaches to resource sharing:

- The Repository Model
- The Client-server Model
- The Abstract Machine Model
- The Component Model

12.3.1.1 Repository Model

In the repository model, all shared resources (data and operation) are stored in a central holding area (library). Each component has access to the repository. The repository may consist of more than one library of shared resources. It may contract or expand as the system is maintained during its useful life.

Examples of software systems that employ this approach include management information systems, CAD systems, and CASE toolsets. Figure 12.5 also provides, by way of example, an overview diagram of the CUAIS project (mentioned in earlier chapters), assuming a repository model. In this approach, the central CUAIS database and controller subsystem interfaces with all other subsystems.

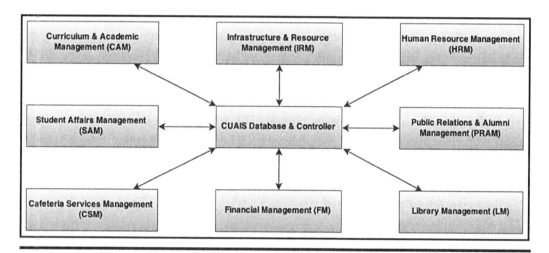

Figure 12.5   OFD for the CUAIS Project, Assuming Repository Model

Advantages of repository-based software include the following:

- Efficient data sharing; no need to explicitly transmit data for sharing
- High level of data independence: subsystems need not be concerned with how other subsystems use data
- Backup and recovery can easily be managed
- Integrating new resources is easy

Disadvantages of repository-based software include the following:

- Since subsystems have to agree on the data model, performance may be compromised.
- Evolution may be difficult, since the data model is standardized. Translating it to a new model can be costly.
- Flexibility in component subsystems may be lost on issues such as backup and recovery.
- It might be challenging to distribute the repository over a number of platforms.

When one considers that the central database may be conveniently partitioned to cater to the needs of each component, these disadvantages can be easily mitigated.

12.3.1.2 Client-Server Model

In the client-server model, a set of independent network servers offers services which may be called on by members of a set of client systems. There is a server version of the software system that typically runs on a designated server machine, and a client version that runs on designated client machines. A network allows communication among clients and servers.

Examples of software systems that employ this approach include management systems with stand-alone components. Also, Figure 8.7 (of Chapter 8) provides a partial overview of the CUAIS project without the central database. It is repeated here (as Figure 12.6) for ease of reference. In this approach, each subsystem would operate independently, but with the capability of communicating with other subsystems comprising the system.

Among the significant advantages of the client-server model are the following:

- A *distributed architecture* is facilitated. The implications of this are very profound and far-reaching, as you will discover in other advanced courses (*electronic communication systems*, *database systems*, and *operating systems*).
- The system can easily grow with additional clients and/or servers.
- The approach provides flexibility in how components communicate (via request, data transfer, or service transfer).
- Accessibility is enhanced.

The model faces three major challenges:

- More sophisticated software systems are required for resource management and communication among the components (but the required technology is readily available).
- Each server must take responsibility for backup and recovery issues.
- In situations where data is replicated in the interest of efficiency, maintaining the integrity of the data is problematic.

12.3.1.3 Abstract Machine Model

In this model (also called the *layered model*), the software is organized into layers, each providing a set of services. Each layer communicates directly only to the layer immediately above or below it.

By way of example, the *open system interconnection model* (OSI model) for communications protocols typifies a layered approach. Additionally, some operating systems and compilers are designed using the layered model. Chapter 25 (Section 25.1.2) provides additional insight on this approach.

The main advantage of the layered approach is that problems can be easily isolated and addressed. The main drawback is that the approach is not relevant to all kinds of system problems.

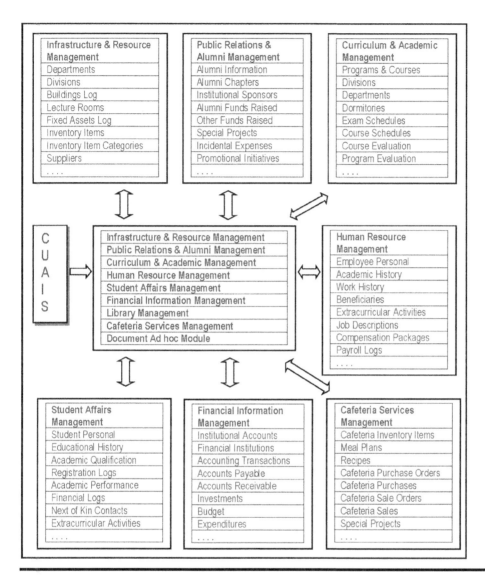

Figure 12.6   Partial Information Topology Chart for the CUAIS Project

12.3.1.4 Component Model

The basic component model describes a pre-client-server approach where a set of component systems resides on a given machine, typically in a network environment. Obviously, this scenario can be easily facilitated in a client-server environment. But observe that it could also be implemented in a non-client-server environment, for example, a minicomputer environment (precisely what used to happen prior to the introduction of client-server technology).

The component-based systems are made to communicate via interface programs and/or the introduction of controlled redundancy. To illustrate, suppose that **System-A** and **System-B** both use a database file, **FileX**. Let us assume that **System-A** owns **FileX** and therefore has update rights. **System-B** has a copy of **FileX**, namely **FileXp**, which it uses for data retrieval only. An interface program could be responsible for periodically copying **FileX** to **FileXp**, so that **System-B** "sees" accurate data.

Multi-component as well as single-user systems (on stand-alone machines) also typify the component model. The approach is relevant whenever a complex software system can be constructed by putting autonomous components together, or the system can be decoupled into identifiable autonomous components. Figure 12.6 would therefore also be applicable if the CUAIS project was being constructed using the component approach.

Two significant advantages of component-based systems as described here are local autonomy and fast response. Two drawbacks are the potential data integrity problems that they pose, and their limited scope of applicability.

With advances in OOM, we have witnessed a resurgence of a revised component model, called component-based software engineering (CBSE). As mentioned earlier in the course (review Section 1.4), CBSE promises significant benefits to the software engineering discipline.

12.3.2 System Controls

System control relates to how executing components are managed. There are two approaches that you should be familiar with—*centralized control* and *event-based control*

In the centralized control approach, one system component is designated as the system controller and has responsibility for execution of the other components. Two models are prevalent:

- The *call-return mode*l is relevant to sequential systems: A component is invoked by a call; after execution, control goes back to the caller.
- The *manager model* is relevant to concurrent systems: One component controls the starting, stopping, and coordination of other components. Components may run sequentially or in parallel.

In the event-driven approach, the events control the execution of system components. Two models are prevalent:

- In *a broadcast model*, an event is broadcasted to all components. A component that can handle the event responds to it. *Ethernet* and *Token Ring* network protocols are good examples of the implementation of broadcast-driven control.
- In an *interrupt-driven model,* an interrupt handler responds to interrupts from various components and arbitrates which component will respond to the intercept. Operating systems software and real-time systems are typically designed with interrupt handlers.

12.4 Integration/Interface Design

System integration is easiest in the waterfall lifecycle model. However, as you are now aware, the reality is that quite often, one of the more pragmatic approaches to software engineering is employed, leading to concerns about integrating each new component into the existing software architecture. Since integration design is very closely related to the architectural design it is not unusual that the former is intercorporate into the latter. Here, the software engineer specifies exactly how the various system components will communicate.

In specifying how new components will communicate with previously existing components, please note the following guidelines:

- Identify and specify all *utility operations* to be employed.
- Design intermediate data structures that may be required.
- Identify and specify system-wide utilities, for example, date validation, error message processors, help retrieval processors, printing aids, etc.

12.5 Software Design and Development Standards

Software standards are very important in the design and development of software. The larger and more complex the project, the greater is the need for (and therefore benefit of) such standards. Every software engineering company should have such standards; standards should also exist in companies where information technology (more specifically, the development of software systems) plays a critical role in the realization of corporate goals.

12.5.1 Advantages of Software Standards

The existence of software standards leads to a number of benefits, some of which are mentioned below:

- Enhancement of consistency in design and coding
- Enhancement of understandability of the system
- Enhancement of the maintainability of the system
- Improvement of the likelihood of reliable software
- Enhancement of good software documentation
- Imposition of some degree of order on software design and coding; programmers have to code to established standards, and not go off in unwieldy explorations
- Enhancement of a high level of efficiency and productivity during software construction
- Facilitation of a higher level of quality of the final product
- Facilitation of better project management

These advantages significantly contribute to the development of software systems of high quality. The converse is also true: failure to develop and observe meaningful software standards is a prescription for spaghetti code and mediocre software systems. It is through standards that the software engineering industry has been able to produce platform-independent software systems that are immune to international borders and cultural barriers. Had there not been software standards, we would not have many things that we sometimes take for granted—the World Wide Web, operating systems, database systems, telephone services, television, radio, aviation, etc. In other words, without standards we'd all still be in the Stone Age, and software engineering as we know it would not exist.

12.5.2 Issues That Software Standards Should Address

The essential issues that software standards should address vary from one organization to another. They also vary with the category of software being developed. However, there are some fundamental ones that should always be treated. These are mentioned below:

Naming of Non-database (Hardware Related) Objects: This includes the naming of workstations, printers, user profiles, output queues, etc. It includes all relevant objects related to the hardware used in the organization.

Naming of Database Objects: This activity includes the naming of physical database files (tables), logical views, database stored procedures and triggers, etc.

Naming of User Interface Objects: This includes the naming of operations (or application programs), menus, and other user interface objects that may be employed in software construction. These objects will vary according to the software development tool being used.

Database Design: Standards must be set in respect of database design. Chapter 13 and your course in *database systems* will elucidate this area.

Screen and Report Layout: It is always a good habit to have standardized screen and report layouts for a given system. Chapter 14 will provide further clarification.

Forms Design: Where the software requires manual inputs, or generates standard forms as output, these forms must be carefully designed. Chapter 14 will elucidate this area.

User Interface Design: There should be guidelines with respect to the user interface (see Chapter 14).

Operation Specification. This relates to the method used to specify the required operations of a system. Chapter 12 will provide further clarification.

Programming standards: This includes generic coding approaches, use of special keyboard keys (e.g. function keys), error handling, etc. Chapters 14 and 15 will provide further clarification.

Software Development Tools: It may be necessary to define guidelines and benchmarks for the use of certain software development tools.

System Documentation: There should be clear guidelines for software documentation in respect of system help, user manuals, and other technical documents.

Figure 12.7 provides an example of an object naming convention that has been used by the current author on several software engineering projects. The inventory management system (IMS) project of Appendix C applies or prescribes this convention for several of the above-mentioned areas.

Object Name: SSSS_XXXXXXX_MMn where interpretations apply.
- SSSS represents the system or subsystem abbreviation (2 – 4 bytes);
- MMn   represents the object mode or purpose (1-3 bytes);
- XXXXXXXXX represents the descriptive name of the object (4-15 bytes).

For example, valid subsystem abbreviations for an Inventory Management System (IMS) are as follows:
- AM: Acquisitions Management Subsystem
- FM: Financial Management Subsystem
- SC: System Controls Subsystem

Valid mode abbreviations include:
- DM: A data model — could represent a whole database, a base relation, or group of base relations
- BR: A base relation (if relational DB model)
- OT: An object type (if OO DB model)
- LVn: A logical view (e.g. LV1, LV2, etc.)
- NXn: An index to a base table or object type (e.g. NX1, NX2, etc.)
- PK: Primary Key
- FKn: Foreign Key (e.g. FK1, FK2, etc.)
- ICn: Integrity Constraint (e.g.IC1, IC2, etc.)
- AO: An ADD operation
- MO: A MODIFY operation
- ZO: A DELETE (Zap) operation
- IO: An INQUIRE operation
- FO. A FORECAST operation.
- RO: A REPORT operation
- XO: A utility operation
- DS: A database synonym or alias of a known database table
- DC: A database constraint
- DT: A database Trigger
- DP: A database procedure or function
- DK: A database package
- MF: A Message file — a special-purpose database table (file) to store the text (and other essential details) for diagnostic error and status messages

The descriptor used for a database base relation or object type is consistently used for other objects that directly relate to that object. For example, the objects related to the management of inventory items may be:
- AM_ItemDef_BR — a base relation to store data on inventory items
- AM_ItemDef_NX1 — an index on the base relation
- AM_ItemDef_AO — an operation to ADD inventory items
- AM_ItemDef_MO — an operation to MODIFY inventory items
- AM_ItemDef_ZO — an operation to DELETE inventory items
- AM_ItemDef_IO — an operation to INQUIRE on inventory items
- AM_ItemDef_RO — an operation to REPORT on inventory items
- AM_ItemDef_XO — a utility operation related to inventory items
- AM_ItemDef_LV1 — a logical view of the base relation

Attribute implementation names are merely abbreviations of their more descriptive names, prefixed by an appropriate abbreviation of the entity.

Figure 12.7   Proposed Object Naming Convention

12.6 The Design Specification

The *design specification* (DS) is another important deliverable for a software engineering project; it signals the end of the design phase (but remember, you may use a reversible life cycle model). Like the requirements specification, it is a formal, comprehensive document, providing further technical insights to the former. If the requirements specification is the initial blueprint of the software, the design specification is the final blueprint; it represents a refinement of the requirements specification, and contains information that is used for the construction of the software product.

Observe: Some textbooks discuss one software blueprint in the form of a highly technical requirements specification. This course favors a less-intimidating requirements specification as a precursor to a more detailed and technical design specification.

12.6.1 Contents of the Design Specification

The design specification includes details such as:

*   Acknowledgments
*   Introductory Notes
*   System Overview
*   Database Design Specification (Chapter 13)
*   User Interface Design Specification (Chapter 14)
*   Operations Design Specification (Chapter 15)
*   Architectural Design Specification
*   Product Documentation Specification (Chapter 16)
*   Message Management Specification (Chapter 16)
*   Software Development Standards (possibly as an appendage)
*   Refined Schedule for Software Development and Implementation (review Chapter 11)

12.6.2 How to Proceed

You proceed to construct the design specification by using the requirements specification as your input. If the requirements specification is as accurate and comprehensive as it should be, then this is all you need. Figure 12.8 provides basic guidelines,

Step	Clarification
1. Overview Refinement	Refine the overview and introductory notes. Assuming the OOD approach, your overview should include a refined information topology chart and possibly an object flow diagram.
2. Define of Software Engineering Standards	If there are no organizational standards for software design and development, develop standards for the project.
3. Preparation of Detailed Specifications	For each subsystem, provide the following: ■ Item 1 ■ Database (or object structure) specification ■ Operations Specification ■ User Interface Specification ■ Message Specification ■ Documentation Specification ■ Security Specification
4. Preparation of Development Schedule	Prepare or refine the schedule for software development and implementation.
5. Specification Refinement	Conduct brainstorming sessions to verify the specifications.
6. Final Refinement	Prepare the acknowledgements and refine the introductory statements.

Figure 12.8   Steps in Constructing the Design Specification

but please note that this is not cast in stone: your approach may vary in some areas, depending on the nature of the project. Also observe that this figure is a refinement of the lower portion of Figure 10.10 (of Chapter 10).

If at this point, you are still not confident about putting a design specification together, do not panic; after the next four chapters, and with practice, you will be in much better shape. The tools available for putting this deliverable together are very important. CASE tools (particularly OO-CASE tools) are excellent (review Chapters 1 and 2). However, even in their absence, you can still be very effective with basic desktop processing tools.

Finally, please note that good software engineering will ensure that the requirements specification and the design specification both inform the final product documentation. Moreover, with experience, you will be able to work on both deliverables in parallel rather than in sequence (review Section 12.1). An excellent product is a credit to excellent design, not a coincidence.

12.7 Summary and Concluding Remarks

It is time once again to summarize what has been covered in this chapter:

- The software design process consists of architectural design, interface design, object structure design, operations design, user interface design, documentation design, message design, and security design.
- The design may proceed as FOD or OOD. OOD relies on OO methodologies and is the preferred approach for contemporary software systems.
- Architectural design addresses the issue of integrating the various components that comprise the software system. It addresses issues such as resource sharing and system controls. Alternate strategies for resource sharing include the repository model, the client-server model, the abstract machine model, and the component model. Alternate strategies for system control include the centralized control and the event-driven control. In many cases, interface design is combined with architectural design.
- Software design and development standards are absolutely necessary if software products of a high quality are to be developed.
- The design specification (DS) is the software engineering deliverable that results from the software design process. This becomes the blueprint for the software system. The software engineer must be clear on what goes into the DS and how to prepare it.

The next few chapters discuss important components of the design specification (DS). As you proceed through these chapters, please reserve the liberty to periodically examine Appendix C as this provides excerpts from a DS for the inventory management system of earlier mention.

12.8 Review Questions

1. Outline the software design process, explaining each aspect.

2. Compare function oriented with object-oriented design.

3. Discuss the four approaches to resource sharing as covered in the chapter. For each approach, cite advantages and disadvantages.

4. Examine Figure 12.5. What conclusions can you draw about the system represented? Also examine Figure 12.6. What conclusions can you draw about the system represented?

5. Discuss the importance of software design and development standards. Describe six important issues that these standards must address.

6. Which deliverable comes after software design? What is this deliverable comprised of? How would you proceed to construct such a deliverable?

References and Recommended Readings

[Lee 2002] Lee, Richard C. and William M. Tepfenhart. 2002. *Practical Object-Oriented Development With UML and Java*. Upper Saddle River, NJ: Prentice Hall. See chapter 4.

[Martin 1993] Martin, James, and James Odell. 1993. *Principles of Object-Oriented Analysis and Design*. Eaglewood Cliffs, NJ: Prentice Hall. See chapters 1 and 3.

[Peters 2000] Peters, James F. and Witold Pedrycz. 2000. *Software Engineering: An Engineering Approach*. New York, NY: John Wiley & Sons. See chapter 7.

[Pfleeger 2006] Pfleeger, Shari Lawrence. 2006. *Software Engineering Theory and Practice*, 3rd ed. Upper Saddle River, NJ: Prentice Hall. See chapters 5 and 6.

[Pressman 2015] Pressman, Roger. 2015. *Software Engineering: A Practitioner's Approach*, 8th ed. New York, NY: McGraw-Hill. See chapters 9–14.

[Schach 2011] Schach, Stephen R. 2011. *Object-Oriented & Classical Software Engineering*, 8th ed. New York, NY: McGraw-Hill. See chapters 7, 11–14.

[Sommerville 2016] Sommerville, Ian. 2016. *Software Engineering*, 10th ed. Boston, MA: Pearson. See chapters 4–6.

[Zhu 2005] Zhu, Hong. 2005. *Software Design Methodology: From Principles to Architectural Styles*. Boston, MA: Elsevier.

Chapter 13

Database Design

If you review the OO modeling hierarchy (Figure 10.11) of Chapter 10, you will notice that the left-hand side is characterized by the term *object structure analysis* (OSA). This chapter focuses on OSA, or more precisely, *object structure design*. In this chapter, we shall relax any distinction between *database design* and *object structure design*, for the following reason: As established in the previous chapter, irrespective of the software engineering paradigm employed, data and object structure design are of paramount importance. By way of observation, most software engineering environments embrace the idea of a relational database, upon which an OO user interface is superimposed. This chapter presumes that convention, and provides you an overview of the database design experience. For a more comprehensive coverage of database systems, please refer to the recommended readings.

The chapter proceeds under the following captions:

- Introduction
- Approaches to Database Design
- Overview of File Organization
- Summary and Concluding Remarks

13.1 Introduction

A *database* is the record-keeping component of a software system. Database design is critical part of software engineering. Underlying most software products is a database that stores data that is critical to the successful operation of the software. Figure 13.1 provides you with some examples of this. In most cases, the database is superimposed by a user interface, and may therefore sometimes not be obvious to the end user. However, whether it is obvious or not the database component is real and potent. A full discussion of database design is beyond the scope of this course; it is best done in a course in database systems. This chapter provides you with a useful overview of the territory. Much of the information presented here is really a summary of more elaborate details available in the author's work on database systems (see [Foster 2016]).

Database design is very crucial as it affects what data will be stored in and therefore accessible from the software system. Hence, it affects the success of the system. Poor design leads to the following software flaws:

- Poor response time, hence
- Poor performance
- Data omissions
- Inappropriate data structures
- Redundancies
- Modification anomalies
- Integrity problems
- Lack of data independence
- Difficulty in system maintenance
- Inflexibility
- Lack of clarity
- Security and reliability problems Pressure on the programming effort to compensate for the poor design

Software Category	Database Need
Operating Systems	A sophisticated internal database is needed to keep track of various resources of the computer system including external memory locations, internal memory locations, free space management, system files, user files, etc. These resources are accessed and manipulated by active jobs. A job is created when a user logs on to the system, and is related to the user account. This job can in turn create other jobs, thus creating a job hierarchy. When you consider that in a multi-user environment, there may be several users and hundreds to thousands of jobs, as well as other resources, you should appreciate that underlying an operating system is a very complex database that drives the system.
Compilers	Like an operating system, a compiler has to manage and access a complex dynamic database consisting of syntactic components of a program as it is converted from source code to object code.
Information Systems	Information systems all rely on and manipulate internal databases, in order to provide mission-critical information for organizations. All categories of information systems (DSS, EIS, MIS, WIS and SIS) are included.
Expert Systems	At the core of an expert system is a knowledge base containing cognitive data which is accessed and used by an inference engine, to draw conclusions based on input fed to the system.
CAD, CAM and CIM Systems	A CAD, CAM or CIM system typically relies on a centralized database (repository) that stores data that is essential to the successful operation of the system.
Desktop Applications	All desktop applications (including hypermedia systems and graphics software) rely on resource databases that provide the facilities that are made available to the user. For example, when you choose to insert a bullet or some other enhancement in an MS Word document, you select this feature from a database containing these features.
CASE and RAD Tools	Like desktop applications, CASE and RAD tools rely on complex resource databases to service the user requests and provide the features used.
DBMS Suites	Like CASE & RAD tools, a DBMS also relies on a complex resource database to service the user requests and provide the features used. Additionally, a DBMS maintains a very sophisticated meta database (called a data dictionary or system catalog) for each user database that is created and managed via the DBMS.

Figure 13.1   Illustrations of the Importance of Database

Poor database design puts pressure on the software development team to program its way out of the poor design. By contrast, good design leads to the exact opposite of these conditions induced by poor design.

Some objectives of database design include the following:

- Comprehensive data capture
- Efficiency
- Flexibility and Reliability
- Control of Redundancy
- Security and Protection
- Consistency and Accuracy
- Ease of access and ease of change
- Availability of information on demand
- Desirable data integrity
- Data independence—immunity of application programs to structural, storage, or hardware changes of the database
- Clarity and multi-user access

13.2 Approaches to Database Design

Broadly speaking, there are two approaches database design:

- Conventional files
- Database approach which includes
  o Relational model
  o Object-oriented model
  o Hierarchical model
  o Network model
  o Inverted List model

13.2.1 Conventional Files

Figure 13.2 illustrates the idea of conventional file approach. Application programs exist to update files or retrieve information from files.

This is a traditional approach to database design that might still abound in very old *legacy systems* (to be discussed in Chapter 22). You may use this approach if the software system is already designed using this approach, and the task is to maintain it.

Note: Do not attempt to redesign the software system without management consent. Also, be aware that people (including managers) sometimes get annoyed with a software engineer who walks around looking for every problem to fix. Ironically, fixing problems often created by human ineptitude or limitations is an integral part of your job. Just be discreet in the execution of your job (Chapters 18 and 19 provide more guidelines on how to conduct yourself on the job).

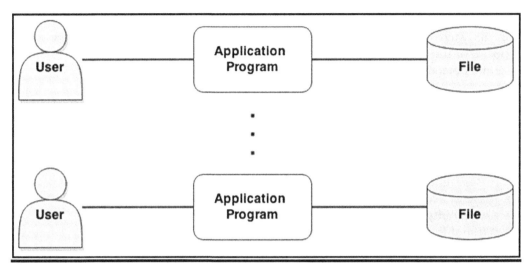

Figure 13.2   Conventional File-Based Design

13.2.2 Database Approach

In the database approach, a database is created and managed via a database management system (DBMS) or CASE tool. A user interface, developed with appropriate application development software, is superimposed on the database, so that end users access the system through the user interface. Figure 13.3 illustrates the basic idea.

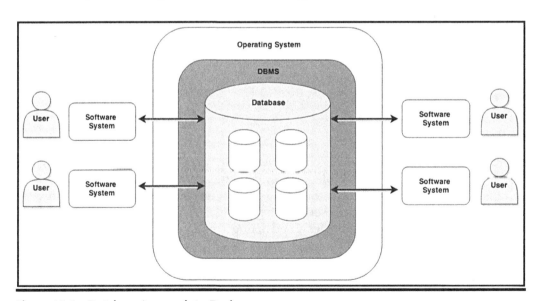

Figure 13.3   Database Approach to Design

Of the five methodologies for database design, the relational model and the OO model are the two that dominate contemporary software engineering; this is expected to continue into the foreseeable future. The other three approaches are traditional (from the 1960s and 1970s), but occasionally show up in legacy systems. They will not be discussed any further; for more information on them, see the recommended readings.

Following are the steps involved in designing a relational or OO database:

1. Identify data entities or object types
2. Identify relationships
3. Eliminate unnecessary relationships
4. Develop an *entity-relationship diagram* (ERD) or an *object-relationship diagram* (ORD)
5. Prepare the database specification
6. Develop and implement the database

Note: Step 6 belongs to the field of database systems and will not be explored any further in this course. However, you should appreciate the close nexus between database design and software engineering.

13.2.3 *Identifying and Defining Entities or Object Types*

Identifying information entities or object types requires skills, experience, and practice. In this regard, the techniques discussed in Chapter 7 are applicable. As you will recall, an information entity is a concept or object about which data is (to be) stored. An object type is concept or object about which data is (to be) stored, and upon which a set of operations are defined. If you relax the latter part of the definition of an object type, you will readily see that information entities and object types are similar.

Of equal importance is the structure of each information entity (object type). We define structure in terms of *attributes:* An entity (object type) is made up of *attributes* (also called elements or properties) which describe the entity (object). Attributes are non-decomposable (atomic) properties about the entity (object), as illustrated in the example below (Figure 13.4).

Entity	Attributes	Primary Key
Department [MSPDEPT]	Dept Number [DeptNo, N4], Department name [DeprName A30]	DeptNo
Location [MSPLOCN]	Location [LocCode, A7], Location Name [LocnName, A30], Distance from Head Office [LocDist, N5,2]	LocCode
Employee [MSPEMPL]	Employee Number [EmpNum, N7], Employee Name [EmpName, A15], Date of Birth [EmpDOB, N7], ...	EmpNum
Projects [MSPPROJ]	Project Number [ProNum, N5], Project Name [ProNam, A30] ...	ProjNum
Inventory Item [MSPItem]	Item Number [ItemNum , A8], Item Name [ItemName, A20], Item Unit [ItemUnit, N5,2] ...	ItemNum
Supplier [MSPSUPLR]	Supplier Number [SupplNum, A7], Supplier Name [SuplName, A30]	SupplNum
Warehouse [MSPWHOUS]	Warehouse Number [WHNum, A3], Warehouse Name [WHName, A30], Warehouse Size [WHSize, N5,2] ...	WHNum

Note:
1. It is a good habit to indicate the proposed system same (based on an established naming convention) for each database object (in the figure, system names for data entities are indicated in square brackets).
2. For each attribute, assign a system name as well as some indication of the physical characteristics of the attribute. In the figure, this information is also enclosed in square brackets next to the respective attributes. The simple convention used here, is N for numeric data and A for alphanumeric data, with the length and decimal positions indicated.
3. For a more comprehensive coverage, it is advisable to allow for a row for each attribute, so that additional information about the attribute can be specified.
4. The process can be automated.

Figure 13.4   Partial Entity-Attributes List for a Manufacturing Environment

13.2.4 *Identifying Relationships*

Identifying *relationships* among entities (object types) also requires skills, experience, and practice, but can often be intuitively recognized by observation. To identify relationships, you have to know what a relationship is and what types of relationships there are. Your course in database systems will elucidate these issues to some level of detail. For now, you may consider a relationship as a mapping involving two or more information entities (or object types) so that a data item (an object) in one relates in some way to at least one data item (object) in the other(s) and vice versa. There are seven types of relationships:

- One-to-one (1:1) relationship
- One-to-many (1:M) relationship
- Many-to-one (M:1) relationship
- Many-to-many (M:M) relationship
- Component relationship (if OO database)
- Aggregation relationship (if OO database)
- Super-type-sub-type relationship (if OO database)

The first four types of relationships are referred to as *traditional relationships* because up until object model (for database design) gained preeminence, they were essentially the kinds of relationships that were facilitated by the relational model. Observe also, that the only difference between a 1:M relationship and an M:1 relation is a matter of perspective; thus, a 1:M relationship may also be described as an M:1 relationship (so that in practice, there are really three types of traditional relationships). Put another way:

> If E1 and E2 are two information entities (or object types) and there is a 1:M relationship between E1 and E2, an alternate way of describing this situation is to say that there is an M:1 relationship between E2 and E1.

For traditional relationships, to determine the type of relationship between two entities (object types) E1 and E2, ask and determine the answer to the following questions:

- How many data items (objects) of E1 can reference a single data item (object) of E2?
- How many data items (objects) of E2 can reference a single data item (object) of E1?

To test for a component relationship between any two relations (object types) E1 and E2, ask and determine the answer to the following questions:

- Is (a data item of) E1 composed of (a data item of) E2?
- Is (a data item of) E2 composed of (a data item of) E1?

For a subtype relationship, the test is a bit more detailed; for entities (or object types) E1 and E2, ask and determine the answer to the following questions:

- Is (a data item of) E1 also an (a data item of) E2?
- Is (a data item of) E2 also an (a data item of) E1?

The test is identical for object types, except that in the object-oriented paradigm, the term *instance* is preferred to "data item." Possible answers to the questions are always, sometimes, or never. The possibilities are shown in Figure 13.5.

Possibility	Implication
E1 always E2, E2 always E1	E1 and E2 are synonymous
E1 always E2, E2 sometimes E1	E1 is a subtype of E2
E1 always E2, E2 never E1	Makes no sense
E1 sometimes E2, E2 always E1	E2 is a sub-type of E1
E1 sometimes E2, E2 sometimes E1	Inconclusive
E1 sometimes E2, E2 never E1	Makes no sense
E1 never E2, E2 always E1	Makes no sense
E1 never E2, E2 sometimes E1	Makes no sense
E1 never E1, E2 never E1	No subtype relationship exists

Figure 13.5 Testing for Sub-type Relationship

Example 13.1 Identifying Relationships for a Manufacturing Environment

Referring to the sample manufacturing subsystem of Figure 13.4, several relationships can be identified, as indicated in Figure 13.6.

Name of Relationship	Participating Entities	Type	Optional or Mandatory
Supplies	Suppliers, Inventory Items	M:M	M
P-uses	Projects, Inventory Items	M:M	M
Assigned	Projects, Employees	M:M	M
Belongs	Employee, Department	M:1	M
Hosts	Location, Department	1:M	M
Situated – W	Warehouse, Location	1:1	M
Situated – S	Supplier, Location	1:M	M
Supplier to Projects	Supplier, Projects	M:M	M
SPJ	Supplier, Inventory Item, Project	M:M	O
Composed of	Inventory Item, Inventory Item	M:M	M
Stocks	Warehouse, Inventory Item	M:M	M

Figure 13.6 Relationships List for a Manufacturing Environment

13.2.5 Developing the ERD or ORD

An *entity-relationship diagram* (ERD or E-R diagram) is a graphical illustration of entities and their relationships in the database. In the OO paradigm, the equivalent diagram is called an *object-relationship diagram* (ORD or O-R diagram). For small and medium-sized projects, it is a very useful modeling technique. However, as the size and complexity of the system increases, the ORD/ERD tends to become unwieldy. In these circumstances, unless the software engineering team is using a CASE tool that facilitates generation and maintenance of the ERD/ORD, other pragmatic approaches are recommended. One such approach is to tabulate as illustrated above. Another approach is to construct for each information entity (object type), an *object/entity specification grid* (O/ESG). This will be discussed shortly.

The symbols used in an ERD are as shown in Figure 13.7. Figure 13.8 shows the ERD for the manufacturing system of Figure 13.4, but also includes additional relationship; the ERD shown employs the Crows-foot notation. In the diagram, the convention to show attributes of each entity has been relaxed. Note also that relationships are labeled as verbs so that in mapping one entity (or object type) to another, one can read an object-verb-object formation. If the verb is on the right or above the relationship line, the convention is to read from top-to-bottom or left-to-right. If the verb is on the left or below the relationship line, the convention is to read from bottom-to-top or right-to-left.

It is customary to indicate on the ERD, the *multiplicity* (also called the *cardinality*) of each relationship. By this we mean, how many occurrences of one entity can be associated with one occurrence of the other entity. This information is particularly useful when the system is being constructed. Moreover, violation of multiplicity constraints could put the integrity of the system is question, which of course is undesirable. Usually, the DBMS does not facilitate enforcement of multiplicity constraints at the database level. Rather, they are typically enforced at the application level by the software engineer.

Several notations for multiplicity have been proposed, but the Chen notation (first published in 1976, and reiterated in [Chen 1994]) is particularly clear; it is paraphrased here: Place beside each entity, two numbers [x,y]. The first number (x) indicates the minimum participation, while the second (y) indicates the maximum participation.

An alternate notation is to use two additional symbols along with the Crow's Foot notation: an open circle to indicate a participation of zero, and a stroke (|) to indicate a participation of 1. The maximum participation is always indicated nearest to the entity box. The Chen notation is preferred because of its clarity and the amount of information it conveys. Figure 13.9 provides an illustrative comparison of the two notations.

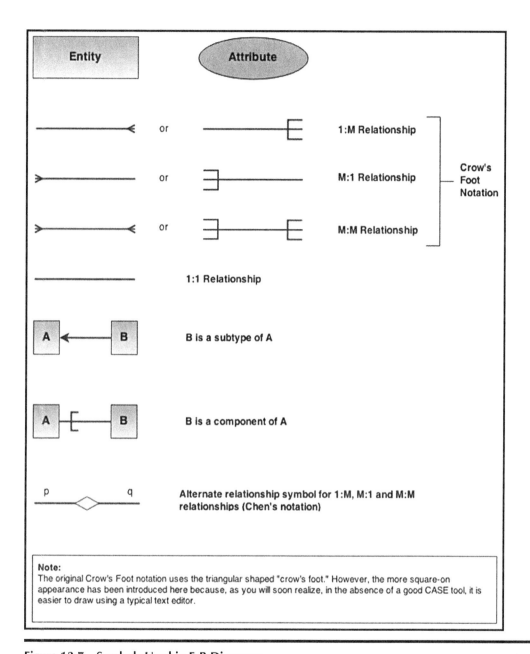

Figure 13.7   Symbols Used in E-R Diagrams

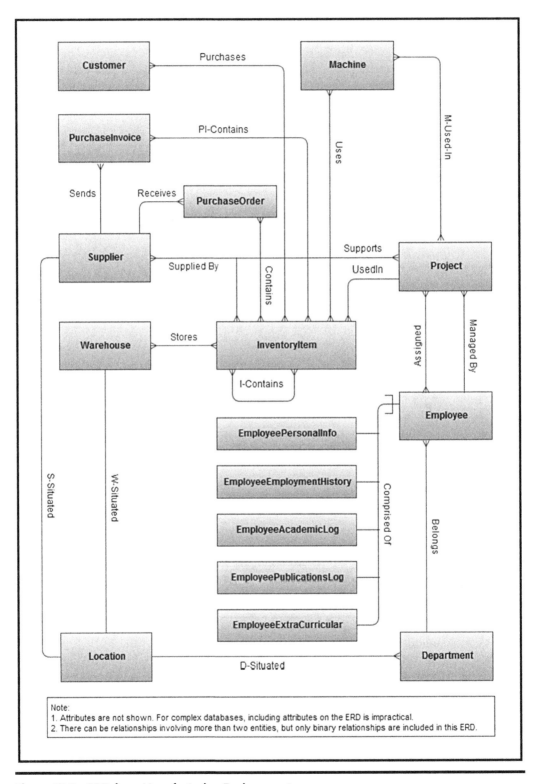

Figure 13.8   ERD for a Manufacturing Environment

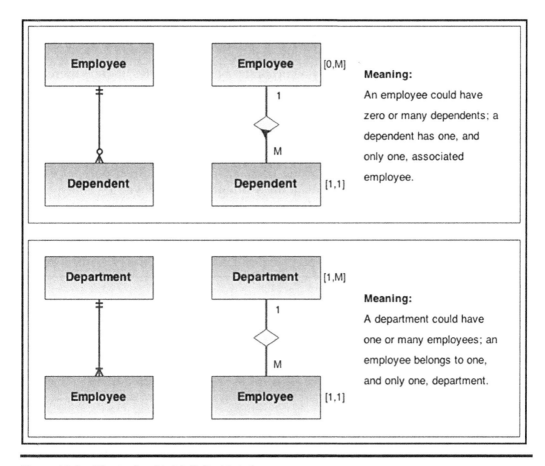

Figure 13.9   Illustrating Multiplicity Notations

Let us now turn our attention to the O-R diagram. The symbol for object type (mentioned in Chapter 12) is also used in the ORD, and replaces the entity symbol (as you will soon see, they are actually similar). Assuming the UML notation, the following guidelines apply:

- Similar to the information entity, a box (square or rectangle) represents an object type. The object-type box has two additional compartments: one for attributes, and the other for operations.
- A triangle or arrowhead (pointing towards the super-type) represents an inheritance relationship.
- An open diamond represents an aggregation relationship (the parts existing independent of the whole).
- A filled in diamond represents a composition relationship (the parts only exist as part of the whole).
- A line connecting two object types represents a traditional relationship; the *multiplicity* (also called cardinality) of this relationship is indicated by a pair of integers next to each object type; the lower value is indicated first, and an asterisk is sometimes used to mean "many." The multiplicity of a relationship is the level of

participation of each object type (or entity) in the relationship. The role that each object type plays in the relationship is also indicated next to the object type.

Since the object symbol automatically incorporates object attributes, there is therefore no need for an attribute symbol. In any event, including attributes and operations on the ORD tends to clutter the diagram rather quickly. It is therefore recommended that you omit this detail from the diagram for very large and/or complex systems. The upcoming subsection will suggest a creative and elegant way to represent attributes and related operations for object types comprising a system.

Figure 13.10 illustrates an ORD (using UML notation) depicting aspects of a college environment. According to the diagram, **Student** and **Employee** are subtypes of **College Member**. Additionally, **Employee** is a composition **of Employee Personal Info**, **Employee Work History**, **Employee Academic Log**, **Employee Publication**, **Employee Extra Curricular,** and **Employee Dependents Log**.

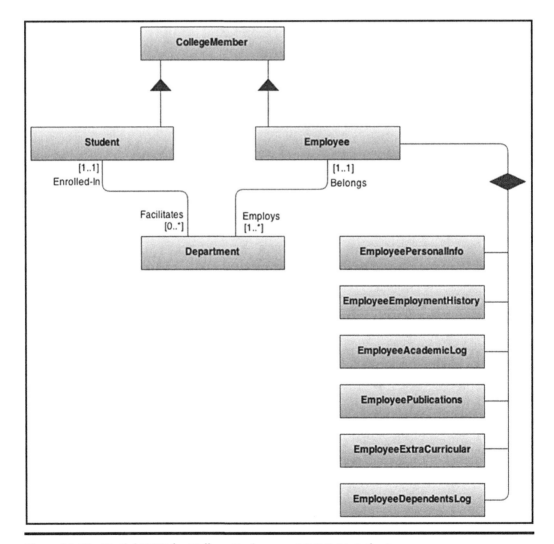

Figure 13.10   Partial ORD for College Environment (UML Notation)

Once, created, the ERD/ORD (or its equivalent) must be maintained for the entire life of the software system. If you are fortunate to be using a sophisticated CASE tool that automatically generates the diagram from the current database, then this will not be a problem for you.

13.2.6 Implementing Relationships

Once the ERD has been developed, the next logical step is to figure out how to account for each relationship of the ERD in the actual database that must be designed and implemented. Assuming the E-R model, relationships can be implemented by following a set of guidelines as outlined Figure 13.11.

Relationship	Recommended Implementation Strategy
1:M	To implement a 1:M relationship, store the primary key of one as a foreign key of the other (foreign key must be on the "many side"). To illustrate this strategy, carefully compare figure 13-8 with figure 13-12. Figure 13-8 includes several 1:M relationships for which there are implementation proposals in figure 13-12.
M:M	To implement an M:M relationship, introduce a third intersecting relation (also called an *associative entity*). The original relations/entities form 1:M relationships with the intersecting relation. The new relation is usually keyed on all the foreign keys (or a *surrogate #*). Figure 13-13 illustrates this principle; to gain additional insights, compare figure 13-8 with figure 13-13. Figure 13-8 includes several M:M relationships for which there are implementation proposals in figure 13-12.
Subtype	To implement a subtype relationship, introduce a foreign key in the subtype, which is the primary key in the referenced super-type. Further, make the foreign key in the subtype, the primary key of that subtype. In the case of *multiple inheritance* (where a subtype has more than one super-type), make the introduced foreign keys in the subtype, candidate keys, one of which will be the primary key. Figure 13-14 includes several subtype relationships for which there are implementation proposals in figure 13-15.
Component	To implement a component relationship, introduce in the component relation, a foreign key which is the primary key in the summary relation. This foreign key will form part of the primary key (or a candidate key) in the component relation. Figure 13-14 includes several component relationships that were first introduced in figure 3-3, and for which there are implementation proposals in figure 13-15.
1:1	To implement a 1:1 relationship, introduce a foreign key in one relation (preferably the primary relation) such that the primary key of one is an attribute in the other. Then enforce a constraint that forbids multiple foreign keys referencing a single primary key. Alternately, treat the 1:1 relationship as a subtype relationship (but ignore enforcing inheritance). Figure 13-8 includes two 1:1 relationships for which there are implementation proposals in figure 13-12.
# **Note:** A surrogate is an atomic attribute introduced either automatically by the DBMS, or manually by the database designer. It is used to uniquely identify records in a relation and simplify its design.	

Figure 13.11   Guidelines for Implementing Relationships

Take some time to carefully study these strategies outlined in Figure 13.11; you will gain more experience with them (or probably have already done so) in your database systems course. It is imperative that you learn them and know how to apply them; with practice, you will. Figures 13.12 through 13.15 provide illustrations of the application of these strategies but your course in database system will provide more opportunities for applying them. Here is a summary of each of these figures:

• Figure 13.12 illustrates a methodology called the *relations-attributes list* (RAL)—a summarized list of all entities comprising the database and the essential attributes of each entity.
• Figure 13.13 provides an example of how to treat an M:M relationship.
• Figures 13.14 and 13.15 illustrate how to treat subtype and component relationships.

Relation	Properties (Attributes)	Comment
Core Entities (Including Resolution of 1:M Relationships):		
MES_Customer_BR	*Cust#*	Primary Key (PK)
	CustName ...	
MES_Department_BR	*Dept#*	PK
	DeptName ...	
MES_Employee_BR	*Emp#*	PK
	EmpName ...	
	EmpDep#	Refers to **Department.Dept#**
MES_Supplier_BR	*Supp#*	PK
	SupplName ...	
	SupplLoc	Refers to **Location.LocCode**
MES_Project_BR	*Proj#*	PK
	ProjName	
	ProjManagerEmp# ...	Refers **Employee.Emp#**
MES_InventoryItem_BR	*Item#*	PK
	ItemName ...	
MES_Dependent_BR	*DepnEmp#*	Refers to **Employee.Emp#**; K1
	DepnRef	K2
	DepnName	
MES_Location_BR	*LocCode*	PK
	LocName ...	
MES_Machine_BR	*Mach#*	PK
	MachName ...	
MES_Warehouse_BR	Whouse#	PK
	WhouseLoc	Refers to **Location.LocCode**
	WhouseName ...	
Resolution of Component and/or Subtype Relationships		
MES_EmpPersonalInfo_BR	*EPI_Emp#*	Refers to **Employee.Emp#**; PK
	EPI_Specialty ...	
MES_EmpWorkHistory_BR	*EWH_Emp#*	Refers to **Employee.Emp#**; K1
	EWH_JobSeqNo	K2
	EWH_Organization	
	EWH_JobTitle ...	
MES_EmpAcademicLog_BR	*EAL_Emp#*	Refers to **Employee.Emp#**; K1
	EAL_SeqNo	K2
	EAL_Institution	
	EAL_StartDate	
	EAL_ExitDate	
	EAL_Award ...	

Note: Primary key and foreign keys are *italicized*. K1 .. K4 represent composite candidate keys.

Each of these relations would typically encompass several other attributes; only the essential ones are shown here.

Figure 13.12   Relations-Attributes List for the Manufacturing Environment

(*Continued*)

Relation	Properties (Attributes)	Comment
Resolution of Component and/or Subtype Relationships (continued)		
MES_EmpPublicationsLog_BR	*EP_Emp#*	Refers to **Employee.Emp#**; K1
	EP_PubSeqNo	K1
	EP_Title	
	EP_PubType . . .	
MES_EmpExtraC_BR	*EX_Emp#*	Refers to **Employee.Emp#**; K1
	EX_ActivityCode	K2
	EX_ActivityDesc . . .	
Other relations that might have been missed or to be added . . .		
. . .		
Resolution of M:M Relationships		
MES_MachineUsageMap_BR	*MU_Mach#*	Refers to **Machine.Mach#**; K1
	MU_Item# . . .	Refers to **InventoryItem.Item#**; K2
MES_MachProjectMap_BR	*MP_Mach#*	Refers to **Machine.Mach#**; K1
	MP_Proj# . . .	Refers to **Project.Proj#**; K2
MES_SuppItemMap_BR	*SI_Supp#*	Refers to **Supplier.Suppl#**; K!
	SI_Item#	Refers to **InventoryItem.Item#**; K2
	SI_Ref#	Surrogate PK
MES_ItemProjMap_BR	*IP_Item#*	Refers to **InventoryItem.Item#**; K1
	IP_Proj#	Refers to **Project.Proj#**; K2
	IP_Ref#	Surrogate PK
MES_ProjSuppMap_BR	*PS_Proj#*	Refers to **Project.Proj#**; K1
	PS_Supp#	Refers to **Supplier.Suppl#**; K2
	PS_Ref#	Surrogate PK
MES_ProjWorkSchedu;e_BR	*PW_Emp#*	Refers to **Employee.Emp#**; K1
	PW_Proj#	Refers to **Project.Proj#**; K2
	PW_Ref#	Surrogate PK
MES_ItemStruct_BR	*IS_ThisItem#*	Refers to **InventoryItem.Item#**; K1
	IS_ComplItem#	Refers to **InventoryItem.Item#**; K2
	IS_Ref#	Surrogate PK
MES_PurchOrdSummary_BR	*OrderRef#*	Surrogate PK
	OrderSupp#	Refers to **Supplier.Suppl#**; K1
	Order#, OrderDate	K2, K3
	OrderStatus . . .	
MES_PurchOrdDetail	*POD_OrderRef#*	Refers to **PurchOrdSummary.OrderRef#**; K1
	POD_Item#	*References* **InventoryItem.Item#**; K2
	POD_Quantity	
	POD_UnitPrice	
Note: Primary key and foreign keys are *italicized*. K1 . . K4 represent composite candidate keys.		

Each associative entity represents the intersecting relation used to implement a M:M relationship.

Figure 13.12 (Continued)

Relation	Properties (Attributes)	Comment
Resolution of M:M Relationships (continued)		
MES_PurchInvoiceSummary_BR	*PI_Ref#*	Surrogate PK
	PI_Supp#	Refers to **Supplier.Suppl#**; K1
	PI_Invoice#	K2
	PI_OrderRef#	Refers to **PurchOrdSummary.OrderRef#**; K3
	PI_Date	K4
	PI_Amount	
	PI_Status . . .	
MES_PurchInvoiceDetail_BR	*PID_PIRef#*	Refers to **PurchInvoiceSummary.PI_Ref#**; K1
	PID_Item#	Refers to **InventoryItem.Item#**; K2
	PID_Quantity	
	PID_UnitPrice . . .	
MES_SaleInvoiceSummary_BR	*SI_Ref#*	Surrogate PK
	SI_Cust#	Refers to **Customer.Cust#**; K1
	SI_Invoice#	K2
	SI_Date	K3
	SI_Amount	
	SI_Status . . .	
MES_SaleInvoiceDetail_BR	*SID_SIRef#*	Refers to **SaleInvoiceSummary.PI_Ref#**; K1
	SID_Item#	Refers to **InventoryItem.Item#**; K2
	SID_Quantity	
	SID_UnitPrice . . .	
MES_StockPile_BR	*SP_Whouse#*	Refers to **Warehouse.Whouse#**; K1
	SP_Item#	Refers to **InventoryItem.Item#**; K2
	SP_Quantity . . .	
Note: Primary key and foreign keys are *italicized*. K1 . . K4 represent composite candidate keys.		

Each associative entity represents the intersecting relation used to implement a M:M relationship.

Figure 13.12 (Continued)

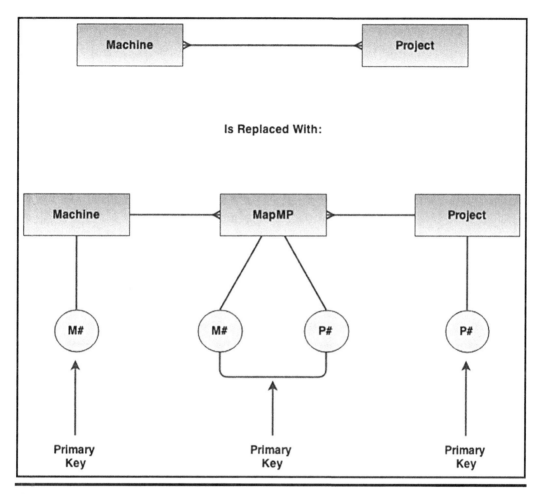

Figure 13.13   Implementing M:M Relationships

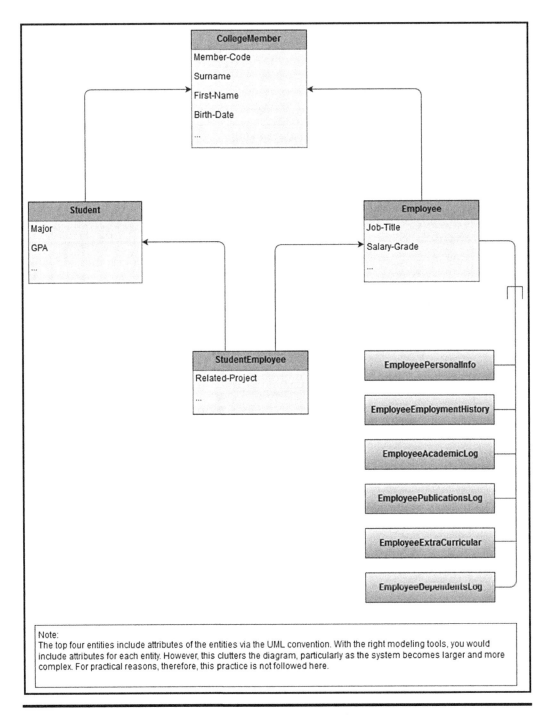

Figure 13.14   Illustrating Subtype and Component Relationships

Relation	Attributes	Primary Key
College Member	*MemberCode*, Surname, First-Name, BirthDate, ...	MemberCode
Student	*MemberCode*, Major, GPA, ...	MemberCode
Employee	*MemberCode*, JobTitle, SalaryGrade, ...	MemberCode
StudentEmployee	*MemberCode*, Related-Project, ...	MemberCode
EmployeePersonalInfo	*MemberCode*, Address, Telephone, ...	MemberCode
EmployeeEmploymentHistory	*MemberCode, JobSeqNo*, Organization, ...	[MemberCode, JobSeqNo]
EmployeeAcademicLog	*MemberCode, LogSeqNo*, Institution, Period-Attended, Award, ...	[MemberCode, LogSeqNo]
EmployeePublicationsLog	*MemberCode, PubCode*, Title, Book-Journal-Flag, ...	[MemberCode, PubCode]
EmployeeExtraCurricular	*MemberCode, ActvCode*, Activity-Description, ...	[MemberCode, ActvCode]

Each relation and each attribute will need additional clarification prior to database construction and table creation.

Note the following:
1. Primary key attributes and foreign key attributes are in italics.
2. This is not a comprehensive RAL. For several of the relations included, there are additional attributes to be added (indicated by the three periods in the attributes column).

Figure 13.15   Illustrating the Implementation of Subtype and Component Relationships

13.2.7 *Preparing the Database Specification*

The database specification may be in different forms, depending on the available resources. In an OO environment where you have the use a CASE tool that supports UML, it may simply be a detailed ORD where each object type is represented as explained and illustrated in Section 12.2.2 of the previous chapter. In an FO environment, it may simply be a detailed ERD where the attributes of each entity are included on the diagram.

For large, complex projects (involving huge databases with tens of information entities or object types), unless a CASE or RAD tool which automatically generates the ERD/ORD is readily available, manually drawing and maintaining this important aspect of the project becomes virtually futile. In such cases, an *object/entity specification grid* (O/ESG) is particularly useful. The grid contains the following components:

- Descriptive name of the entity (object type)
- Implementation name of the entity (object type)—typically indicated in square brackets
- Reference identification for each entity, to facilitate easy referencing
- Descriptive name, implementation name (in square brackets), and characteristics (in square brackets) for each attribute
- References (implying relationships) to other entities in the system (indicated in curly braces)
- Comments on the entity and selected attributes

- Indexes (including primary key or candidate keys) to be defined on the entity
- Operations to be defined on each entity (object type)
- Optionally, implementation names of operations are be indicated in square brackets next to respective operations

The convention for specifying attribute characteristics is to use a letter to represent the nature of the data (A for alphanumeric, N for numeric, and M for memo) followed by numbers representing the length and precision (for decimals). Figure 13.16 provides an illustration of an O/ESG for the manufacturing environment, or the college/university environment of earlier discussions. The ESG for three entities is included in the figure. In actuality, there would be one for each entity (object type) comprising the system. Also note the special data attributes that reference other entities in the figure (E1.3, E2.6, E2.8). In order to determine when to introduce such references, you need to apply principles of database design. These principles are best discussed in a course on database design (see the recommended readings).

E1 – Department [HR_Department_BR]

Attributes:
1. Department Number [DeptNo] [N4]
2. Department Name [DeptName] [A35]
3. Department Head Employee Number [DeptHead] [N7] {**Refers to E2**}

Comments:
This table stores definitions of all departments in the organization.

Indexes:
1. Primary Key: [1] – constraint HRDepartment_PK
2. HRDepartment_ NX1 on [2]

Valid Operations:
1. Maintain Departments [HRDepartment_MO]
   1.1. Add Departments [HRDepartment_AO]
   1.2. Update Departments [HRDepartment_UO]
   1.3. Delete Department [HRDepartment_ZO]
2. Inquire on Departments [HRDepartment_IO]

E2 – Employee [HR_Employee_BR]

Attributes:
1. Employee Identification Number [EmpNo] [N7]
2. Employee Last Name [EmpLName] [A20]
3. Employee First Name [EmpFName] [A20]
4. Employee Middle Initials [EmpMInitl] [A4]
5. Employee Date of Birth [EmpDOB] [N8]
6. Employee's Department [EmpDepNo] [N4] {**Refers to E1**}
7. Employee Gender [EmpGender] [A1]
8. Employee Marital Status [EmpMStatus] [A1]
9. Employee Social Security Number [EmpSSN] [N10]
10. Employee Classification Code [EmpClass] [A3] {**Refers to E3**}
....

Comments:
This table stores standard information about all employees in the organization.

Indexes:
1. Primary Key: [1] – constraint HREmployee_PK
2. HREmployee_ NX1 on [2, 3, 4]
3. HREmployee_ NX2 on [7]

Valid Operations:
2. Manage Employees [KREmployee_MO]
   1.1. Add Employees [HREmployee_AO]
   1.2. Update Employees [HREmployee_UO]
   1.3. Delete Employees [HREmployee_ZO]
2. Inquire on Employees [HREmployee_IO]

E3 – Employee Classification [HR_Classif_BR]

Attributes:
1. Classification Code [ClsCode] [A3]
2. Classification Description [ClsDesc] [A30]

Comments:
This table stores definitions of all employee classifications.

Indexes:
1. Primary Key: [1] – constraint HRClassif_PK
2. HRClassif_ NX1 on [2]

Valid Operations:
1. Maintain Classifications [HRClassif_MO]
   1.1 Add Classifications [HRClassif_AO]
Delete Classifications [HRClassif_ZO]
2. Inquire on Departments [HRDepartment_IO]

Figure 13.16   Sample Object/Entity Specification Grid

13.3 Overview of File Organization

In planning the underlying database for software system, it is important that the software engineer understands the different types of file organization techniques and the rationale and benefits of each. From your earlier programming courses, you should recall that there are four types of file organization:

- Sequential File organization
- Relative (direct) File Organization
- Indexed Sequential File Organization
- Multi-Access File Organization

13.3.1 Sequential File Organization

In a sequential file, records are arranged in arrival sequence usually stored on systematic tape. Accessing of records must also be done sequentially (the file may be sorted in a particular order).

Sequential file organization is useful when a large volume of records is to be added or updated in bulk or batch mode. Figure 13.17 illustrates what a sequential file of student records may look like. Accessing the Nth record means first accessing N-1 records.

Sequential file organization is not suited for interactive processing. This is so because records have to be accessed sequentially in arrival sequence. If you have a file with thousands of records, attempt to provide interactive processing on a sequential file would produce very poor results, and would therefore be counterproductive.

	Student ID	Last Name	First Name	Address
1	93010101	Foster	Bruce	Fox Lane …
2	93060101	Ming	Rose	Rose Lane …
3	92120101	Mano	Howard	Mano Lane …
4	91010101	Henry	Adrian	Abbey Court …
…				
n	99120101	Foxley	Sharon	Fox Lane …

Figure 13.17   Illustration of Sequential File

13.3.2 Relative or Direct File Organization

In relative (direct) file organization, records are arranged in some logical order where there is a relationship between the key used to identify a particular record and the record's address on the storage medium. This translates to the possibility of accessing the file randomly via the access key, or sequentially via arrival sequence. In computer science, this is often represented as follows:

$$F(key) \rightarrow Address$$

Use of a superimposed linked list is one method of implementing relative file. Each record has a pointer to the next logical record. Access is improved but additional data has to be stored with each record. Figure 13.18 illustrates a relative file implemented by linked list (on surname). Records may be accessed sequentially (via arrival sequence or Last-Name pointer), or randomly via Student ID.

Other more desirable methods of implementation of relative files are direct mapping, table lookup, hash functions, and open addressing with buckets (you should be familiar with these methods from your course in data structures and algorithms). These methods facilitate random (direct) access of the file(s). Consequently, interactive processing is facilitated.

	Student ID	Last Name	First Name	Address	Last-Name Pointer
1	93010101	Foster	Bruce	Fox Lane …	4
2	93060101	Ming	Rose	Rose Lane …	END
3	92120101	Mano	Howard	Mano Lane …	2
4	91010101	Henry	Adrian	Abbey Court …	3
…					

Figure 13.18   Illustrating Relative Access via Linked List (based on Last-Name) or Direct Access via Student ID

13.3.3 Indexed Sequential File Organization

In an indexed sequential file organization, records are ordered sequentially, but can also be accessed randomly via some key.

Indexed sequential access method (ISAM) is suitable for batch processing as well as interactive processing. ISAM files are typically implemented as B-trees (or some derivative of the B-tree). It is the most widely used method of file organization (again, please review your data structures).

13.3.4 Multi-Access File Organization

In multi-access file organization, a record can be accessed by any key order. An enhancement of ISAM, multi-access file organization is typically implemented by sophisticated DBMS suites and CASE tools. The DBMS maintains the index(es) that may be defined on the file in a manner that is transparent to the user. These indexes are typically B-tree or bitmap implementations.

13.4 Summary and Concluding Remarks

Let us summarize what we have covered in this chapter:

- A database is the record-keeping component of a software system. It must be properly designed. Failure to do so will seriously compromise the quality of the software system.

- Contemporary databases are designed to be relational or object-oriented, but mostly relational.
- Database design involves five steps: identifying the information entities (or object types), identifying relationships among the entities, eliminating unnecessary relationships, developing the ERD or ORD, and preparation of the database specification.
- There are four types of file organizations: sequential, direct, indexed-sequential, and multi-access. In sequential file organization, the records are organized in arrival sequence. The file can only be accessed sequentially. In direct file organization, each record has a specific address, thereby allowing random access. Indexed-sequential file organization supports both sequential access and random access. In a multi-access file, the records can be accessed sequentially or randomly, as well as via alternate access paths.

This is merely an introduction to database design from the context of software engineering. Study of database systems is a field of computer science, so a full discussion is beyond the scope of this course (please see the recommended readings). Appendix C provides you with a real example of the database specification for the Inventory Management System of earlier mention. The next chapter discusses design of the software user interface.

13.5 Review Questions

1. How important is database design? Cite four concrete examples of database playing an important role in computer software.

2. Identify the problems that are likely to occur due to poor database design.

3. Identify six objectives of good database design.

4. Outline the steps involved in the design of a database.

5. The following is an excerpt from the requirements for a college academic administration system:
   - Courses are offered by various departments without any overlap (a department offers between 5 and 30 courses).
   - The courses make up academic programs, in some instances a course may occur in more than one program. Academic programs are offered by departments (no overlap allowed).
   - A faculty typically consists of several departments.
   - A lecturer is scheduled to lecture at least two courses. Each course is lectured in a specific lecture room.
   - A student may register for several courses; typically, a course is pursued by a minimum of fifteen students.
   - Each student is registered to one department only.
   From the information given, develop an ERD for the system.

6. The Inventory Management Information System (IMIS) of a marketing company has the following database specification:

• The company has several warehouses, each storing certain inventory items without overlap.

• The company has a cadre of suppliers, each supplying various items of inventory, with possible overlap.

• The company purchases items by first sending a purchase order to a supplier (of course, the supplier could receive several orders). Each purchase order details the items required. In responding to the purchase order, the supplier submits an invoice, detailing the items supplied, along with other relevant information.

• The company may also sell items from its inventory. In such a case, a sale-invoice is submitted to the customer, which details the items sold, along with other relevant information.

• A sale-invoice is usually with respect to a sale-order, received from a customer. A sale-order is essentially a purchase order, coming into the company, from one of its customers.

• Each inventory item belongs to a category.

• A department may make a requisition for inventory items. In response, inventory items may be issued to department(s).

From this information, develop an ERD for the system.
By conducting a brainstorming session (or otherwise), and using your E-R diagram as well as guidelines illustrated in Figure 13.10, construct an initial entity specification grid (ESG) for the IMIS project.

References and Recommended Readings

[Chen 1994] Chen, Peter. 1994. "The Entity-Relationship Model – Toward a Unified View of Data," In *Readings in Database Systems*, 2nd ed., pp. 741–754. San Francisco, CA: Morgan Kaufmann.

[Date 2004] Date, C. J. 2004. *An Introduction to Database Systems*, 8th ed. Menlo Park, CA: Addison-Wesley. See chapters 1, 3, 5, 6, 11, and 12.

[Foster 2016] Foster, Elvis C. with Shripad Godbole. 2016. *Database Systems: A Pragmatic Approach*, 2nd ed. New York, NY: Apress.

[Hoffer 2015] Hoffer, Jeffrey A., Ramesh Venkataraman, and Heikki Topi. 2015. *Modern Database Management*, 12th ed. Boston, MA: Pearson. See chapters 3 and 4.

[Kendall 2014] Kendall, Kenneth E., and Julia E. Kendall. 2014. *Systems Analysis and Design*, 9th ed. Boston, MA: Pearson. See chapter 13.

[Lee 2002] Lee, Richard C. and William M. Tepfenhart. 2002. *Practical Object-Oriented Development With UML and Java*. Upper Saddle River, NJ: Prentice Hall. See chapter 8.

[Martin 1993] Martin, James, and James Odell. 1993. *Principles of Object-Oriented Analysis and Design*. Eaglewood Cliffs, NJ: Prentice Hall. See chapters 6 and 7.

[Rumbaugh 1991] Rumbaugh, James, Michael Blaha, William Premerlani, Frederick Eddy and William Lorensen. 1991. *Object Oriented Modeling and Design*. Eaglewood Cliffs, NJ: Prentice Hall. See chapter 4.

Chapter 14

User Interface Design

User interface management is a field of computer science that has been developed and given much attention in recent years. Indeed, courses in this area have become prevalent in many curricula of Computer Science and/or other related fields. Several texts have been written on the subject. It is therefore impossible to cover all that is entailed in one chapter. However, since user interface design is a very important aspect of software design, a brief summary of the subject matter is provided here.

The chapter proceeds under the following subheadings:

- Introduction
- Types of User Interfaces
- Steps in User Interface Design
- Overview of Output Design
- Output Methods versus Content and Technology
- Guidelines for Designing Output
- Overview of Input Design
- Guidelines for Designing Input
- Summary and Concluding Remarks

14.1 Fundamental Concepts

A user interface is the portion of computer software that facilitates *human–computer interaction* (HCI). Moreover, it is the window through which end users access the software system. As you are aware, most software systems have user interfaces. User interface design is therefore applicable to all software systems that require user interaction. It is through the user interface that users communicate with the software and with each other. Many good software products have suffered neglect in the marketplace, due to poor user interface design. On the other hand, many mediocre products have managed to survive market competition, due to attractive user interface design and aggressive marketing. Proper user interface design is therefore critical to

DOI: 10.1201/9780367746025-17

the success of a software engineering project, since this could determine the user acceptance and by extension, the success of the software system in the marketplace.

There are five main aspects of a user interface. These are:

- User Needs
- Human Factors
- Interface Design
- Interface Programming
- Environment

A well-designed user interface will meet the requirements in all of these areas. It will enhance effective use of the software, thus promoting end-user confidence in the product. Let us briefly look at each aspect.

14.1.1 User Needs

Among the basic user needs that the user interface must address are the following:

- **Functionality**—the capacity provided the user to carry out desired tasks and activities.
- **Flexibility**—the provision of alternative approaches to solving a problem.
- **Effective Control**—users like to feel that they are in control and not the software system.
- **Reliability**—the software must offer the user some assurance that it will facilitate solution to certain problems in a consistent manner.
- **Security**—controlled access to the overall system, specific resources of the system, and data managed by the system.
- **Consistency of Design** —the user must be able to anticipate system behavior; also, information must be presented to the user in a consistent manner.
- **Standardization**—the user interface must conform to established standards for the software.
- **Intelligibility**—the user interface must promote easy learning and understanding of the system.

14.1.2 Human Factors

Good user interface design must be guided by the following human factors:

- **Minimal Memory Taxation:** Taxation on human memory should be kept at a minimum.
- **Minimal Skilled Activities:** Required number of skilled tasks should be minimized; it is better to have a few skilled tasks and several operational tasks, than vice versa.
- **Shortcuts:** Shortcuts should exist for experienced users.
- **Help Facility:** There should be a help facility for all users.

- **Good Color Scheme:** The color scheme must not create pressure on the eye, but must be welcoming.
- **Minimal Assumptions:** The number of blanket assumptions about users should be kept at a minimum.

14.1.3 Design Considerations

The following considerations should be factored into the design and construction of a user interface:

- **Command Alignment:** Are the commands used appropriately aligned to typical human thinking?
- **Understandability:** How easily learned and understood are the system rules?
- **Semantic Alignment:** How aligned are the terminologies and other semantics to typical human thinking in that problem domain?
- **User Categories:** Are all user categories (experts, knowledgeable intermittent, and novices) catered for?
- **Screen Design:** How adequate are the user panels? Is there consistency among the panels?
- **System Documentation:** How adequate is the system documentation (including the help system)? Is the design appropriate?
- **System Menu:** How well-structured and adequate is the menu system?
- **User Interactions:** What kinds of user interaction are allowed?
- **System Messages:** How are feedback and diagnostic error messages handled?
- **Reversibility of Actions:** Can users reverse their actions if they need to?
- **Change Confirmation:** Do users get a chance to confirm their requests for changes before these changes take effect?
- **Locus of Control:** Who has control (or the perception of it)—the system or the user?
- **Responsiveness:** How responsive is the system?
- **Complexity Hiding:** Are users shielded from complexity details, or are they overwhelmed by the system complexity?
- **Usability:** How easy is it to learn and use the system?

14.1.4 User Interface Preparation

Preparation of the user interface involves actual development (programming) as well as preparation of the operating environment. Actual user interface development will be discussed further in Chapter 18. As for preparation of the operating environment, this involves various issues related to office preparation and computer site preparation.

Against this background, various user interface theories and models have been developed. Your course in user interface management explores these; they will therefore not be explored any further in this chapter (for additional insights, see [Schneiderman 2017]). Rather, we will focus our attention on user interface design alternatives.

14.2 Types of User Interfaces

User interfaces can be put into three broad categories—m*enu-driven interface, command interface,* and *graphical user interface* (GUI). Command interface is the oldest category. Traditionally, this is how software systems were written. In order to use a command-based system, one had to first be familiarized with the command language for the system. Older operating systems such as Unix and System i (formerly OS-400) are still predominantly command-based. Next are menu-driven systems. They are characterized by menu(s) of options from which the user selects the desired option. The System i operating system is also menu-driven—you can access each system command from a menu. Many legacy systems that run on mainframes and mini-computers are menu-driven. The GUI is the newest type of user interface, and it represents the contemporary trend. You are able to use a mouse to select items (also from a menu), to drag and drop objects, and perform several other activities that we often take for granted. Figure 14.1 compares the approaches in terms of relative complexity of design (COD), response time (RT), and ease of usage (EOU) for each user interface category.

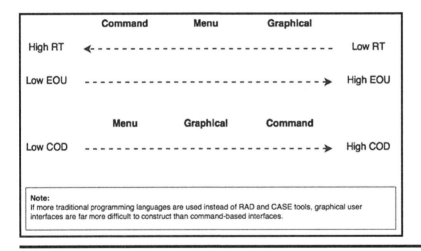

Figure 14.1   Comparison of User Interface Categories

Up until the mid-1990s, menu-driven interfaces were the most frequently used, dominating the arena of business information and application systems. Since the late 1980s, GUIs have become very popular, and clearly dominate user interfaces of the current era. Of course, the approaches can be combined. An excellent example of a software system that combines all three user interface categories is the System i operating system. Being traditionally command-based, modern versions of the operating system fully support all three user interaction categories. Another example of this hybrid approach is the Windows operating system. Though predominantly GUI-based, you are allowed to key in specific system commands if you so choose. The operating systems Linux and Unix tend to be the opposite of Windows. Though predominantly command-based, each supports various GUI-based systems that can be superimposed on the underlying command-based system.

The interaction styles facilitated depend to a large extent on the type of interface supported. There are seven possible interaction styles, summarized in Figure 14.2:

Interaction Style	Type(s) of Interface
Menu Selection	Menu, Graphical
Form Fill-ins	Menu, Graphical, Command
Command Language	Command
Natural Language	Command
Direct Manipulation	Graphical
Function Key	Menu, Command
Question - Answer	Menu, Command, Graphical

Figure 14.2   Interaction Styles for User Interfaces

14.3 Steps in User Interface Design

How you design the user interface will depend to a large extent on the type of user interface your software requires, It will also depend on the intended users of the software (experts, knowledgeable intermittent, or novices).

14.3.1 Menu or Graphical User Interface

If the user interface is to be menu driven or graphical, the following steps are recommended (assuming OOD):

1. Put system objects (structures and operations) into logical groups. At the highest level, the menu will contain options pointing to summarized logical groups.
2. For each summarized logical group, determine the component sub-groups where applicable, until all logical groups have been identified.
3. Let each logical group represent a component menu.
4. For each menu, determine options using an object-oriented strategy to structure the menu hierarchy (object first, operation last).
5. Design the menus to link the various options. Develop a menu hierarchy tree or a user interface topology chart (UITC) as discussed in Chapter 8.

The partial UITC of chapter six has been repeated in Figure 14.3 for ease of reference. This chart relates to the CUAIS project of earlier discussions. Also recall from Section 8.4.2 that the UITC is comparable to Schneideman's *Object-Action Interface* (OAI) model for user interfaces [Schneiderman 2017].

CUAIS Main Menu
1. Infrastructure & Resource Management System
2. Curriculum & Academic Management System
3. Financial Information Management System
4. Student Affairs Management System
5. Public Relations & Alumni Management System
6. Human Relations Management System
7. Cafeteria Services Management System
8. Library Information Management System
9. Other Ad Hoc Services

2. Infrastructure & Resource Management System
1.1 Department Definitions
1.1.1 Add Department Definitions
1.1.2 Modify Department Definitions
1.1.3 Delete Department Definitions
1.1.4 Inquire/Report on Department Definitions
1.2 School/Division Definitions
1.2.1 Add School/Division Definitions
1.2.2 Modify School/Division Definitions
1.2.3 Delete School/Division Definitions
1.2.4 Inquire/Report on School/Division Definitions
1.3 Building Definitions
1.3.1 Add Building Definitions
1.3.2 Modify Building Definitions
1.3.3 Delete Building Definitions
1.3.4 Inquire/Report on Building Definitions
1.4 Lecture Room Definitions
1.4.1 Add Lecture Room Definitions
1.4.2 Modify Lecture Room Definitions
1.4.3 Delete Lecture Room Definitions
1.4.4 Inquire/Report on Lecture Room Definitions
1.5 Fixed Asset Logs
1.5.1 Add Fixed Assets
1.5.2 Modify Fixed Assets
1.5.3 Delete Fixed Assets
1.5.4 Inquire/Report on Fixed Assets
1.6 Inventory/Resource Items
1.6.1 Add Inventory/Resource Items
1.6.2 Modify Inventory/Resource Items
1.6.3 Delete Inventory/Resource Items
1.6.4 Inquire/Report on Inventory/Resource Items

1. Infrastructure & Resource Management System
1.7 Supplier Definitions
1.7.1 Add Supplier Definitions
1.7.2 Modify Supplier Definitions
1.7.3 Delete Supplier Definitions
1.7.4 Inquire/Report on Supplier Definitions
1.8 Purchase Order Summaries
1.8.1 Add Purchase Order Summaries
1.8.2 Modify Purchase Order Summaries
1.8.3 Delete Purchase Orders
1.8.4 Inquire/Report on Purchase Order Summaries
1.9 Purchase Order Details
1.9.1 Add Purchase Order Details
1.9.2 Modify Purchase Order Details
1.9.3 Delete Purchase Order Details
1.9.4 Inquire/Report on Purchase Order Details
1.10 Purchase Invoice Summaries
1.10.1 Add Purchase Invoice Summaries
1.10.2 Modify Purchase Invoice Summaries
1.10.3 Delete Purchase Invoices
1.10.4 Inquire/Report on Purchase Invoice Summaries
1.11 Purchase Invoice Details
1.11.1 Add Purchase Invoice Details
1.11.2 Modify Purchase Invoice Details
1.11.3 Delete Purchase Invoice Details
1.11.4 Inquire/Report on Purchase Invoice Details
1.12 Purchase Return Summaries
1.12.1 Add Purchase Return Summaries
1.12.2 Delete Purchase Returns
1.12.3 Inquire/Report on Purchase Return Summaries
1.13 Purchase Return Details
1.13.1 Add Purchase Return Details
1.13.2 Modify Purchase Return Details
1.13.3 Delete Purchase Return Details
1.13.4 Inquire/Report on Purchase Return Details
1.14 Supplier-Resource Mappings
1.14.1 Add Supplier-Resource Mappings
1.14.3 Delete Supplier-Resource Mappings
1.14.4 Inquire/Report on Supplier-Resource Mappings
. . . .

Figure 14.3 Partial User Interface Topology Chart for the CUAIS Project

1. Curriculum & Academic Management System	3. Financial Information Management System
2.1 Academic Program Definitions	3.1 Chart of Accounts
2.1.1 Add Academic Program Definitions	3.1.1 Add Account Definitions
2.1.2 Modify Academic Program Definitions	3.1.2 Modify Account Definitions
2.1.3 Delete Academic Program Definitions	3.1.3 Delete Account Definitions
2.1.4 Inquire/Report on Academic Program Definitions	3.1.4 Inquire/Report on Account Definitions
2.2 Course Definitions	3.2 Financial Institutions
2.2.1 Add Course Definitions	3.2.1 Add Financial Institution Definitions
2.2.2 Modify Course Definitions	3.2.2 Modify Financial Institution Definitions
2.2.3 Delete Course Definitions	3.2.3 Delete Financial Institution Definitions
2.2.4 Inquire/Report on Course Definitions	3.2.4 Inquire/Report on Financial Institution Definitions
2.3 Academic Department Definitions	3.3 Financial Transactions
2.3.1 Add Academic Department Definitions	3.3.1 Add Financial Transactions
2.3.2 Modify Academic Department Definitions	3.3.2 Modify Financial Transactions
2.3.3 Delete Academic Department Definitions	3.3.3 Delete Financial Transactions
2.3.4 Inquire/Report on Academic Department Definitions	3.3.4 Inquire/Report on Financial Transactions
2.4 Dormitory Definitions	3.4 Purchase Orders — Summaries & Details
2.4.1 Add Dormitory Definitions	3.4.1 Add Purchase Orders
2.4.2 Modify Dormitory Definitions	3.4.2 Modify Purchase Orders
2.4.3 Delete Dormitory Definitions	3.4.3 Delete Purchase Orders
2.4.4 Inquire/Report on Dormitory Definitions	3.4.4 Inquire/Report on Purchase Orders
2.5 Course Schedules	3.5 Purchase Invoices — Summaries & Details
2.5.1 Add Course Schedules	3.5.1 Add Purchase Invoices
2.5.2 Modify Course Schedules	3.5.2 Modify Purchase Invoices
2.5.3 Delete Course Schedules	3.5.3 Delete Purchase Invoices
2.5.4 Inquire/Report on Course Schedules	3.5.4 Inquire/Report on Purchase Invoices
2.6 Examination Schedules	3.6 Sale Orders — Summaries & Details
2.6.1 Add Examination Schedules	3.6.1 Add Sale Orders
2.6.2 Modify Examination Schedules	3.6.2 Modify Sale Orders
2.6.3 Delete Examination Schedules	3.6.3 Delete Sale Orders
2.6.4 Inquire/Report on Examination Schedules	3.6.4 Inquire/Report on Sale Orders
2.7 Course Evaluations	3.7 Sale Invoices — Summaries & Details
2.7.1 Add Course Evaluations	3.7.1 Add Sale Invoices
2.7.2 Modify Course Evaluations	3.7.2 Modify Sale Invoices
2.7.3 Delete Course Evaluations	3.7.3 Delete Sale Invoices
2.7.4 Inquire/Report on Course Evaluations	3.7.4 Inquire/Report on Sale Invoices
2.8 Academic Program Evaluations	3.8 Investments
2.8.1 Add Academic Program	3.8.1 Add Investment Entries
2.8.2 Modify Academic Program	3.8.2 Modify Investment Entries
2.8.3 Delete Academic Program	3.8.3 Delete Investment Entries
2.8.4 Inquire/Report on Academic Program	3.8.4 Inquire/Report on Investments
. . . .	. . . .

Figure 14.3   (Continued)

4. Student Affairs Management System
4.1 Student Personal Records
4.1.1 Add Student Personal Records
4.1.2 Modify Student Personal Records
4.1.3 Delete Student Personal Records
4.1.4 Inquire/Report on Student Personal Records
4.2 Student Educational History
4.2.1 Add Student Educational History
4.2.2 Modify Student Educational History
4.2.3 Delete Student Educational History
4.2.4 Inquire/Report on Student Educational History
4.3 Student Academic Qualification
4.3.1 Add Student Academic Qualification
4.3.2 Modify Student Academic Qualification
4.3.3 Delete Student Academic Qualification
4.3.4 Inquire/Report on Student Academic Qualification
4.4 Student Next of Kin Contacts
4.4.1 Add Student Next of Kin Contacts
4.4.2 Modify Student Next of Kin Contacts
4.4.3 Delete Student Next of Kin Contacts
4.4.4 Inquire/Report on Student Next of Kin Contacts
4.5 Student Extracurricular Activities
4.5.1 Add Student Extracurricular Activities
4.5.2 Modify Student Extracurricular Activities
4.5.3 Delete Student Extracurricular Activities
4.5.4 Inquire/Report on Student Extracurricular Activities
4.6 Student Registration Logs
4.6.1 Add Student Registration Logs
4.6.2 Modify Student Registration Logs
4.6.3 Delete Student Registration Logs
4.6.4 Inquire/Report on Student Registration Logs
4.7 Student Academic Performance Logs
4.7.1 Add Student Academic Performance Logs
4.7.2 Modify Student Academic Performance Logs
4.7.3 Delete Student Academic Performance Logs
4.7.4 Inquire/Report on Student Academic Performance
4.8 Student Financial Logs
4.8.1 Add Student Financial Logs
4.8.2 Modify Student Financial Logs
4.8.3 Delete Student Financial Logs
4.8.4 Inquire/Report on Student Financial Logs
. . . .

5. Public Relations & Alumni Management System
5.1 Alumni Information
5.1.1 Add Alumni Information
5.1.2 Modify Alumni Information
5.1.3 Delete Alumni Information
5.1.4 Inquire/Report on Alumni Information
5.2 Alumni Chapters
5.2.1 Add Alumni Chapter Definitions
5.2.2 Modify Alumni Chapter Definitions
5.2.3 Delete Alumni Chapter Definitions
5.2.4 Inquire/Report on Alumni Chapters
5.3 Institutional Sponsors
5.3.1 Add Institutional Sponsors
5.3.2 Modify Institutional Sponsors
5.3.3 Delete Institutional Sponsors
5.3.4 Inquire/Report on Institutional Sponsors
5.4 Alumni Funds Raised
5.4.1 Add Alumni Funds Raised
5.4.2 Modify Alumni Funds Raised
5.4.3 Delete Alumni Funds Raised
5.4.4 Inquire/Report on Alumni Funds Raised
5.5 Other Funds Raised
5.5.1 Add Other Funds Raised
5.5.2 Modify Other Funds Raised
5.5.3 Delete Other Funds Raised
5.5.4 Inquire/Report on Other Funds Raised
5.6 Special Project Definitions
5.6.1 Add Special Project Definitions
5.6.2 Modify Special Project Definitions
5.6.3 Delete Special Project Definitions
5.6.4 Inquire/Report on Special Project Definitions
5.7 Special Project Details
5.7.1 Add Special Project Details
5.7.2 Modify Special Project Details
5.7.3 Delete Special Project Details
5.7.4 Inquire/Report on Special Project Details
5.8 Promotional Initiatives/Activities
5.8.1 Add Promotional Initiatives
5.8.2 Modify Promotional Initiatives
5.8.3 Delete Promotional Initiatives
5.8.4 Inquire/Report on Promotional Initiatives
. . . .

Figure 14.3 (Continued)

6. Human Resource Management System	7. Cafeteria Services Management System
6.1 Employee Personal Records	7.1 Cafeteria Inventory Items
6.1.1 Add Employee Personal Records	7.1.1 Add Cafeteria Inventory Items
6.1.2 Modify Employee Personal Records	7.1.2 Modify Cafeteria Inventory Items
6.1.3 Delete Employee Personal Records	7.1.3 Delete Cafeteria Inventory Items
6.1.4 Inquire/Report on Employee Personal Records	7.1.4 Inquire/Report on Cafeteria Inventory Items
6.2 Employee Academic History	7.2 Cafeteria Meal Plans
6.2.1 Add Employee Academic History	7.2.1 Add Cafeteria Meal Plans
6.2.2 Modify Employee Academic History	7.2.2 Modify Cafeteria Meal Plans
6.2.3 Delete Employee Academic History	7.2.3 Delete Cafeteria Meal Plans
6.2.4 Inquire/Report on Employee Academic History	7.2.4 Inquire/Report on Cafeteria Meal Plans
6.3 Employee Work History	7.3 Cafeteria Recipes
6.3.1 Add Employee Work History	7.3.1 Add Cafeteria Recipes
6.3.2 Modify Employee Work History	7.3.2 Modify Cafeteria Recipes
6.3.3 Delete Employee Work History	7.3.3 Delete Cafeteria Recipes
6.3.4 Inquire/Report on Work History	7.3.4 Inquire/Report on Cafeteria Recipes
6.4 Employee Beneficiaries	7.4 Cafeteria Purchase Order Summaries
6.4.1 Add Employee Beneficiary Information	7.4.1 Add Purchase Order Summaries
6.4.2 Modify Employee Beneficiary Information	7.4.2 Modify Purchase Order Summaries
6.4.3 Delete Employee Beneficiary Information	7.4.3 Delete Purchase Order Summaries
6.4.4 Inquire/Report on Employee Beneficiaries	7.4.4 Inquire/Report on Purchase Order Summaries
6.5 Employee Extracurricular Activities	7.5 Cafeteria Purchase Order Details
6.5.1 Add Student Extracurricular Activities	7.5.1 Add Purchase Order Details
6.5.2 Modify Student Extracurricular Activities	7.5.2 Modify Purchase Order Details
6.5.3 Delete Student Extracurricular Activities	7.5.3 Delete Purchase Order Details
6.5.4 Inquire/Report Employee Extracurricular Activities	7.5.4 Inquire/Report on Purchase Order Details
6.6 Employee Job Definitions	7.6 Cafeteria Sale Order Summaries
6.6.1 Add Employee Job Definitions	7.6.1 Add Sale Order Summaries
6.6.2 Modify Employee Job Definitions	7.6.2 Modify Sale Order Summaries
6.6.3 Delete Employee Job Definitions	7.6.3 Delete Sale Order Summaries
6.6.4 Inquire/Report on Employee Job Definitions	7.6.4 Inquire/Report on Sale Order Summaries
6.7 Employee Compensation Packages	7.7 Cafeteria Sale Order Details
6.7.1 Add Employee Compensation Packages	7.7.1 Add Sale Order Details
6.7.2 Modify Employee Compensation Packages	7.7.2 Modify Sale Order Details
6.7.3 Delete Employee Compensation Packages	7.7.3 Delete Sale Order Details
6.7.4 Inquire/Report Employee Compensation Packages	7.7.4 Inquire/Report on Sale Order Details
6.8 Employee Payroll Logs	7.8 Special Catering Projects
6.8.1 Add Employee Payroll Logs	7.8.1 Add Special Projects
6.8.2 Modify Employee Payroll Logs	7.8.2 Modify Special Projects
6.8.3 Delete Employee Payroll Logs	7.8.3 Delete Special Projects
6.8.4 Inquire/Report on Employee Payroll Logs	7.8.4 Inquire/Report on Special Projects
. . . .	. . . .

Figure 14.3   (Continued)

14.3.2 Command-Based User Interface

If the user interface is command driven, the following steps are recommended:

1. Develop an *operations-set* i.e. a list of operations that will be required.
2. Categorize the operations—user operations as opposed to system operations.
3. If an underlying database is involved, develop a mapping of operations with underlying database objects.
4. Determine required parameters for each operation.
5. Develop a list of commands (may be identical to operations set). If this is different from the operations set, each command must link to its corresponding system operations.
6. Define a syntax for the command language—how users will communicate with the commands (operations).
7. Develop a user interface support for each command (and by extension each operation). This interface support must be consistent with the defined command syntax.
8. Program the implementation of each operation.

Note that construction of a command-based user interface requires much more effort than a menu or graphical interface. This was not always the case: In the early days of visual programming, it was quite arduous to construct a GUI using traditional procedural programming languages. However, with the advent of object-oriented CASE and RAD tools, constructing GUIs is much easier than before.

14.4 Overview of Output Design

An important aspect of the user interface is the output—the results users obtain from the software. Output may be in the form of printed copy, VDU display, tape, diskette, CD or audio; or in the case of expert systems and CAM/CIM systems, output may be in the form of produced motion and/or products. Through output, the software communicates with the user; it is therefore an important aspect of the user interface of computer software.

Let us concentrate on the more traditional and prevalent form of output, namely information. The system must produce outputs according to user requirements. In a way, output is one fundamental test of the usefulness of a system. Some information may require a little processing before output; in other cases, much processing may be required before output.

Below are some important objectives of software output:

1. The output must serve the intended purpose: It must not be redundant and it must be in the required form, if it is to fulfill its desired purpose.
2. The output must fit the user and the situation: For decision support systems, it must fit individual user needs; for management information systems, it must fit functional needs; for expert systems, it must reveal expert analysis as required by the human expert; for CASE and RAD tools, it must generate accurate and accessible code; and so on.

3. The output must be in correct quantity: With the proliferation of distributed systems, this matter is not as critical as it used to be when centralized systems dominated.
4. The output must be on time: This is particularly important in mission-critical systems, real-time systems, and traditional centralized information systems. It is not as critical in distributed system where the user decides when to generate/access system outputs.
5. The method of output (VDU, print, tape storage, disk storage, CD storage, or audio) must be appropriate to the need.

14.5 Output Methods versus Content and Technology

The output method affects output content and presentation. For example, a screen display or a printed report will have screen heading, body and footnote; a tape or diskette file will just have raw data. The output method also affects technology chosen to yield the output. For example, sound cards and speakers are required for audio output; printers are required for hard copy; special cameras are required for microfilm; etc. External output leaves the organization and must adhere to standards, for such outputs. Internal output stays within the organization and must adhere to standards for such output. In choosing output, the technological alternatives and quality factors (*e.g.* reliability, compatibility, and portability) are useful.

14.5.1 Printed Output

Printed output is one of the most common output methods (the other being monitor display). It is inexpensive and can serve a wide and varied user population.

For printed output, required volume also affects choices with respect to the related technology. To illustrate, the type of printer depends on the print volumes required, the speed required, frequency required, quality required.

Reports must be properly formatted, according to established standards. Usually, each operation spec (for system output) has associated output layouts for the programmer.

14.5.2 Monitor Display

Monitor display is the other most common output method. Apart from an initial cost of acquiring a visual display unit (VDU) or work station, it is very fast and economical. Additionally, thanks to GUIs, monitor display is very attractive and effective in enhancing the user's understanding of system outputs.

Monitor display is also very convenient: the user can assess information before transferring to more expensive output medium e.g. printing. One limitation of monitor display is that the number of users that can benefit is to some extent, constrained by the available number of monitors. Of course, monitor display must conform to established standards.

14.5.3 Audio Output

Audio output is usually in the form of voice to a single user or multiple users in a building. This method of output has become quite common. It is useful in situations where users must be free of encumbrances, or where a message on an intercom is adequate communication to the end users.

Increasingly, individuals who have physical challenges and executives with very busy schedules are finding audio-based systems very convenient and helpful.

14.5.4 Microfilm/Microfiche

Microfilm/microfiche is traditionally used to store large volumes of data. It takes up approximately 1% of space a printed copy would take. Special machines are required to create microfilm files. Projector-like machines are then used to magnify the images so that they can be read.

Traditionally, microfilm technology has been used in legal offices, civil engineering, colleges, universities, and banks. The technology comes with a high price tag, since special equipment are required. It can also be time consuming to access information from microfilm machine.

14.5.5 Magnetic and Optical Storage

Traditionally, magnetic storages devices (disks and tapes) have been used for storing data for future usage. System backups are typically done onto disks or tapes; these devices are then stored safely until they are required. Also, before computer networks were as prevalent and sophisticated as they currently are, disks and tapes were used to transfer information between different organizations, or offices of an organization.

Optical storage devices (DVDs, CD-ROMs, CD-RWs) have become commonplace as storage media. These devices store much more information than their traditional predecessors. DVDs (digital versatile disks) are expected to dominate the future, due to their storage capacity and flexibility.

14.5.6 Choosing the Appropriate Output Method

Figure 14.4 provides a comparison of the output methods. Bear in mind that there is no right or wrong output method; what is required is that you prescribe the most appropriate output method for the situation at hand.

Additionally, the following factors should be considered when making decisions about system output:

- Who will use the output?
- How many users will access the output?
- What is the purpose of the output?
- What response speed is required?
- What volumes are required?
- How frequently is the output required?
- What are the environmental requirements?

Method	Advantages	Disadvantages
Printer	1. Affordable 2. Reasonable flexibility in type, quality, location. 3. Can reach large user population inexpensively. 4. Reliable on down time. 5. Handles large volumes of output	1. May be noisy 2. May have compatibility problems with software. 3. May require special expensive supplies. 4. Requires operator intervention. 5. May be slow, depending on model.
Visual Display Unit (VDU), also called the Monitor	1. Interactive 2. Can serve indefinite no. of users, depending on number of terminals. 3. Transmission may occur over widely dispersed Network. 4. Fast, on-the-spot response. 5. Good for frequently accessed transitory information e.g. messages, mail, notices. 6. User has chance to analyze information and decide what to do with it - whether print is required.	1. User may still require printed output. 2. Can be expensive if required for many users.
Audio	1. Good for individual user. 2. Good for transitory information. 3. Good if output is highly repetitive, or transmitted in an intercom or headset. 4. Ideal where worker needs to have hands free for other tasks.	1. Expensive 2. Need to ensure that output does not interfere with other activities. 3. Limited applications.
Microfilm/Microfiche	1. Traditionally handles large volumes of information. 2. Reduced space required for storage. 3. Preserves fragile but frequently used materials. 4. Avoids problems of physically paging through physically cumbersome reports.	1. Requires special hardware & software. 2. May therefore be an expensive initial investment. 3. Can be effectively replaced by DVDs, CD's, Diskettes or tapes.
Magnetic/Optical Disk	1. Traditionally handles large volumes of information. 2. Reduced space required for storage. 3. Avoids problems of physically paging through physically cumbersome reports. 4. Requires no special hardware different from computer system. May be cheaper than Microfilm.	1. Special hardware is required. 2. Software required to present output in a form that the user understands. 3. May therefore be more expensive than the VDU alone or printer alone.

Figure 14.4   Comparison of Output Methods

14.6 Guidelines for Designing Output

Whatever the output medium, it must follow some basic guidelines. Let us briefly focus on these guidelines as they relate to printed output and screen output:

14.6.1 Guidelines for Designing Printed Output

The following guidelines relate to printed outputs:

1. Use information gathered during investigation phase to design reports according to user requirements.
2. Adhere to established software standards as they affect output design.
3. Use standard output design forms—headings, body, and footnotes should conform to established standards.
4. Have a convention for representing variable information. Constant information is usually typed or written on output design forms. Variable information is

usually indicated via some convention. For example, a string of Xs (e.g. XXXXXX...) is used to represent alphanumeric data; a string of 9s (e.g. 99999) is used to represent numeric data.

5. Decide on paper width (80, 132 or 198), quality, and type.
6. Involve the user in the decision-making exercises.
7. The prototype and actual output should be well balanced and attractive.

14.6.2 Guidelines for Designing Screen Output

The above guidelines for printed output all apply to screen (monitor) output. Additionally, the following guidelines may be useful:

1. Where possible, use interactive prototypes to help the user to conceptualize the output being designed at an early stage. This may be supported by the software design tool being used (e.g. CASE tools or presentation software with hyperlinks).
2. Keep screens simple and attractive.
3. Keep screen presentation consistent.
4. Facilitate user movement among screens.
5. Control the duplication of data on screens (duplicate only when necessary).

Screen movement is usually facilitated by one of the following:

- Scrolling
- Calling up detail (e.g. position cursor and press <Enter>)
- On-screen dialog
- Function Keys

Figure 14.5 illustrates these strategies. Screen movement could also be prototyped (review Section 7.6) and used as a good source of feedback from prospective users of the system.

14.7 Overview of Input Design

Input may be in the form of manual input, audio, magnetic, and optical storage devices (output for magnetic or optical storage media is typically used as input for some other system). The arguments presented in the previous sub-section on magnetic and optical storage devices also apply here. We will therefore concentrate on manual inputs.

Desirable objectives of software inputs are as follows:

1. Input forms must capture all required data.
2. Input screens must be well-designed screens.
3. There must be well-designed forms and screens in order to achieve effectiveness, accuracy, ease of use, consistency, simplicity, and attractiveness.
4. If input is automatic, the input files must be in the required formats.

a. Scrolling:

Student Information Query/Update
Starting Point: 2001001_____   Order by: · ID Number

ID Number	Last Name	First Name	Middle Name	Program
2001012	Jones	Bruce	Farnsworth	Computer Science
2001015	Barnaby	Carlos	Kane	Biology
...	...	...	...	...
2006540	McBean	Irene	Isbeth	Mathematics

F1: Exit   F2: Previous Screen   Scroll via the scrolling buttons or the PageUp/PageDown key
F5: Clear Search Argument   PageUp: Previous Page   PageDown: Next Page

b. Calling Up Detail:

Student Information Query/Update

Starting Point: 2001001_____   Order by: · ID Number

ID Number	Last Name	First Name	Middle Name	Program
2001012	Jones	Bruce	Farnsworth	Computer Science
2001015	Barnaby	Carlos	Kane	Biology
...	...	...	...	...
2006540	McBean	Irene	Isbeth	Mathematics

F1: Exit   F2: Previous Screen   Position cursor and press <Enter> for more detail
F5: Clear Search Argument   F7: Page Down   F8: Page Up

c. Using Function Keys:

Student Information Query

Starting Point: 2001001_____   Order by: · ID Number

ID Number	Last Name	First Name	Middle Name	Program
2001012	Jones	Bruce	Farnsworth	Computer Science
2001015	Barnaby	Carlos	Kane	Biology
...	...	...	...	...
2006540	McBean	Irene	Isbeth	Mathematics

F1: Exit   F2: Previous Screen   F7: Page Down   F8: Page Up   F5: Clear

Figure 14.5   Illustrating Screen Movements

d. Using On-Screen Dialog:

Student Information Entry

Student ID#:	930101012
Name:	Bruce Jones
Program:	B.Sc. Computer Science
Lives on Hall? (Y/N)	___

Student Information Entry

Student ID #:	930101012
Name:	Bruce Jones
Address Line 1:	_____
Address Line 2:	_____
State/Province:	_____
Zip Code:	_____

Note: *This screen appears when the answer on the previous screen is "N"*

Figure 14.5   (Continued)

14.8 Guidelines for Designing Input

All the points and guidelines about the design of output screens (Section 14.6) are applicable here.

In the interest of clarity, they are repeated here along with some additional guidelines:

1. Use information gathered during investigation phase to design inputs according to user requirements.
2. Adhere to established software standards as they affect input design.
3. Use standard input design forms—headings, body and footnotes should conform to established standards.
4. Have a convention for representing variable information. Constant information is usually typed or written on output design forms. Variable information is usually indicated via some convention. For example, a string of Xs (e.g. XXXXXX...) to represent alphanumeric data; a string of 9s (e.g. 99999, 999.99, etc.) to represent numeric data.
5. Involve the user in the decision-making exercises.
6. The prototype and actual input should be well balanced and attractive.
7. Where possible, use interactive prototypes to help the user to conceptualize the output being designed at an early stage. This may be supported by the software design tool being used (e.g. CASE tools, or a presentation software product that supports hyperlinks).
8. Keep screens simple and attractive.
9. Keep screen presentation consistent.

10. Facilitate easy user movement among screens and reversibility of actions.
11. Control the duplication of data on screens (duplicate only when necessary).
12. As much as possible, the screen must match the associated form from which data will be entered.
13. For color monitors, avoid outrageous color schemes.
14. Input screen design must conform to established standards of the organization.
15. The principles of forms design, discussed in Chapter 2, must apply to the associated form. In particular:
    a. The form must fulfill its intended purpose i.e. it must collect the data required.
    b. It must be trivially easy to fill out the form with negligible or no error.
    c. The form must be attractive.
    d. The form may be designed via an appropriate software product and reviewed before implementation.

14.9 Summary and Concluding Remarks

Here is a summary of what we have covered in this chapter:

- The user interface is the window through which end users access the software system. It is critical that this component of the software system is properly designed as failure to do so could compromise the success of the software product.
- In planning the user interface, the software engineer must dive due consideration to the end user needs, human factors, design factors, environmental factors, and actual programming of the interface.
- There are three common types of user interfaces: command interface, menu-driven interface, and graphical user interface (GUI). Your approach in designing the interface will be influenced by the type of interface that is required.
- The software engineer must observe guidelines for designing system input as well as system output. Of course, the steps taken will be constrained by the type of input/output that is required.

There is much more that could be said about user interface design, but this chapter has given you an overview that you should find useful. In fact, many computer science programs include a course in this area. Before moving on, take a look as the user interface design section of the Inventory Management System project of Appendix C. The next chapter will discuss operations design.

14.10 Review Questions

1. What is a user interface? Why is user interface design important?

2. What are the five aspects of user interface design? For each of these areas, discuss the factors to be considered when a user interface is being designed.

3. Compare the three categories of user interfaces in terms of response time, ease of usage, and complexity of design.

4. Construct a grid that shows how various interaction styles apply to the different categories of user interface.

5. Outline the steps to be followed when designing a menu-based user interface or a GUI. Also outline the steps to be followed when designing a command-based user interface.

6. Construct a grid that compares the various output methods with respect to advantages and disadvantages of each.

7. State four basic guidelines for designing printed output. State four basic guidelines for designing screen output. State four basic guidelines for designing software input.

References and Recommended Reading

[Kendall 2014] Kendall, Kenneth E., and Julia E. Kendall. 2014. *Systems Analysis and Design*, 9th ed. Boston, MA: Pearson. See chapters 11, 12, and 14.

[Pressman 2015] Pressman, Roger. 2015. *Software Engineering: A Practitioner's Approach*, 8th ed. New York, NY: McGraw-Hill. See chapter 15.

[Schneiderman 2017] Schneideman, Ben, et al. 2017. *Designing the User Interface*, 6th ed. Boston, MA: Pearson.

[Van Vliet 2008] Van Vliet, Hans. 2008. *Software Engineering*, 3rd ed. New York, NY: John Wiley & Sons. See chapter 16.

Chapter 15

Operations Design

This chapter discusses operations design as an integral part of the software design experience. The chapter proceeds under the following captions:

- Introduction
- Categorization of Operations
- Essentials of Operations Design
- Informal Methods for Specifying Operation Requirements
- Formal Specification
- Summary and Concluding Remarks

15.1 Introduction

Whether the functional approach or the object-oriented approach is employed, ultimately, software systems will necessarily have operations. An important aspect of software engineering is the preparation of *operation specifications* for the operations of a system. Hence, operations design forms a very important component of software design.

The spin-off from operations design is a set of operation specifications: each operation has an operation specification (commonly abbreviated as operation spec), which can be pulled by a programmer and used in developing the required operation (program). The more thorough the spec, the easier is the required programming effort.

In OOD, the operations are implementation of the verbs that link objects and allow communication among these objects. A common practice is to make the operations singular (monolithic) in nature, thus further simplifying the subsequent development and maintenance processes. Further, complex operations are made to employ the services of other (simpler) operations, thus promoting code reuse and efficiency.

As established from earlier chapters, we refer to the analysis of operations (and other related issues) as object behavior analysis (OBA). You will recall that in Chapter 13, we used the corresponding term object structure analysis (OSA) to describe analysis relating to data structures of object types comprising the software system. Take some time to review the OO modeling hierarchy (Figure 10.11) of Chapter 10. What you need to remember is that OOD boils down to two things—OSA and OBA.

On the other hand, in FOD, the functions typify (functional) activities in the organization and facilitate management and retrieval of information; verbs may be combined in a single function. This sometimes leads to more complex functions, with a significantly lower level of reusability of code.

**Example 15.1   Contrasting the FOD Approach with the OOD Approach
for Operations DesIGN**

For a typical information entity in a software system, the operations may be defined as follows:

In OOD for a given database entity, the legitimate operations may be: ADD, DELETE, MODIFY, RETRIEVE, QUERY, and LIST. In FOD, for the same entity, the operations (typically referred to as functions) may be MAINTAIN, QUERY, LIST, but may involve other entities.

15.2 Categorization of Operations

Categorization of the operations is a useful strategy, particularly during subsequent software development:

- Some operations will be more complex than others, and will therefore need longer development time.
- Also, by knowing about the relative complexity of operations making up a system, a project manager can make prudent decisions about work schedule.
- Some operations will be more crucial to the overall software product than others; if this information is known, important prioritization decisions can be made.

Thus, by carefully categorizing operations the software designer can significantly contribute to the success of subsequent software development and maintenance. Figure 15.1 provides a four-step approach to categorizing operations.

---

1. Rank the major quality factors (mentioned in chapters 1 and 3) in order of importance based on requirements of the software.
2. Rank each operation, based on the quality factors ranking.
3. Prioritize the operations as mandatory, important, or optional but desirable (i.e. nice to have).
4. Apply a *relative complexity index* to each operation. This may be done mathematically by considering factors such as the number of other system objects (including database objects) that the operation has to communicate with, the anticipated nature of these interactions, the nature of algorithms to be employed, and the anticipated code length. However, in practice, this is usually estimated by an experienced software engineer, after considering these factors.

Figure 15.1   Categorizing Operations

While the first two steps of Figure 15.1 may be considered optional, the latter two are essential, as they help in guiding decisions about the project during the development phase.

15.3 Essentials of Operations Design

Each operation must be assigned a unique system (implementation) name. Preferably, these system names should conform to the established naming convention that is in vogue (review Chapter 12).

For each operation, the following should be clearly stated: what the operation will do; the inputs and outputs; the algorithm(s) to be implemented; any required validation rules; and/or calculation rules. These things are normally provided in an *operation specification* (often abbreviated as operation spec or op spec).

In the interest of clarity and efficiency, operation specs are usually expressed in a standard format, according to the software development standards of the organization (review Chapter 12). Over the years, several formal methods as well as informal methods for software specification have been proposed. The next two sections will summarize some of the commonly used approaches, as well as a pragmatic approach, proposed by the author.

As you proceed, bear in mind that an operation will be implemented by code in a particular programming language. How it is implemented depends on the development environment: In an FO environment, it is typically implemented as a program. In a purely OO environment, the operations may be implemented as a method of a class, a class consisting of several related methods, or a set of related classes. This is also the likely case in a hybrid environment (consisting of an OO user interface superimposed on a relational or object database).

15.4 Informal Methods for Specifying Operation Requirements

Remember, in operations specification, the software engineer is attempting to define and express the requirements of system operations in a manner that promotes comprehension and efficiency, while avoiding the hazard of being ambiguous or too verbose.

In this section, we shall briefly review some traditional approaches to operations specification. We will then look at use of the Warnier–Orr diagram and the UML notation. The section closes with a discussion of a methodology introduced by the author.

15.4.1 Traditional Methods

Traditional informal methods of operations specification include program flow-charting, the use of IPO charts, decision techniques, and pseudo-coding (review Chapters 8 and 10). The main advantages of these approaches are:

- Visual aid (in the case of IPO chart and program flow chart)
- Flexibility and creativity in treating unanticipated or complex situations

Notwithstanding the advantages, these approaches also have inherent flaws, some of which are stated below:

- None of these approaches lends itself to comprehensive coverage of all the requirements of operations of a software product. To illustrate, a flowchart does not readily provide information about the inputs to and outputs from an operation as well as an IPO chart. An IPO chart does not represent logical decisions as well as a flowchart or decision table. Neither does a pseudo-code.
- Pseudo-codes and flowcharts can be difficult to maintain, especially for large, complex systems.
- The use of pseudo-code does not facilitate easy analysis; it does not avoid the possibility for ambiguity, since there is no established standard vocabulary; neither does it facilitate easy automation.

In view of the foregoing, the IPO chart of Figure 15.2 should illustrate the limitations of the technique. The figure shows an IPO chart for an operation that facilitates addition, modification, or deletion of employee records in a human resource management system (HRMS), or some other system requiring employee information. Traditionally, operations that provide such functionalities (addition, update, and deletion) were called MAINTAIN operations, and appeared frequently in software systems designed in the FO paradigm.

Operation Name: HR_Employee_MO		
Operation Description: Allows maintenance of the employee personal information.		
Input	Processing	Output
HR_Employee_BR	1. Accept employee number	Edit List, HR_Employee_BR
Employee Information Form	2. If this is a new employee number, allow **addition** of a new employee	
	3. If this is a preexisting employee number, find out if the user desires **modification** or **deletion**	
	4. If modification is chosen, allow modification of the employee record; otherwise, allow deletion of the employee record	

Figure 15.2 IPO Chart for a MAINTAIN Operation

15.4.2 Warnier Orr Diagram

In the Warnier–Orr approach, the operation is consistently broken down into constituent activities, in a hierarchical manner. Component activities are numbered in a manner similar to the sections of the chapters of this book. The basic idea is as shown in Figure 15.3.

A Warnier–Orr diagram provides two main advantages:

- It shows the relationships among activities of an operation.
- If activities are logically arranged it can replace the flowchart/pseudo-code.

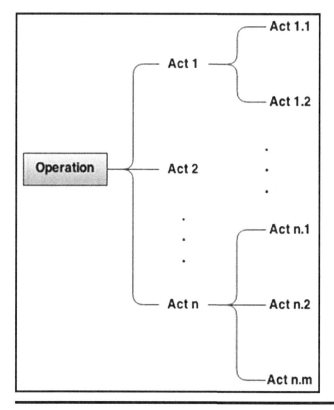

Figure 15.3   Illustration the Warnier–Orr Diagramming Methodology

Disadvantages of the Warnier–Orr diagram are:

- It does not explicitly show inputs to or outputs from an operation.
- It does not include explicit representation of logical decisions.
- The diagram can grow indefinitely, so that in the absence of a CASE tool that supports the technique, maintaining Warnier–Orr diagrams for system operations could be quite time consuming.

Although the technique might appear to have a function-oriented connotation, it is in fact also applicable in an object-oriented scenario. Figure 15.4 provides an example of a Warnier–Orr diagram for a MAINTAIN operation for employee records. The operation name (**HR_Employee_MO**) is indicated in the first activity box. Still referring to the figure, the item **HR_Employee_BR** may be implemented as an object type in an OO environment, but more typically as a relational table in a relational database that is accessed by an appropriately constructed user interface (which may very well be based on an OO software development environment). This operation would form part of the user interface.

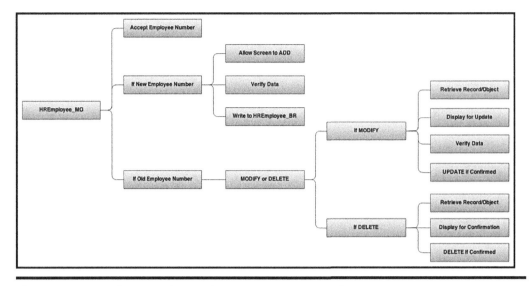

Figure 15.4   Warnier–Orr Diagram for a MAINTAIN Operation

15.4.3 UML Notations for Object Behavior

The unified modeling language (UML) employs a comprehensive set of notations for specifying the behavior of objects when using the object-oriented approach to software design. In conducting OBA, the UML facilitates the following techniques:

- Use-case Diagrams
- State Diagrams
- Activity Diagrams
- Sequence Diagrams and Collaboration Diagrams

These techniques were covered in earlier Chapters 5, 8, and 9. Please review Figures 13.10 and 13.16 of Chapter 13. In those figures, some of the information entities that would be found in the CUAIS project (for a generic college or university) were represented. One such entity (object type) was the **Employee**. The following figures (Figures 15.5 through 15.8) provide examples of a use-case diagram or a typical **Employee** object, a state diagram for a typical **Employee** object, (copied from Chapter 8), an activity diagram for creating (i.e. adding) an **Employee** object, and a collaboration diagram for querying (i.e. inquiring on) an **Employee** object.

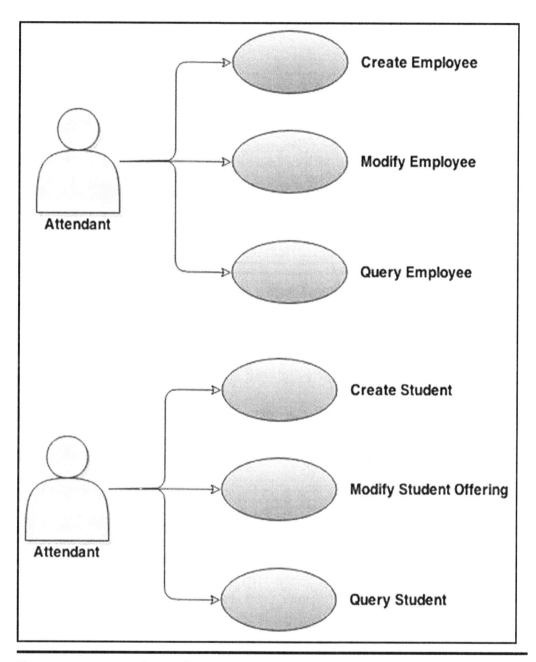

Figure 15.5   Use-case for Employee Processing and Student Processing

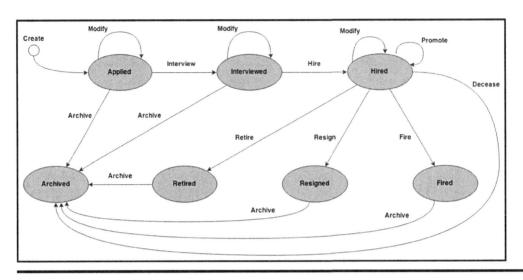

Figure 15.6   State Diagram for Employee Object

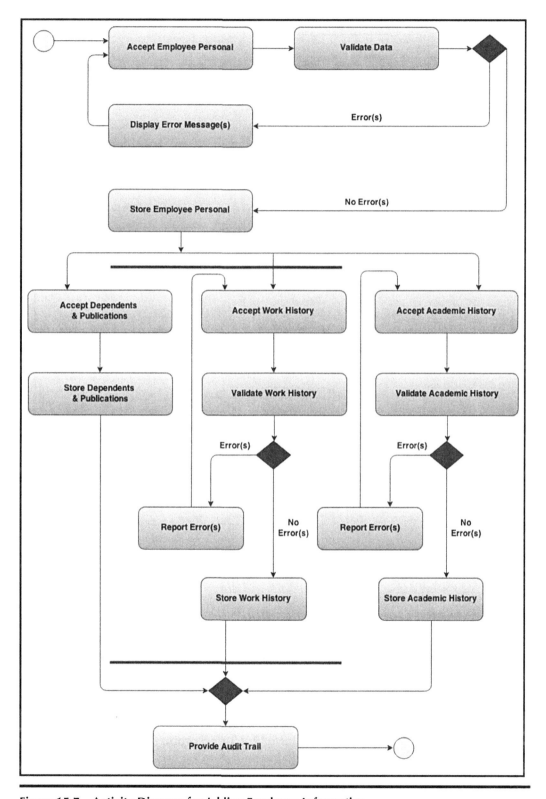

Figure 15.7   Activity Diagram for Adding Employee Information

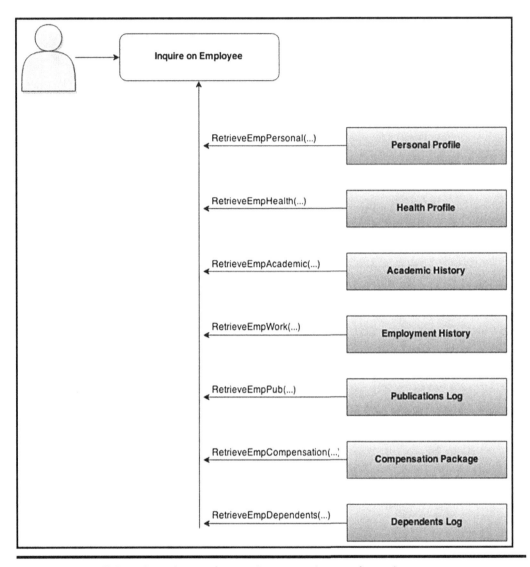

Figure 15.8   Collaboration Diagram for Inquiry on Employee Information

15.4.4 *Extended Operation Specification*

The *extended operation specification* (EOS) approach was developed by the current writer and has been successfully used on various software systems designed by him. In this approach, the software engineer records important requirements about an operation in such a manner as to offset the disadvantages of the previously discussed methods.

The basic idea is to provide enough detail about the required operation, so that a programmer that pulls the spec should have little or no problem in writing the operation (program). The format recommended is shown in Figure 15.9. Observe that the specification also includes categorization information, which may be used during software development. The technique is applicable to both FOD and OOD.

```
System:
Sub-system:
Operation Name:
Operation Description:
Operation Categorization: _____    (Mandatory, Important, Nice)
Complexity Rank [ ] of  Total Possible Rating [ ]
Spec. Author: _____             Spec. Date: __/__/__


Inputs:          _____
                 . . .
                 _____

Outputs:         _____
                 . . .
                 _____

Validation Rules: _____
                 . . .
                 _____

Special Notes:   _____
                 . . .
                 _____

Outline:         _____
                 . . .
```

Figure 15.9   Components of the Extended Operation Specification

Here are a few additional guidelines for the EOS:

1. The required output formats and/or screen formats could be (and are usually) designed and attached.
2. Validation rules are itemized and special notes are itemized.
3. The outline could be in the form of a pseudo-code (as illustrated in Figures 15.11 through 15.14), a flowchart, a Warnier–Orr diagram, or an activity diagram.

Figure 15.10 provides an excerpt from the object/entity specification grid (O/ESG) of Chapter 13, repeated here for convenience; in the current context, we shall concentrate on the **Employee** object type (entity), but recall from Figure 13.16 in Chapter 13, that two referenced entities are **Department** and **Classification**. Figures 15.11 through 15.14 provide sample EOSs for an ADD operation, an UPDATE operation, a DELETE operation, and an INQUIRE operation, respectively.

Three additional points to note from the illustrations (Figures 15.11 through 15.14):

1. The algorithms for adding, modifying and deleting data items (records) have been standardized and can be used for different scenarios. How they are implemented will vary for different software development tools.
2. The term *Virtual Data Collection Object*, as used in Figure 15.14, refers to any user interface widget that may typically be employed by a RAD or CASE tool (for instance, in Delphi, we could us a DB-Grid or a DB-Image; in Team Developer, we could use a Child-Table).

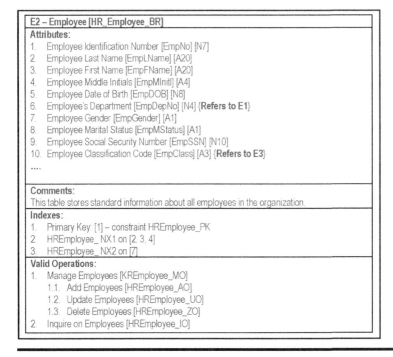

E2 – Employee [HR_Employee_BR]
Attributes:
1. Employee Identification Number [EmpNo] [N7]
2. Employee Last Name [EmpLName] [A20]
3. Employee First Name [EmpFName] [A20]
4. Employee Middle Initials [EmpMInitl] [A4]
5. Employee Date of Birth [EmpDOB] [N8]
6. Employee's Department [EmpDepNo] [N4] {**Refers to E1**}
7. Employee Gender [EmpGender] [A1]
8. Employee Marital Status [EmpMStatus] [A1]
9. Employee Social Security Number [EmpSSN] [N10]
10. Employee Classification Code [EmpClass] [A3] {**Refers to E3**}
....

Comments:
This table stores standard information about all employees in the organization.
Indexes:
1. Primary Key: [1] – constraint HREmployee_PK
2. HREmployee_ NX1 on [2, 3, 4]
3. HREmployee_ NX2 on [7]
Valid Operations:
1. Manage Employees [KREmployee_MO]
   1.1. Add Employees [HREmployee_AO]
   1.2. Update Employees [HREmployee_UO]
   1.3. Delete Employees [HREmployee_ZO]
2. Inquire on Employees [HREmployee_IO]

Figure 15.10   Sample Object/Entity Specification Grid for Employee

3. The term *logical view* has also been introduced in Figure 15.14. A logical view is a virtual database object which stores an access path to data contained in persistent database tables. In the example, we have a join logical view, based on the fact that there is a relationship between **E2 (HR_Employee_BR)** and **E1 (HR_Department_ BR)** on the one hand, and **E2 (HR_Employee_BR)** and **E3 (HR_Classif_BR)** on the other (see Figure 15.10 and review Figure 13.16 of Chapter 13). You will spend much more time working with logical views in your database systems course.

Among the advantages of EOS are the following:

- It allows the software engineer to pack all the relevant information about an operation into one spec so that development is easy.
- It provides information that allows the project manager to make intelligent work assignments during software development.
- Under the operation outline section, the software engineer has the flexibility of using a program flowchart, a pseudo-code, a Warnier–Orr diagram, or an activity diagram.
- Important information such as I/O requirements, categorizations, etc. can be included in the spec.
- The whole process of specifying an EOS for an operation can be automated by developing a software system for that purpose.

In terms of disadvantages of the EOS, no major drawback of the methodology has been identified so far. However, with time it is anticipated that constructive criticisms will inspire further refinement. Automation of this and other related novel methodologies introduced in the course, is the focus of an ongoing research.

Operation Biography:

System:	CUAIS
Subsystem:	Human Resource Management
Operation Name:	HR_Employee_AO
Operation Description:	Facilitates addition of records to the employee file.
Operation Category:	Mandatory
Complexity Rank:	6 of 10
Spec. Author:	E. Foster
Date:	24-10-2004

Inputs:
Employee Personal Information — **HR_Employee_BR**
Department — **HR_Department_BR**
Employee Classification — **HR_Classif_BR**
New Employee Profile Form

Outputs:
Employee Personal Information — **HR_Employee_BR**
Audit Trail (edit listing)

Validation Rules:
1. Employee # must not previously exist.
2. Employee classification code must exist in file **HR_Classif_BR**.
3. Employee's Department Number must exist in file **HR_Department_BR**
4. Blank or null name and address must be rejected.
5. Date of birth must be a valid date in the 20th century.

Special notes: None

Operation Outline (Pseudo-code):
```
START
WHILE (User wishes to continue)
       Accept Key Field(s);
       Check Record Absence or Existence in file HR_Employee_BR;
       IF      (Record Absent)
               Accept Non-key Fields;
               Validate Non-key Fields based on Validation Rules;
               WHILE (Any Error Exists),
                       Re-display Non-key Fields for possible Update;
                       Display appropriate error message(s);
                       Validate Non-key Fields based on Validation Rules;
               END-WHILE;
               Re-display full Record for confirmation;
               IF      (Confirmation Obtained)
                       Write New Record to file HR_Employee_BR;
                       Write New Record to audit file for Additions;
               ENDIF;
               ELSE   Inform the User that nothing was saved;
               END-ELSE;
       ENDIF;
       ELSE   Display Message ('Record already exists');
       Check if User wishes to quit and set an exit flag if necessary;
END-WHILE;
Generate Edit-List;
STOP
```

Figure 15.11   Sample EOS for ADD Operation

Operation Biography:

System:	CUAIS
Subsystem:	Human Resource Management
Operation Name:	HR_Employee_UO
Operation Description:	Facilitates update of employee records in the employee file.
Operation Category:	Mandatory **Complexity Rank:** 6 of 10
Spec. Author:	E. Foster **Date:** 24-10-2004

Inputs:
Employee Personal Information — **HR_Employee_BR**
Department — **HR_Department_BR**
Employee Classification — **HR_Classif_BR**
New Employee Profile Form

Outputs:
Employee Personal Information — **HR_Employee_BR**
Audit Trail (edit listing)

Validation Rules:
1. Employee # must previously exist.
2. Employee classification code must exist in file **HR_Classif_BR**.
3. Employee's Department Number must exist in file **HR_Department_BR**.
4. Blank or null name and address must be rejected.
5. Date of birth must be a valid date in the 20th century.

Special notes: None

Operation Outline (Pseudo-code):
```
START
WHILE (User wishes to continue)
        Accept Key Field(s);
        Check Record Absence or Existence in file HR_Employee_BR;
        IF      (Record Present)
                Retrieve Record and update Audit Log Fields (with before-values);
                Display Non-key Fields for possible Update;
                Validate Non-key Fields based on Validation Rules;
                WHILE (Any Error Exists),
                        Re-display Non-key Fields for possible Update;
                        Display appropriate error message(s);
                        Validate Non-key Fields based on Validation Rules;
                END-WHILE;
                Re-display full Record for confirmation;
                IF      (Confirmation Obtained)
                        Update Audit Log Fields (with current-values);
                        Write New Record to audit file for Updates;
                        Update Record in file HR_Employee_BR;
                ENDIF;
                ELSE   Inform the User that nothing was saved;
                END-ELSE;
        ENDIF;
        ELSE   Display Message ('Record does not exist');
        Check if User wishes to quit and set an exit flag if necessary;
END-WHILE;
Generate Edit-List;
STOP
```

Figure 15.12   Sample EOS for UPDATE Operation

Operation Biography:
System: CUAIS
Subsystem: Human Resource Management
Operation Name: **HR_Employee_ZO**
Operation Description: Facilitates deletion of employee records from the employee file.
Operation Category: Mandatory
Complexity Rank: 6 of 10
Spec. Author: E. Foster   **Date:**   24-10-2004

Inputs:
Employee Personal Information — **HR_Employee_BR**
Department — **HR_Department_BR**
Employee Classification — **HR_Classif_BR**
New Employee Profile Form

Outputs:
Employee Personal Information — **HR_Employee_BR**
Audit Trail (edit listing)

Validation Rules:
1.   Employee # must previously exist.

Special notes: None

Operation Outline (Pseudo-code):
```
START
WHILE (User wishes to continue)
      Accept Key Field(s);
      Check Record Absence or Existence in file HR_Employee_BR;
      IF      (Record Present)
              Retrieve Record;
              Display full Record for confirmation;
              IF      (Deletion Confirmation Obtained)
                      Update Audit Log Fields (with current-values);
                      Write New Record to audit file for Deletions;
                      Delete Record from file HR_Employee_BR;
              ENDIF;
              ELSE   Inform the User that nothing was saved;
              END-ELSE;
      ENDIF;
      ELSE   Display Message ('Record does not exist');
      Check if User wishes to quit and set an exit flag if necessary;
END-WHILE;
Generate Edit-List;
STOP
```

Figure 15.13   Sample EOS for DELETE Operation

Operation Biography:
System: CUAIS
Subsystem: Human Resource Management
Operation Name: HR_Employee_IO
Operation Description: Facilitates inquiry on employee information.
Operation Category: Mandatory
Complexity Rank: 9 of 10
Spec. Author: E. Foster   Date: 24-10-2004

Inputs:
Employee Personal Information — **HR_Employee_BR**
Department — **HR_Department_BR**
Employee Classification — **HR_Classif_BR**

Outputs: Monitor Display
Validation Rules: None

Special notes:
1.   It will be possible to query employee information via any of the following access paths:
1.1 By Identification Number or Social Security Number
1.2 By Name
1.3 By Department, Classification and Name
1.4 By Department, Gender and Name
1.5 By Department and Name
1.6 By Classification, Gender and Name
1.7 By Classification and Name
2.   Each option will invoke one of seven sub-operations (HREmployee_IO1, HREmployee_IO2,   ... HREmployee_IO7).
3.   Utilizes the logical view HREmployee_LV1 which joins HREmployee_BR,
        HRDepartment_BR and HRClassif_BR.

Operation Outline:
START: /* Update */
  While User Wishes to Continue
    Present the User With the Options mentioned above;
    Depending on the User's Choice, Invoke one of sub-operations HREmployee_IO1, ...
      HREmployee_IO7;
  End-While
STOP.

/* Each sub-operation will allow the user to specify appropriate search criteria; this will then be used to retrieve records from the system and display them on the monitor. For instance, */
...
Outline for HREmployee_IO5:
START
While User Wishes to Continue
  Prompt user for Starting Department and Name;
  Starting at that point in **HREmployee_BR**, Load and Display a *Virtual Data Collection Object* with
  all records until End-of-File;
End-While;
STOP.
...

Figure 15.14   Sample EOS for INQUIRE Operation

15.5 Formal Specifications

A formal specification of software is a specification that is expressed in a language whose syntax and semantics have been formally defined.

Formal specification methods have been developed and are widely used in software engineering. Several *program description languages* (PDLs) have been proposed and used in software engineering. Some examples are mentioned below:

- PSL—Problem Statement Language
- Ada PDL
- Z-specifications (pronounced zed specifications)
- Larch Specifications
- B Specifications
- Lotos Specifications

The prediction that by the twenty-first century, a significant proportion of software systems would be developed by formal methods has not been realized, due to a number of reasons:

1. Successful informal methods such as structured methodologies and OO methodologies have been on the increase.
2. Market dynamics puts pressure on the software engineering industry to produce software at much faster rates than formal methods would allow.
3. Formal methods are not well suited for some scenarios, e.g. user interface development.
4. Formal methods provide limited scope for quality factors such as scalability and portability. To some extent, maintainability and flexibility are also negatively affected. As you are aware, these are crucial requirements for contemporary software.

One significant advantage of formal methods is that they force precise, unambiguous specification of software. Because of this, formal methods are widely used in areas of software engineering where precision is required. Two examples of such areas are hardware synthesis and compilation. However, to do justice to the field, further exploration of formal specifications is best treated in a course on formal methods.

15.6 Summary and Concluding Remarks

It's time once more for us to summarize what we have covered in this chapter:

- Operations design is an integral part of the OBA process (assuming the OOD paradigm). The spin-off from operations design is a set of operation specifications: each operation has an operation specification that outlines the blueprint for the operation.
- Categorizing operations is very useful particularly during the development of the software system. Each operating can be categorized by considering its

alignment with quality factors or importance to the software system, the level of importance of the operation, and its relative complexity.

- Each operation spec must have a unique name, followed by unambiguous guidelines that will help a programmer to easily write the actual operation.
- Informal methods of operations specifications include (but are not confined to) the following: traditional methods (program flow charts, pseudo-code, and IPO charts); Warnier–Orr diagrams; UML diagrams (use-case diagrams, state diagrams, activity diagrams, sequence diagrams, and collaboration diagrams); EOS formulations.
- Formal methods of operations specifications include (but are not confined to) PSL, ADA/PDL, Z-specifications, Larch specifications, B specifications, and Lotos specifications.

Armed with the software development standards, architectural specification, the database specification, the user interface specification, and the operation specification, you are almost ready to embark on the actual software development with confidence. Bear in mind that if you are using an OO-CASE tool, actual design and development may be merged into one modeling phase, since many of the diagrams may be executable diagrams. We still have a few design issues to cover, and these will be discussed in the next chapter.

15.7 Review Questions

1. How important is operations design? Explain how it affects the development of a software system.

2. Outline an approach for categorizing operations comprising a software system.

3. The **Student** entity would be an important component of the CUAIS project. It contains attributes **StudentID, Name, Gender, DateOfBirth, Major, Dept#**, among others. Each **Student** object has a unique identification number. The department (**Dept#***)* to which a student is assigned must previously exist in the **Department** entity. Also, the student's major must reside in the **AcademicProgram** entity. **Gender** must be male or female and **DateOfBirth** must be a valid date in the twentieth or twenty-first century.
   3a. Propose an O/ESG for the **Student** entity.
   3b. Propose an EOS for the operation to allow addition of valid student records. Your outline may include a Warnier–Orr diagram, a pseudo code, or an activity diagram.
   3c. Propose an EOS for an operation to allow users to run interactive query of student information.
   3d. Propose a collaboration diagram for interactive query of student information.

4. Examine Figure 15.6 and thoroughly explain all the state transitions and the operations that trigger them.

5. Examine Figure 15.7 and thoroughly explain the behavior of the **ADD Employee** operation.

6. Examine Figure 15.8 and thoroughly explain what is represented therein.

7. When is formal specification relevant? Give three examples of formal specification languages.

References and Recommended Reading

[Peters 2000] Peters, James F. and Witold Pedrycz. 2000. *Software Engineering: An Engineering Approach*. New York, NY: John Wiley & Sons. See chapter 5.

[Pfleeger 2006] Pfleeger, Shari Lawrence. 2006. *Software Engineering Theory and Practice*, 3rd ed. Upper Saddle River, NJ: Prentice Hall. See chapters 4.

[Schach 2011] Schach, Stephen R. 2011. *Object-Oriented & Classical Software Engineering*, 8th ed. New York, NY: McGraw-Hill. See chapters 13 and 14.

[Van Vliet 2008] Van Vliet, Hans. 2008. *Software Engineering*, 3rd ed. New York, NY: John Wiley & Sons. See chapter 10–12.

Chapter 16

Other Design Considerations

We have looked at all the major aspects of software design. However, there are a few outstanding areas that need some attention. This chapter covers these areas as outlined below:

- System Catalog
- Product Documentation
- User Message Management
- Design for Real-time Software
- Design for Reuse
- System Security
- The Agile Effect
- Summary and Concluding Remarks

16.1 The System Catalog

The system catalog (also called data dictionary) is a very useful and effective management and analysis tool for the software engineer or system manager. It stores critical control information that will be useful to anyone managing or working on the development/maintenance of the software system.

Typically, the catalog commences its existence during the design phase, but may be introduced as early as during the preparation of the requirements specification. Once introduced, it serves the project for the rest of its useful life.

16.1.1 Contents of the System Catalog

The system catalog contains database related information as well as operational control information. Essential database-related contents of the system catalog are the following:

DOI: 10.1201/9780367746025-19

- Name, description, proper coding, and aliases of each information entity (or object type) in the system
- Name, description, proper characteristics (e.g. type and length), and coding of data element in each information entity
- Required editing and integrity checks on data elements
- Indication as to whether a data element references another element in another entity (object type) and the type of reference (relationship)

Operational contents include:

- Name and description of sub-systems (and/or modules)
- Name and description of each operation (or function)—including super-operations as well as sub-operations
- For each operation, specification of entities (object types) used, and/or impacted

System rules include:

- Relationships among entities (object types)
- Data integrity rules (if not already specified above)
- Calculation rules for operations
- Other operational rules

Observe: In purely OO environments, the approach is somewhat different: For each object type, a set of allowable operations is specified. The reason for this is that purely OO software tools will allow the software to encapsulate the object's related operations with its data structure into a class.

16.1.2 *Building the System Catalog*

The system catalog may be constructed in any of three of ways—via static tables, via dynamic files, or via automatic creation.

Via Static Tables: In the absence of a sophisticated software development environment, the catalog can be built using static table(s) via some word processor e.g. Word Perfect, Microsoft Works, Microsoft Word, etc. Maintenance will be particularly challenging, since any change to the model will necessitate manual changes to the tables.

Via Dynamic Files: Again, assuming an environment devoid of sophisticated software development tools, an alternative to static tables would be to create data file(s) that can store the pertinent information, then load and manage the information via some entry/update application program(s). This could be done in any programming language.

Automatic Creation: A few business-oriented operating systems (e.g. System i) have the facility to allow the software engineer to define and maintain a data dictionary as an important component of software developed in that environment. Additionally, the more sophisticated software development products (CASE tools and DBM suites e.g. Oracle, DB2, Informix, etc.) have built-in features to automatically build and maintain a system catalog, while a software product is being developed.

The catalog can then be accessed by the software engineer. This approach has become mainstream in contemporary software engineering.

Whichever of the first two methods is employed, a top-down approach to defining the system catalog is recommended:

- Specify all sub-systems of the software system
- Specify all entities (object types)
- Specify all elements of entities
- Specify all operations

16.1.3 Using the System Catalog

The system catalog is an excellent means of system documentation:

- You can obtain assorted views of the system e.g. entity-operations lists; operation-entities lists; entity-elements lists; terms and aliases; list of system operations; list of system entities; etc.
- You can also obtain information about relationships. In the case of automatic creation, you might even be able to obtain an ERD/ORD from the catalog.
- You will also be able to obtain assorted views of the systems business rules on a system-wide basis, by operation, or by entity (object type).

Information Entities / Object Types

Entity/OT Name	Descriptive Name	Subsystem/Module	Comment
AMProgram_BR	Academic Program	Academic Management	Stores definition of all academic programs
AMCourse_BR	Course	Academic Management	Stores definition of all courses offered
...			

Operations

Operation Name	Operation Title	Subsystem/Module	Inputs	Outputs	Comment
AMProgram_AO	Add Academic Program	Academic Management	Keyboard Entry	AMProgram_BR	Allows addition of academic programs
...					

Operational Business Rules

Operation Name	Rule(s)
AMProgram_AO	1. Program Code must not previously exist on file.
	2. Program Title is mandatory.
	3. Program must be offered by an existing academic department.
...	

Database Business Rules (Relationships)

Entity/OT Name	Referenced Entity / Object Type	Ref-Type	Comment
AMProgStruuct_BR	AMProgram_BR	M:1	
	AMCourse_BR	M:1	
...			

Note:
1. The entity names and operation names used here are based on the naming convention described in appendix 10. 2. The catalog can also be used to provide information on the properties (elements) comprising each entity (or object type).

Figure 16.1   Illustration of System Catalog

The system catalog is very effective in guiding the design phase (particularly with respect to database, operations, menu interface, and business rules). It can also be used as a management tool in design, development, and maintenance phases. Figure 16.1 provides a very basic illustration of the kind of information that a system catalog could provide.

16.2 Product Documentation

Product documentation typically involves three main areas: the *system help facility*, the *users' guide*, and a more technical document often referred to as *the system guide*. Of course, there are variations according to the product, as well as the organization of responsibility. For instance, the users' guide and system guide may be merged as one document or set of documents. Also, if the software system is very large and complex, there may be a *product overview* document.

16.2.1 The System Help Facility

In help design, the software engineer specifies exactly how the help system will be structured and managed for the software in question. Two important considerations here are:

- The structure of the help system
- Content of the help system

16.2.1.1 Structure of the Help System

There are three alternatives for structuring the help system: it may be *panel-by-panel*, *context sensitive*, or *hypermedia-based*.

Panel-by-panel Help: Each panel displayed by the software has a corresponding help panel, invoked when the user makes a request for it. The user request is typically affected by pressing a given function key on the keyboard, or clicking at a help button on the screen.

Context-Sensitive Help: Depending on the cursor position at the time of user request, help information, specific to that locality is provided. This approach is difficult to develop and maintain, since there must be close synchronism between cursor location and help information provided. For example, the operating system called System i (formerly OS-400) has a help facility that is completely context sensitive.

Hypermedia-based Help: The help facility is developed as a hypermedia system, invoked by the user making an explicit request (clicking help menu/button or pressing a function key). The user accesses information based on the choices made from a menu, an index, or a hyperlink. This approach has become very popular, because it comes close to offering the same conveniences as the context-sensitive help, but with much less complexity of design.

16.2.1.2 Content of the Help System

The content of the help system must be relevant to the software it is designed for. Quite often, the help system is developed by practicing system documenters, and not the software engineers who design or develop the product. In such circumstances, close coordination is required, if a high-quality product is to be the outcome.

Unfortunately, the content of the help system has been a problem area for some software systems. Very often, the product documentation that is marketed with the software is voluminous, but inadequate. Different vendors try to document, each producing documents with their own inadequacies. The end result is an inundation of documentation that the user has to scan through in order to get a good handle of the product. The situation is particularly worrying when the product is a software development tool that software engineers use is order to be more proficient on the job.

Ideally, the software engineers should be closely involved in the design and development of the help system (but due to market dynamics, this is often impractical). Alternately, system documenters should have sound appreciation of software engineering.

16.2.2 The User's Guide and System Guide

Traditionally, when a business application is developed for the organization, there used to be a distinction between the user documentation (*user's guide*) and the system documentation (*system manual*). The former was a nontechnical document for end users of the system, the latter, a more technical document for the information systems professional. This approach is still relevant for such categories of software. However, as more sophistication is added to the business applications, this distinction is becoming more nebulous.

The user's guide is a nontechnical document, suited for the end users. It contains step-by-step instructions on how to use the software, preferably on a module-by-module basis. The system manual is a technical document that is ideally suited for system managers and software engineers, whose responsibility it will be to maintain the system. It is a technical summary of the system, outlining:

* The main components
* The operating constraints and configuration issues
* Security issues
* Syntax and explanations of system commands (for command interface)
* Explanations of system commands (for menu or graphical interface)

For other categories of software such as development tools for software engineering, this distinction (between technical and nontechnical) is not relevant. What is required here is a comprehensive set of documents that the user (in this case the software engineer) can use. Typically, the set begins with a *product overview*, then depending on the complexity and scope of the product, there is a document or set of documents for different aspects of the software. This may also be supported by comprehensive (set of) system manual(s). In the era of command-based user interfaces, system manuals were very voluminous. Nowadays (in the era of GUIs), they are not as bulky.

Another contemporary trend is to provide the system documentation in electronic form rather than via large volumes as used to be the case. This cuts down on the marketing cost of software engineering companies, allowing for easier packaging, shipping, and handling.

Ideally, the software engineers should be closely involved in this aspect of software documentation. Failure to do so often results in poor quality in the documentation of the product. There is no scarcity of software products that have been poorly documented.

Rational Software (now a division of IBM), the company responsible for Rational product line and the Unified Modeling Language (UML), provides a positive example of good software documentation. Most of the product documentation was done by the chief software engineers behind the products—Grady Booch, Ivar Jacobson, and James Rumbaugh (see [Rumbaugh 1999]). The product documentation for Oracle (despite being quite voluminous) also provides an excellent example of good product documentation: it is comprehensive, nonintimidating, easy to use, well organized, and packed with good illustrations.

16.3 User Message Management

Undoubtedly, it will be necessary for the operations of the software system to provide user messages to guide the user along. There are three kinds of user messages that a software system may provide:

- **Error Messages** inform the user that an attempted activity is invalid, or an entry is invalid. For example, keying in an invalid date, or attempting to access a data item (record) that does not exist, should each elicit a software response that that activity is not permissible at that point in time.
- **Status Messages** inform the user on the current state of an ensuing activity, for example displaying the number of records read from a database file.
- **Warning Messages** alert the user that an attempted activity could result in problems.

There should be some standard as to how and where on the screen, user messages will be displayed. Two possibilities are, on the last line of the screen, or in a pop-up window.

In message management, the software engineer specifies how messages will be stored, retrieved, and displayed to users of the software. In many scenarios, message management is addressed in predetermined software development standards (review Chapter 12).

16.3.1 Storage and Management of Messages

Two approaches to storing and managing user messages are possible:

- Each application operation stores and manages its own user messages. This is the easy way out. The main problems are:
  o It promotes inconsistencies.

o It makes the software difficult to maintain.
o There is no independence between system errors and system applications.
- Store all user messages in a system-wide message file. Messages are given unique identification codes and can be accessed by any application operation. This is the preferred approach and avoids the problems of the first approach.

Obviously, the second approach is preferred to the first, since it provides more flexibility and control, particularly as the size and complexity of the project increases. It also leads to a more maintainable software product.

16.3.2 Message Retrieval

If the operation-confined approach described above is employed, message retrieval is not an issue. However, if the system-wide approach is employed, then it might be prudent to define and specify an operation to retrieve user messages and return them to calling operations. This retrieval operation would accept an input argument of the message identification code and return it with an additional argument containing the message text. The calling operation would determine how and where that message is displayed, preferably based on established standards (Chapter 12).

16.4 Design for Real-Time Systems

Not only are computers used to manage information, but also complex manufacturing processes. In many instances, computers are required to interact with hardware devices. The hardware designed in such circumstances is embedded real-time software. Real-time software must react to events generated by hardware and issue control responses that will determine the behavior of the system.

The stages in the design of real-time systems are mentioned below:

1. **Stimulus Identification:** Identify the stimuli that the system must respond to and the appropriate response to each stimulus.
2. **Timing Constraints:** For each stimulus and response, establish a timing constraint.
3. **Aggregation:** Aggregate stimuli into classes (categories). Define a process for each class of stimuli, with allocation for concurrent processes.
4. **Algorithm Design:** Design the required algorithm for each stimulus-response combination. By aggregation, derive algorithms for processes.
5. **Scheduling:** Design a scheduling system that will synchronize processes according to established (time) constraints.
6. **Integration:** Establish a method of integrating the system into a larger system if necessary, via a *real-time executive*.

16.4.1 Real-Time System Modeling

In modeling real-time systems, the software engineer indicates precisely, all the state transitions (the causes and effects). Techniques used include (review state transition diagrams of Chapter 8):

- Finite state machines (state diagrams)
- State transition diagrams

Since these techniques were introduced earlier (Chapter 8) and will be explored in other advanced courses (such as Compiler Construction), they will not be discussed further.

16.4.2 Real-Time Programming

Real-time programming remains an exciting (and sometimes lucrative) arm of software engineering. Real-time programming is typically done using:

- Assembly language, where the programmer has full control
- Intermediate-level languages such as C, C++, and Ada
- OOPLs such as Java and C#

A *real-time executive* is a software component that manages processes and resource allocation in a real-time system. It determines when processes start and stop, and what resources the processes access. The essential components of a real-time execution are:

- A real-time clock
- An interrupt handler
- A scheduler
- A resource manager
- A dispatcher (responsible for starting and stopping processes)

16.5 Design for Reuse

Like in other engineering disciplines, software must be designed with reuse as a given necessity. Hence:

- Principles of inheritance and polymorphism must be deliberately ingrained into the software design
- Where possible, software must be constructed by using tested and proven components.
- New software must be designed so that they can be reused in constructing other software products.

In order for this to be achieved:

- Software engineers must be adequately trained.
- The industry must strive towards a zero-tolerance level for blatant software errors.
- Standardization must take more preeminence than it has taken in the past.

- Software (and components) documentation must be an integral part of software engineering.

Software reuse may be considered at different levels:

- Application systems and subsystems
- System components such as object types (classes) and operations
- Ubiquitous algorithms e.g. sort algorithms, forecasting algorithms, date validation, etc.
- Methodologies

Advantages to reusable software include:

- Improved software quality
- Reduction of development time and cost
- Enforcement of software engineering standards
- Improved reliability of software
- Reduction of risks for new software engineering projects

16.6 System Security

System security has always been and will continue to be an integral part of software design. There are three levels of security that should be addressed:

- Access to the system
- Access to the system resources
- Access to system data

16.6.1 Access to the System

Access to the system typically involves a login process. Each legitimate user is provided with an account, without which they cannot access the system. Figure 16.2 provides some details that are stored in the user account.

Element	Categorization
Account (login) name	Essential
User name	Essential
User password (usually encrypted and/or disguised)	Essential
User group(s) or class(es)	Optional
User logo or picture	Optional

Figure 16.2   Details Stored About a User Account

Among the categories of software that employ a system-level user account are the following:

- Management Support Systems
- Data Warehouses
- Operating Systems
- Web Information Systems
- Computer-Aided Design
- Computer-Aided Manufacturing
- Computer-Integrated Manufacturing
- Database Management Systems
- CASE Tools and RAD Tools

For these systems, the user accounts are stored in an underlying database file. When the user attempts to log on, this file is accessed to determine whether this is a legitimate user. If the test is successful, the user is admitted in; otherwise, an error message is returned to the user. Of course, the database file must be appropriately designed to store all required details about the user.

Of course, not all software products require the use of system-level user accounts as described above. Some products make use of user accounts already defined and stored in the underlying operating system. Others use no system-level security at all. Desktop applications and multimedia enabling software are two categories of software that embody this latter approach.

16.6.2 Access to System Resources

Once a user gains access to the system, the next level of security to be addressed is access to system resources. By system resources, we mean database files, source files, commands, programs, services, etc. Depending on how complex the system is, this could be quite extensive, involving the storage and (often transparent) management of a user-resource matrix (also called an *access matrix*). When the user attempts to access a resource, the access matrix is checked to determine whether the attempted access is legitimate. If it is, the operation is allowed; otherwise, an error message is returned to the user.

One way to implement the resource access matrix is to incorporate its design into the underlying database for the software system, and include related operations for managing it. This approach will become much clearer to you when you study database systems (and learn more about the system catalog) as well as operating systems. Alternately, the matrix may be implemented as an independent system and integrated in the current software system. While the first approach is easier, the latter approach provides more flexibility, and the ability to reuse the resource access matrix in multiple projects (see [Foster 2011] for more details).

16.6.3 Access to System Data

If your system involves access of user data in an underlying database, then it might be desirable to manage access to actual data contained in the database. One very effective and widely used methodology for this is the use of logical views.

As mentioned earlier, a logical view stores the definition of the virtual database file (table), but stores no physical data. It is simply a logical (conceptual/external) interpretation of data stored in core database files. Figure 16.3 provides an example of a situation requiring logical views. You will learn more about, and develop a better appreciation of logical views after you have completed a course in database systems.

Referring to the CUAIS project, your CUAIS database would most definitely include a **Student** entity that stores basic information about student. Suppose that your college/university has twenty academic departments. You could create a logical view of the **Student** entity for each department. Each view will filter and reorganize data stored in the **Student** entity so that the department chair for that department has read-only access to records of all students in his/her department, but cannot access the information for students in any other department. Rather than giving department chairs unfettered access to the **Student** entity, it provides them with controlled access to information that they need to know, and nothing more.

Figure 16.3   Example of a Situation Requiring Logical Views

16.7 The Agile Effect

Recall from Section 1.4, that one of the main flaws of the traditional waterfall life cycle model is its inability to adjust to changes in a graceful manner. Of the other six pragmatic life cycle models that have been proposed, the agile model has emerged as the most widely used approach among contemporary software engineering enterprises.

One of the huge benefits of the agile approach is its nimble capacity to anticipate and respond to changes. At its core, the agile methodology calls for an incremental approach to software engineering. With this in mind, and in the context of Section 12.6, the design specification for the software system may be constructed in one of two ways:

- Have a *single design specification* that is updated whenever enhancements are made to the original product
- Have a *series of design specifications*, each representing successive iterations of the product

Single Document: Having a single design specification ensures that the software system is kept close to its original mission. The software engineer(s) will carry the responsibility of crafting and preserving a design specification that comprehensively and accurately represents the software system at each stage of its existence. Here are two related considerations:

- The first iteration of the design specification may be as comprehensive as possible. This would minimize the need for multiple subsequent iterations.
- Alternately, the first iteration of the design specification may be minimalistic. This would necessitate an expansion of the original blueprint, with each expansion of the software system.

Series of Design Specifications: Having a series of design specifications, each representing the current iteration of the software system, is another way of preserving the integrity of the software system with respect to its blueprint. Following are two related considerations:

- The first iteration of the design specification would likely be minimalistic. However, in the interest of professional integrity, it should also clearly state the scope and areas for subsequent improvements.
- Each subsequent iteration of the software system should preserve the original integrity by stating what expansions are being addressed, and providing a revised scope and areas for subsequent improvements.

16.8 Summary and Concluding Remarks

Let us summarize what has been discussed in this chapter:

- The system catalog is a data dictionary of the software system. It contains information such as the name, description, and characteristic details of the main system components including information entities and operations. It may also contain definition of system rules.
- With a sophisticated DBMS suite or CASE tool, the system catalog is automatically created and maintained by the software development tool. In a more primitive software development environment, the catalog can be created and maintained as static word processing documents or dynamic files maintained by utility programs.
- Once created, the system catalog should be carefully maintained as it contains useful information about the software system.
- Software documentation typically includes a help facility, a user's guide, and a system guide.
- The help facility may be panel-by-panel, context sensitive, or hypermedia-based. Whichever approach is used, the help facility must be carefully planned.
- The system guide is a technical document, for software engineers and/or managers of the system. The user's guide is a nontechnical document for end users. Both should be carefully planned.
- User messages may be error messages, status messages, or warning messages. They are designed to assist the user in successfully using the software system.
- User messages may be managed on an operation-by-operation basis, or via a system-wide message file. The latter approach is preferred as it provides more flexibility to the software engineer, and leads to fewer problems during software maintenance.
- Real-time software systems are hardware-intensive systems that operate in real time based on hardware signals and responses, rather than human intervention. A real-time system passes through the stages of stimulus identification, timing constraints, aggregation, algorithm design, scheduling, and integration.

- Real-time system modeling involves extensive use of diagramming techniques such as finite state machines, state transition diagrams, and activity diagrams. Real-time system programming involves low-level programming.
- Code re-use is not magic; it must be carefully planned and managed.
- There are three levels of software system security: access to the system, access to system resources, and access to system data. Thoughtful design of the security mechanism is paramount to the success of the software system.

By following the principles and methodologies in this and the previous four chapters, you are now in a position where you can confidently put together a design specification for your software system. Remember, we are assuming reversibility between phases of the SDLC as well as between stages within any given phase of the SDLC. The alternative to this assumption is to assume the waterfall model; however, as was mentioned in Chapter 1, this model is particularly problematic, especially for large, complex systems.

Take some time to review the ingredients of the design specification (see Chapter 12), and the various issues discussed in this division of the text. Then examine the sample design specification of Appendix C with fresh eyes. You have been armed with the basic skills needed to design quality software!

16.9 Review Questions

1. What details might be stored in a system catalog? How might it be constructed? How might it be used?

2. Briefly discuss three approaches to structuring the help system of a software product.

3. What information is normally provided in the user's guide of a software product?

4. Describe how you would manage error and status messages for a large software engineering project.

5. Describe the six stages in the design of real-time systems.

6. Describe the three levels of security that many software products are required to address. For each level, outline an effective approach for dealing with security at that level.

References and Recommended Readings

[Foster 2016] Foster, Elvis C. with Shripad Godbole. 2016. *Database Systems: A Pragmatic Approach*, 2nd ed. New York, NY: Apress. See chapters 13 and 14.
[Foster 2011] Foster, Elvis C. 2011. Dynamic Menu Interface Designer. In Papadopoulos, Y. and Petratos, P. *Enterprise Management Information Systems*. Athens: Athens Institute for Education and Research. See chapter 23. Retrievable from https://www.elcfos.com/papers-in-cs.

[Kendall 2014] Kendall, Kenneth E., and Julia E. Kendall. 2014. *Systems Analysis and Design*, 9th ed. Boston, MA: Pearson. See chapters 8 and 15.

[Rumbaugh 1999] Rumbaugh, James, Ivar Jacobson and Grady Booch. 1999. *The Unified Modeling Language Reference Manual*. Reading, MA: Addison-Wesley.

[Schneiderman 2017] Schneideman, Ben, et al. 2017. *Designing the User Interface*, 6th ed. Boston, MA: Pearson. See chapters 13 and 14.

[Sommerville 2016] Sommerville, Ian. 2016. *Software Engineering*, 10th ed. Boston, MA: Pearson. See chapters 15 and 20.

Chapter 17

Putting the Pieces Together

The last 16 chapters of the course have been spent on discussion of software engineering as discipline, as well as on various aspects of software engineering investigation, analysis, and design. Indeed, we have discussed several methodologies that may be applied during these stages of a software engineering project. Before moving on to latter stages of the SDLC, let us pause to put into perspective, what have been covered so far. This brief chapter advances under the following captions:

- How a Software Engineering Project Begins
- The First Deliverable
- The Second Deliverable
- The Third Deliverable
- Other Subsequent Deliverables
- Summary and Concluding Remarks

17.1 How a Software Engineering Project Begins

As mentioned in Chapter 3, a software system concept typically emerges after a needs assessment. This assessment may be conducted for an organization, a group or related organization, an individual endeavor, or a marketplace. The assessment may be requested by an organization, or initiated by an insightful software engineer.

The needs assessment identifies a problem set, i.e., a set of related problem(s), followed by a set of related software systems to address the problem(s). In most cases, it will not be feasible to pursue all the identified projects at the same time, so a selection process is conducted to select the project(s) to be pursued. Figure 17.1 provides a summary. Project selection is typically based on factors such as backing from management; appropriate timing; alignment to corporate goals; required resources and related cost; internal constraints of the organization; and external constraints of the environment/society.

DOI: 10.1201/9780367746025-20

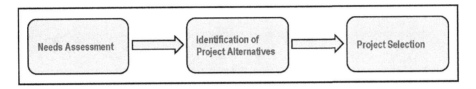

Figure 17.1   Summary of the Software Engineering Project Selection Exercise

17.2 The First Deliverable

Chapter 3 also articulates that the first deliverable for a software engineering project is the initial system requirement (ISR), also called the project proposal. This deliverable introduces the project to decision makers in an organization, or prospective sponsors of the initiative. Among the important ingredients of this deliverable are the following: problem definition; proposed solution; system scope; system objectives; system justification; feasibility analysis report; summary of system inputs and/ or information entities; summary of main outputs; initial project schedule; profile of the project team; and concluding remarks. Figure 17.2 provides a summary to assist you in remembering these elements.

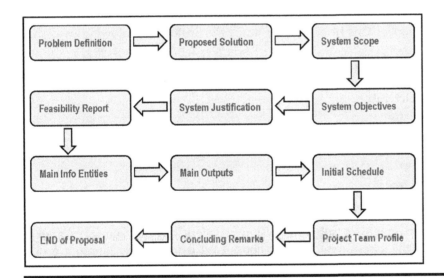

Figure 17.2   Main Ingredients of a Software Engineering Project Proposal

The project proposal is typically prepared and delivered in two formats—a detailed document as well as an oral presentation (aided via an appropriate presentation software such as PowerPoint, Publisher, Prezi, or some other product).

Among the diagramming techniques recommended for the project proposal are the following:

- For system scope: the object flow diagram (OFD) is excellent for the main system components and how they interact; the information topology chart (ITC)

is great for depicting the main information entities and how they relate to the various subsystems.

- For feasibility analysis report, evaluation grids are highly recommended for comparing alternate solutions.
- For the initial project schedule, an activity table will suffice at this early stage.
- Other techniques that may be considered include amortization schedules to illustrate projected cash flow or NPV analysis; bar charts to illustrate time-based comparisons; and process-oriented flowchart (POF) to provide a summarized global perspective of the intended software system.

17.3 The Second Deliverable

As enunciated in Chapter 6, the second deliverable of a software engineering project is the requirements specification. This deliverable uses the project proposal as input to a more elaborate series of activities, and leading to the specification of detailed guidelines for the software system. Among the requisite activities are the following: detailed information-gathering campaign (as covered in Chapter 7); meticulous analysis and synthesis of the information collected; and system modeling via various diagramming techniques (covered in earlier chapters). These activities are summarized in Figure 17.3. As the software engineer on a project, you have the liberty of choosing what methodologies you will employ. However, bear in mind that it is not unusual for a software engineering enterprise to have a set of standards that they require their engineers to follow (review Chapter 12).

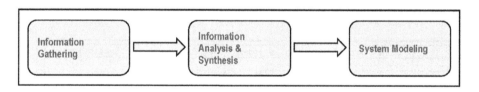

Figure 17.3 Summary of the Main Activities During Preparation of the Requirements Specification

This course recommends that the requirements specification include the following constituents: acknowledgments; problem synopsis; system overview; storage requirements; operational requirements; system rules; interface specification; system constraints; security requirements; revised schedule; concluding remarks; and appendices. For ease of memory, these are depicted in Figure 17.4.

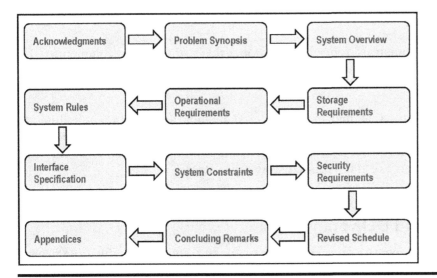

Figure 17.4   Main Ingredients of the Requirements Specification

17.4 The Third Deliverable

The latter part of Chapter 12 (Section 12.6) describes the design specification as the third main deliverable in a software engineering project. This deliverable comes as the result of taking the requirements specification as input, and conducting a series of analysis and/or brainstorming activities as outlined in Chapters 12 and 16. Among the requisite activities are the following: architecture design, interface design, database design, operations design, user interface design, documentation design, message design, and security design. These activities have been summarized in Figure 12.1 of Chapter 12, repeated here in Figure 17.5 for convenience. The figure also specifies the main components of the design specification, connecting each component with its related activity.

Design Activity	Resultant Design Specification Component	Suggested Chapter in Refined Deliverable
Architectural Design	System Architecture Specification	System Overview
Interface Design	System Interface Specification	
Database Design	Database Specification	Database Specification
User Interface Design	User Interface Specification	User Interface Specification
Operations Design	Operations Specification	Operations Specification
Documentation Design	System Documentation Specification	
Message Design	Message Specification	Other Design Considerations
Security Design	Security Specification	

Figure 17.5   Important Software Design Activities

The design specification is perhaps the most important deliverable for a software engineering project; for this reason, it is sometimes called the blueprint for the software system. The main components of this deliverable are highlighted in Figure 17.6. In short, the design specification synthesizes the information from the requirements specification to produce a blueprint for the software system. Here are two points worth remembering:

- It is possible for a substandard software system to emerge from an outstanding design; reckless or inept development will ensure this.
- However, it is not possible for an excellent software system to emerge from a poor or inept design specification.

Given the importance of the design specification, it therefore behooves the conscientious software engineer(s) to be diligent and unremittingly careful in crafting a deliverable that is accurate and thorough. Missing this benchmark can prove to be very costly. Since as human beings, we are prone to error, it is always a great idea to enlist the input of multiple experts, to employ multiple brainstorming sessions, to advance in iterations, and to cross-check the work before declaring it as a finished deliverable.

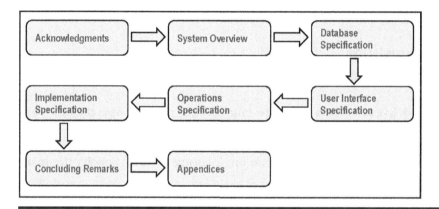

Figure 17.6   Main Ingredients of the Design Specification

17.5 Other Subsequent Deliverables

If you review Section 1.4 and Figure 1.3 of Chapter 1, you will see that other deliverables of a typical software engineering project are the actual software system and product documentation. Over time, these may be revised based on the changing needs and requirements of end users of the software system.

In order to get to the point of having an actual software system, the imperative stage of software construction (development) must be embarked upon.

17.6 Summary and Concluding Remarks

Here is a summary of what has been covered in this chapter:

- A software engineering project is typically conceptualized after a needs-assessment activity.

- The first deliverable of a software engineering project is the system proposal. This deliverable defines the problem to be addressed, followed by a proposal as to how the problem will be addressed (review Section 17.2).
- The second deliverable is the requirements specification, outlining the essential requirements of the software system. This deliverable contains specific elements that must be included (see Section 17.3).
- The third deliverable is the design specification, which serves as the blueprint for the software system. This deliverable must contain specific elements as summarized in Section 17.4.
- Other deliverables include the actual software system and its related documentation.

It is now time to embark on what is regarded by many as the most exciting part of software engineering—the actual construction (development) of the product. The next section of the book (covering three chapters) will discuss important software development issues.

17.7 Review Questions

1. What scenario typically leads to the conceptualization of a software system?

2. What is the first deliverable in a software engineering project? Specify the main components of this deliverable.

3. What is the second deliverable in a software engineering project? Specify the main components of this deliverable.

4. What is the third deliverable in a software engineering project? Specify the main components of this deliverable.

5. Identify two additional deliverables that succeed the design specification a software system.

SOFTWARE DEVELOPMENT

This relatively short division of the text involves three chapters:

- Chapter 18: Software Development Issues
- Chapter 19: Human Resource Management
- Chapter 20: Software Economics

The brevity of this division is by no means an indication of the level of importance to be given to software development. To the contrary, software construction remains a critical aspect of the software engineering experience; after all, without development, there is no software product.

Traditionally, software engineering projects would dedicate roughly twenty percent of the time to requirements specification and design specification, and the rest of the time to development. As emphasized throughout the text, this has proven to be an inappropriate and imprudent approach to the discipline of software engineering. Rightly so, much focus has been shifted to the earlier phases of the SDLC, in an attempt to have a more balanced approach, thus leading to the production of higher quality software. This course embraces that shift. Software construction then, ought to be a rewarding, exciting, enjoyable experience of building a product that was carefully, thoughtfully, skillfully, and methodically planned. It should not be characterized by guesswork and failure, but confidence and anticipation. After all, you are merely implementing a plan that passed through the rigors of investigation, analysis, design, and several iterations of refinement.

It is against this background that you are invited to embark on the software construction experience. Good software construction requires the observance of some management techniques that have either been discussed in previous chapters (for example PERT/CPM), or are too involved to be thoroughly discussed in this course (for example human resource management and team motivation). It also presumes the mastery of fundamental programming skills. Nonetheless, the chapters pull the critical pieces together to provide you with a comprehensive perspective.

DOI: 10.1201/9780367746025-21

Chapter 18

Software Development Issues

This chapter discusses important software development issues that are critical to the successful construction of a top-quality software product. It proceeds on the presumption that appropriate steps were taken in the earlier phases of the software engineering experience (namely investigation, analysis, and design), and under the following subheadings:

- Introduction
- Standards and Quality Assurance
- Management of Targets and Financial Resources
- Leadership and Motivation
- Planning of the Implementation Strategy
- Summary and Concluding Remarks

18.1 Introduction

If enough thought and planning were put in the design phase, software development should proceed smoothly. In fact, for the meticulous and keen software engineer, software development is fun; it represents the beginning of seeing the fruition of diligent work, invested up front, in the investigation, planning, and design of the software. Indeed, the following statement is worth remembering.

> It is possible to obtain a poor software system from an excellent design—thanks to incompetent or careless development. But it is impossible to arrive at an excellent software system from a poor design.

Software development may employ any of the life cycle strategies discussed in Chapter 1. In the interest of clarity, these strategies are listed and clarified (in the context of development) below:

- **Waterfall Model:** Keep writing code until the software system is finished.
- **Phased Prototype Model:** Develop the software system chunk by chunk.
- **Iterative Development Model:** Develop the software system iteration by iteration.
- **Rapid Prototype Model:** Take a swipe at developing a working model of the software system and hope for the best; then use user feedback to refine.
- **Formal Transformation Model:** Generate provable code for the software system.
- **Component-based Model:** Develop the software system by integrating used and tested components; write code only for the missing pieces and for integrating the components.
- **Agile Development Model:** Merge the ideas of phased prototyping and iterative development, focusing on the construction and delivery of a product with the minimum required essentials features. Then refine/expand in subsequent versions as required.

During software development, a team of software developers uses a set of software development tools to write and/or generate code for the software system, based on the design specification. The design specification is used as input to this phase of the SDLC; the output from this phase is the actual software product (including its documentation).

Management of the project is of paramount importance during this phase. Things may deviate from related project expectations, and it is here that the management skills of the project manager will be challenged. Accordingly, the project manager will need to exhibit good project management skills; he/she will also need to be creative. Recall that Chapter 11 discussed tools such as PERT, CPM, and project management software systems. We will focus on the other related issues in this and the next two chapters.

18.2 Standards and Quality Assurance

In order to ensure that the quality of the software constructed is acceptable, the project manager as well as the software engineers must be concerned about standards and quality. Software quality affects how end users will relate to the product, the level of acceptance the product will receive in the organization, and how the product will perform in the marketplace. As such, the following statement is worth remembering:

Excellent software quality is not magic (though by design, it often appears that way); rather, it's a consequence of sound logic. Excellent software quality will not just happen, if it was not deliberately built into the design, and consistently pursued during development and implementation.

With respect to standards and quality assurance, three issues are worth mentioning:

- The relationship between quality and standards
- The software quality factors
- The quality evaluation

18.2.1 The Relationship between Quality and Standards

During software development, many of the issues addressed in Chapter 12 on software development standards are applicable. To review, the following issues will be of importance:

- Naming of database and non-database objects
- Naming of operations
- Programming standards
- User Interface standards
- Documentation standards
- Database design standards

In this regard, the project manager must be knowledgeable and preferably an expert in order to effectively manage the software engineering project. In complex situations, he/she may have to lead and manage by example.

Quality assurance (QA) involves the process of ensuring that the software meets the requirements and design specifications, and conforms to established standards. QA is applicable to all aspects of business. With regard to software development, Figure 18.1 defines a procedure for maintaining a high software quality during a software engineering project. Essentially, the procedure establishes a loop between software development and assessment against established standards and requirements. Looping continues until the standards and requirements are met.

```
1.  Define and specify requirements
2.  Set standards
3.  Define development strategies and activities to meet standards and requirements
4.  Work to defined standards and requirements
5.  Measure the outcome against defined standards and requirements
6.  If no problem, go to step 8
7.  If problem, identify problems and areas for improvement.  Go back to step 4
8.  If end, stop; else go to step 4
```

Figure 18.1   Procedure for Maintaining Software Quality

In business, a concept that has become very popular since the 1990s, and has remained so, is *total quality management* (TQM). A full discussion of TQM is beyond the scope of this course. Suffice it to say, TQM is a comprehensive philosophy that encompasses all aspects of the organization, and all management functions. Two sound principles of TQM are worth stating here:

- Get it right the first time
- Do it again

The first statement implies that the programmer or software engineer should spend some time to convince himself/herself that a program or module or system works before submitting it for QA evaluation. The QA evaluation then becomes a means of fine-tuning the system rather than identification of major rework.

The second statement implies that the process of proliferation of good standards, models, and methodologies should be pursued. As a practical example, a programmer should spend some time to get his/her first ADD operation correctly. This can then be used as the basis for other ADD operations.

Quality should be proactive rather than reactive. Put another way, quality should seek to avoid errors, rather than respond to them. Quality should focus on all phases of the system life cycle equally rather than the maintenance phase primarily. Quality must characterize all aspects of the software, not just some aspects.

Quality standards must be effectively communicated to all relevant levels of the organization. In particular, the project team must be aware of such requirements. Here is a rhetorical question worth remembering:

What good are standards if no one reads or observes them?

18.2.2 Software Quality Factors

The software quality factors were first introduced in Chapter 1; they were also mentioned in Chapter 3, and again in Chapter 12. They cannot be overemphasized; they affect the success or failure of the software. As a software engineer, when you design software, you must design with these factors constantly in mind. These factors, along with your software standards, if observed during construction, will ensure a high quality in the final product. In the interest of ease of reference, the quality factors are repeated here:

Efficiency, Reliability, Flexibility, Security, User-friendliness, Integrity, Growth Potential, Maintainability, Adequacy of Documentation, Functionality, Cohesiveness, Adaptability, Productivity, Comprehensive Coverage

In the electronic book *DevOps without Measurement is a Fail* (see [New Relic 2015]), the author identifies and elaborates on five essential drivers for the success of a software system; these are summarized below:

- **Business Success:** The software system should be engineered to be aligned to the success of the business that it serves.
- **Customer Experience:** The software system must be engineered to meet the needs and expectations of the intended end users (also called customers). Otherwise, they will become dissatisfied with the product.

- **Application Performance:** The software system must be engineered and managed to ensure and maintain adequate performance levels. In this regard, the database performance is of paramount importance.
- **Speed:** Contemporary consumers of software systems are very demanding about speed—they require quick development, delivery, and response. They typically become impatience (often to the point of intolerance) when these expectations are not met.
- **Quality:** From the conceptualization to delivery and maintenance of the software system, quality needs to be engineered into the project. Important benchmarks to be controlled include (but are not limited to) deployment success rate, application error rates, incident severity, and existence of software bugs.

As mentioned earlier in the course, these drivers and quality factors do not always align with each other. For instance, quality and speed often generate opposing forces. Given this reality, software development often involves pragmatic tradeoff decisions by the software developer(s).

18.2.3 Quality Assurance Evaluation

Operations comprising the software should be tested singly as well as collectively. A *QA evaluation* (traditionally referred to as a structured walk-through) is a comprehensive test of a software system, or component thereof. The evaluation may be done at different levels—for an operation, a group or related operations (a module), a subsystem, or the entire software system.

Important parties to be involved in a major QA evaluation include the following:

- Project manager
- Software engineer or programmer responsible for the component
- A principal user
- Someone to take notes (if not the software engineer)

In some circumstances, a tester or a software engineer, prior to a QA evaluation with all the relevant parties, may test the component. The objective of the QA evaluation is to verify that the operation, module, subsystem, or system meets user requirements and conforms to QA standards. Usually, a standard form is developed to record the result of such evaluations. Bear in mind, however, that the forms may be in hard copy, or stored on the computer. Figure 18.2 illustrates what an evaluation form might look like.

Software Quality Assurance Evaluation Report	
Section A: Basic Information	
Project:	
Project Component:	
Operation:	
Section B: Conformance to Standards	
Observation(s)	Recommended Action(s)
1.	
2.	
. . .	
n.	
Section C: Assessment of User Requirements	
Observation(s)	Recommended Action(s)
1.	
2.	
. . .	
n.	
Section D: Evaluation Summary	
[ ] Accept Work [ ] Revise Work and Resubmit [ ] Reject Work	
Walkthrough Conducted by:	
Signature:	
Date Recorded:	

Figure 18.2   Illustrative QA Evaluation Form

18.3 Management of Targets and Financial Resources

The creativity and management skills of the project manager are challenged most strongly during software development. The critical question to address is how can resources be allocated in order to ensure that targets are met? PERT/CPM is project management technique that is commonly used (review Chapter 11).

Two additional areas of concern are:

- Budget and expenditure
- The value of the software system

18.3.1 Managing Budget and Expenditure

Budget management is an important aspect of software development. There are two components to budget management: planning the budget and monitoring the expenditure. The budget planning would have been completed from the early (planning) stages of the project. What is required during development is monitoring of the targets, the resources, and the expenditure. Following is a summary of what is required in each aspect. However, please note that a comprehensive treatment of budget management is beyond the scope of this course (for additional information, see recommended reading [Morse 2000]).

18.3.1.1 Budget Preparation

The budget typically spans a fiscal year. However, it might be broken down on quarterly basis or a monthly basis. The budget contains *summery items* and *line items* that make up the summary items. The summary items and some of the line items are typically predetermined, based on the organization's *chart of accounts*.

The budget may also be split into broad categories of expenditure items, for instance, *capital expenditures* and *recurrent expenditures*. Capital expenditures relate to investments in fixed assets and/or infrastructure. Recurrent expenditures relate to operational issues (such as salaries, stationary, fuel, heating, electricity, transportation, etc.)

Figure 18.3 illustrates how a budget might be composed. Please note:

1. Each summary item (e.g. Network Upgrade) would have associated detail line items that may be part of the main document, or included as an appendage. The details will show how each summary amount was arrived at.
2. In some instances, supporting documentation (such as quotations from vendors) may be required.

Since one's budget will ultimately affect one's ability to pursue the projects of intent, it is imperative that the budget be comprehensive. Also, to avoid embarrassments due to budget overruns, it is always better to over-estimate rather than under-estimate expenditure.

Division: Information Systems and Services	
Expenditure Areas	**Projected Expenditure**
Recurrent Expenditure	**3,100,000**
Salaries	1,400,000
Office Supplies	400,000
Electricity	500,000
Telephone	150,000
Systems Maintenance	400,000
Travel	150,000
Human Resource Development	100,000
Capital Expenditure	**400,000**
New Information system for Marketing	200,000
Upgrade the Financial Management System	100,000
Network Upgrade	100,000
Total Budget	**3,500,000**

Figure 18.3   Example of a Budget

18.3.1.2 Budget Monitoring

Once a voted budget is in place, the project manager will be aware of this. Software development should proceed according to the budget that is in vogue; ensuring this is the responsibility of the project manager. Issues such as project crashing (review Chapter 11) or recasting of certain targets may become relevant as the project proceeds. However, note that recasting of targets reflects poorly on the project management, and must only be considered when it is clear that the circumstances warranting these changes are beyond the control of the project manager.

18.3.2 Managing Software Cost and Value

Yet another significant consideration is the actual costing of the software product. This is particularly important if the product is to be marketed, or its value included as part of the organization's capital assets. One popular model for estimating software cost is the *COCOMO model*. This will be discussed in more detail in Chapter 20. The basic proposal of the model is that software cost is a function of software size (measured as source code instructions). Of course, other algorithmic cost models have been proposed. Typically, software cost is influenced by the following factors:

- Size (measured in lines of code)
- Complexity (determined by the number of complex calculations and algorithms involved)
- Value added to the organization that acquires the software
- Consumer demand for the product

18.4 Leadership and Motivation

A study of leadership and motivation belongs more appropriately to the field of *organizational theory and behavior* (OTB) and cannot be comprehensively covered in this course. Suffice is to say that the project manager (who might very well be a software engineer) must be appropriately prepared (academically and experientially) to offer effective leadership. He/she must be familiar with various leadership and motivational theories. In passing, a few important points will be stated.

The project manager should know when to apply certain principles and theories of management. Some of the paradigms have been listed in Figure 18.4. These will be clarified in the upcoming chapter.

Autocratic Leadership
Democratic Leadership
Laissez-Faire Leadership
Transformational Leadership
Super Leadership
Task-oriented Leadership
Relations-oriented Leadership
Path-Goal Leadership
Contingency Leadership

Figure 18.4   Management/Leadership Styles

The ability to adjust to situations and manage accordingly is called situational (contingency) management; this is the preferred approach. The project manager must be able to identify the abilities, needs, and perceived values of members of his/

her team and assign activities that will help the respective individuals to realize those needs. This will ensure optimum productivity.

The project manager must be a good motivator. The project manager's most valuable resource is his/her human resource. If you have a team of individuals who are perpetually unmotivated, that team will not achieve much. Software engineering is serious business and must be taken as such.

A full discussion of motivation theory is beyond the scope of this course. Figure 18.5 provides a brief summary of some prominent motivation theories. Since the software engineer may be often called upon to lead project teams, it is in his/her interest to be cognizant of these theories, in order to be effective at project management.

Theory Name	Theory's Main Idea
McGregor's Theory X	This theory purports that workers are only motivated by money. They are basically selfish and lazy and therefore need to be closely controlled and directed. The point to note here is that some people are negative, and are not self-motivated.
McGregor's Theory Y	This theory is the opposite of theory X. It purports that workers are motivated by many different factors apart from money, and that they enjoy their work. The theory argues that given the opportunity, workers will happily take on responsibilities and make decisions for their organization. The point to note here is that some people are self-motivated and welcome challenges.
Vroom's Expectation Theory	Vroom's theory asserts that an employee's motivation to complete a task is a function of the employee's expectation (of the reward), the value placed on such reward, and the instrumentality to accomplish the task. So, to get an employee motivated to complete a job, the project manager needs to ensure that the employee is sold on the reward associated with completion of the job.
Maslow's Theory	Maslow's theory of motivation argues that there is a motivational hierarchy that people seek to have satisfied. The hierarchy consists of physical needs, safety needs, love and acceptance, self-esteem, knowledge and understanding, aesthetics, and self-actualization. The needs are fulfilled in that order. The idea here is that the project manager must understand what motivational needs the team member has in order to best know how to motivate him/her.
Herzberg's Theory	This theory submits that worker satisfaction and performance are influenced by two sets of factors. The *hygiene factors* include work conditions, company policy, salary, inter-personal relationships, and salary. These factors do not motivate, but their absence could lead to de-motivation. The *motivation factors* include achievement, growth potential, recognition, status, and responsibility. These factors inspire the employee to perform well. The idea here is that project managers must ensure the presence of all the factors of production in the workplace.

Figure 18.5   Some Popular Motivation Theories

18.5 Planning of Implementation Strategy

Plans for the implementation must be made and finalized well ahead of the implementation time. This matter will be discussed in Chapter 21. The main implementation issues to be considered are

- Operating environment
- Installation
- Code conversion
- Training
- Change over
- Marketing

As you will soon see, these issues could influence the success or failure of the software product in the organization and/or marketplace.

18.6 Summary and Concluding Remarks

Here is a summary of what has been discussed in this chapter:

- If enough thought and planning were invested in the design of the software, actual construction will be an enjoyable, exciting, and rewarding experience. Prudent project management will also be required of the project manager (who is likely to be a lead software engineer) during this period.
- Excellent software quality is not a miracle; it will not just happen, if it was not deliberately built into the design, and consistently pursued during development and implementation. Procedures for building quality in the software product must be clearly outlined and followed.
- The QA evaluation is used to ensure that established software standards are upheld during the development process.
- During software development, resources, targets, and expenditure must be carefully managed to ensure that the project meets its deadlines.
- Leadership and motivation are also critical factors during software developed. Ideally, the project manager must be an excellent leader and motivator.
- The implementation strategy for the software system must be planned well ahead of the completion of the software development.

The matter of leadership and motivation is extremely important in a software engineering project. For this reason, it is given a bit more attention in the upcoming chapter.

18.7 Review Questions

1. What are the critical issues to be managed during the development phase of a software engineering project?

2. Explain how software standards relate to quality assurance. Outline a procedure for maintaining software quality.

3. Discuss the QA evaluation exercise and propose an instrument for use during this experience.

4. Describe a technique for managing resources, targets, budget, and expenditure during software development.

5. How important is leadership during a software engineering project? Explain.

References and Recommended Readings

[Boehm et al., 1997] Boehm, Barry W., C. Abts, B. Clark, and S. Devnani-Chulani. 1997. *COCOMO II Model Definition Manual*. Los Angeles, CA: University of Southern California.

[Kendall 2014] Kendall, Kenneth E., and Julia E. Kendall. 2014. *Systems Analysis and Design*, 9th ed. Boston, MA: Pearson. See chapter 16.

[Lewis 1993] Lewis, Phillip V. 1993. *Managing Human Relations*. Boston, MA: Kent Publishing.

[Morse 2000] Morse, Wayne, James Davis, and Al L. Hartgraves. 2000. *Management Accounting: A Strategic Approach*. Cincinnati, OH: South-Western College Publishing. See chapter 11.

[New Relic 2015] New Relic. 2015. DevOps without Measurement is a Fail. Accessed July, 2017. https://try.newrelic.com/rs/412-MZS-894/images/MeasuringDevOpsSuccess_eBook_FINAL.pdf.

[Peters 2000] Peters, James F. and Witold Pedrycz. 2000. *Software Engineering: An Engineering Approach*. New York, NY: John Wiley & Sons. See chapters 12–14.

[Pfleeger 2006] Pfleeger, Shari Lawrence. 2006. *Software Engineering Theory and Practice*, 3rd ed. Upper Saddle River, NJ: Prentice Hall. See chapters 7–9.

[Pressman 2015] Pressman, Roger. 2015. *Software Engineering: A Practitioner's Approach*, 8th ed. New York: McGraw-Hill. See chapters 19–26.

[Sommerville 2016] Sommerville, Ian. 2016. *Software Engineering*, 10th ed. Boston, MA: Pearson. See chapters 22-26.

[Van Vliet 2008] Van Vliet, Hans. 2008. *Software Engineering*, 3rd ed. New York, NY: John Wiley & Sons. See chapter 13.

Chapter 19

Human Resource Management

Human Resource Management (HRM) is arguably, the most important aspect of management in general. This argument can be easily supported, since the most important resource in any organization is the human resource.

In most medium-sized and large organizations, there is a human resource director, with responsibilities for the human resource needs of the organization. As indicated in Chapter 1, in progressive organizations, HRM is given the same level of priority as information technology.

At every level, managers have as one of their responsibilities, HRM. Software engineers who operate as project managers are by no means excluded. People work on projects, it requires people to administer software systems; people are needed to administer backup procedures, and conduct preventive maintenance, etc. These people will need direction, and in many cases this direction will come from the lead software engineer.

HRM cannot be fully discussed in one chapter (in fact it is treated as a course in many undergraduate and graduate programs). This chapter should therefore not be construed as a substitute for training and education in this area, but rather an overview of the essential aspects of HRM from a software engineering perspective. The chapter will proceed under the following captions:

- Management Responsibilities
- Management/Leadership Styles
- Developing the Job Description
- Hiring
- Maintaining the Desired Environment
- Preserving Accountability
- Grooming and Succession Planning
- Summary and Concluding Remarks

DOI: 10.1201/9780367746025-23

19.1 Management Responsibilities

The job a manager/leader may be summarized in the following broad objectives:

- To motivate people to be the best they can be.
- To seek out the best interest of his or her employees.
- To create the environment that will ensure the achievement of all organizational objectives to which his/her job applies (directly or indirectly).
- To be an example of accountability, professionalism, and excellence.

In the pursuit of these objectives, the manager uses various strategies and assumes certain responsibilities (some of which would be indicated on a job description). These strategies and responsibilities will vary according to the organization, the nature of the job, and the individual. Nonetheless, some general functions of management have been identified by management scholars:

Planning, Organizing, Coordinating, Commanding, Controlling, Hiring

In contemporary literature on management, the trend is to replace the functions of *coordinating* and *commanding* with a single function—*leading*. The rationale is that *leading* is a more friendly term than *commanding* and *coordinating*. Discussion of these basic functions is left to the reader. Suffice it to say that the software engineer should be cognizant of these responsibilities and functions because as mentioned in Chapter 2, the job of a software engineer is a management position that often requires supervision of a software engineering team.

19.2 Management/Leadership Styles

There is a subtle but important difference between leadership and management: Leadership involves winning the trust and respect of followers, so that they willingly follow. Management relates to control and ruling of subordinates. Ideally, it is desirable that effective managers also be effective leaders but that may not always be the case. Moreover, there appears to be a significant overlap between leadership and management. In the upcoming subsections, the difference between leadership and management will be deemphasized; rather, we will summarize selected theories on leadership/management styles that have been forwarded by social scientists.

19.2.1 Autocratic Management

In *autocratic management*, the manager dictates to and/or commands his/her team members, and requires them to follow his/her instructions.

Two obvious advantages of this approach are:

- The manager gets things done his/her way.
- The manager has the sense of total control.

Two obvious disadvantages can also be identified:

- The manager's way may not always be the most prudent. The approach presumes that team members are not smart enough to be trusted with major responsibilities. This, we know to be fallacious.
- People do not like to be dictated to; experience has shown that treating people like this tends to appeal to their latent propensity to rebellion, and brings dissent to the surface. The manager's sense of control may therefore be false.

19.2.2 Egalitarian (Democratic) Management

In *democratic management* the manager solicits ideas and inputs from his/her team members. The best suggestions are taken and incorporated into planning and operation exercises.

Advantages of this approach are as follows:

- Team members are encouraged to participate in important decision-making. This gives them a sense of belonging and commitment.
- The sense of belonging and commitment motivates team members to share in the corporate mission and consider going the extra mile in pursuit of it.
- Healthy work relationships can be built between managers and team members.

The main disadvantage of this approach is that does not always yield positive results; experience has shown that there are scenarios that do not warrant it.

19.2.3 Laissez Faire Management

In the *laissez faire* approach, the manager assumes little or no control. He/she allows team members to do whatever they please. Although there are situations that warrant this management style (for example, a Christmas party), in most cases it results in chaos, and lack of achievement of significant objectives.

19.2.4 Path-Goal Leadership

In *path-goal leadership*, the leader influences performance, satisfaction, and motivation of his/her team members by

- setting clear achievable goals for team members;
- offering rewards for achieving these goals;
- clarifying paths towards these goals;
- removing obstacles to performance and eventual achievement of goals.

The leader does this by adopting a certain leadership style based on the situation:

- **Directive Leadership:** This can be done by specifying and assigning objectives (goals), strategies, and tasks in pursuit of the established objectives, providing advice and direction to team members.

- **Supportive Leadership:** This is done by building good relationships with team members.
- **Participative Leadership:** Decision-making is based on consultation with team members.
- **Achievement-oriented Leadership:** The leader sets challenging goals and high-performance expectations. Much confidence is expressed in the group's ability. This inspires the members to give their best.

Advantages of the approach are

- It builds confidence of team members and inspires them to contribute their best.
- It can build a healthy working environment.

Disadvantages are

- The environment could become overly competitive, thus creating animosity among team members.
- If team members are not adequately prepared for this, the efforts could be counterproductive.

19.2.5 Transformational Leadership

In transformational leadership, the leader presents himself/herself as an agent of change (presumably for the better). A strong relationship is built between leader and followers. Transformational leadership often involves a vision to forge an organization in a new direction and elicit change. It is characterized by strong ideas, inspiration, innovation, and individual concerns.

The main advantage of this leadership style is that followers are "fired up" to effect the required changes as enunciated by the leader. There is a strong sense of commitment to set goals.

The disadvantages are:

- If change does not come in a timely manner frustration could overtake some of the team members.
- The approach begs the question, what happens after the desired changes are achieved?

19.2.6 The Super Leader Approach

In the *super leader* approach, the leader sets himself/herself up as an icon to be emulated by team members. By example, he/she sets high standards, and challenges team members to emulate them.

The approach also addresses the matter of succession planning (to be discussed later) by grooming selected team members to be super leaders at various levels.

The advantages of this approach are the following:

- The approach promotes the idea of leadership by example, a principle that resonates well with people, and traces back to biblical history (characters such as Jesus Christ, Moses, and Joshua are described in Judeo-Christian literature as super leaders), and forward to current times (you can do doubt identify super leaders in your workplace, social/professional affiliation, or country).
- Like path-goal leadership, the super leader approach motivates people to be the best they can be.

The main challenge of this approach is that the super leader must be well prepared and versed in the activities that he/she desires the team members to engage in.

19.2.7 Task-Oriented Leadership

In *task-oriented leadership*, the leader's primary focus is the conducting of activities, in pursuit of established goals. These activities must be done at whatever cost.

The main advantage of the approach is that it is achievement-oriented, and is therefore likely to produce a high level of productivity.

The main disadvantages of the approach are as follows:

- The leader could become insensitive to the human needs of team members while being absorbed with pursuing his/her objectives.
- This problem could result in a demoralized, demotivated team, thus inhibiting the leader's ability to achieve the very goals being pursued.

19.2.8 Relations-Oriented Leadership

A *relation-oriented leader* seeks, as a primary focus, to build good relationships with team members. The thinking behind this is that if team members have a good relationship with their leader it will inspire them to perform.

The main advantage of this approach is that when it works, the results are very convincing. One possible reason for this is that team members feel a strong sense of belonging, and ownership of the project.

The main disadvantages of the approach are as follows:

- The approach does not always work. In fact, it is possible for team members to enjoy good relationships with their leader, and still not perform well.
- The worst-case scenario of this approach is that the team becomes a social club where there is much fellowship, but little work.

19.2.9 Servant Leadership

Servant leadership is service-oriented leadership; it could also be reasonably described as leadership by example. The leader operates from a place of deep commitment to serve the community of focus (employees, customers, and/or a wider community).

Key aspects of servant leadership are service ahead of self-interest; listening with a view to learning the related requirements; focusing on what is feasible; and always seeking to be helpful. According to its principal proponent, Robert Greenleaf, the main objective of servant leadership in creation of "a more just and caring world." Famous servant leaders throughout history include Jesus Christ, Abraham Lincoln, Mother Teresa, Mahatma Gandhi, Nelson Mandela, et. al.

Significant advantages of servant leadership are as follows:

- The primary objective is caring for and uplifting other human beings.
- This approach results in the creation of strong and enduring relationship bonds.

Like the super leadership style, the main challenge of servant leadership is that the leader must be well prepared and committed to fulfilling the requirements of the role. Also, without an adequate supply of resources, burnout could become an inhibiting factor.

19.2.10 Contingency Leadership

Contingency leadership theory is an argument for pragmatism: Since the individual approaches all have their advantages and disadvantages, the manager should reserve the right to employ different approaches, depending on the scenarios that present themselves.

The main advantage of contingency leadership is that the weaknesses of any approach are avoided while capitalizing on the strengths of the respective approaches.

The main disadvantage of the approach is that inexperienced managers could make bad judgments about which strategy to employ. However, with experience, contingency managers usually make excellent decisions on the average.

19.3 Developing Job Descriptions

As a computer science professional with management responsibilities, you may be called upon develop job descriptions for junior positions in your division or department, from time to time. Different organizations will have different standards regarding how their job description should be written. Figure 19.1 provides a checklist of the essential components of a job description. For an example, please review Figure 2.1 of Chapter 2.

Component	Clarification
Company Heading	Name and address of the company
Job Summary	Overview of the job
Main Functions	Itemized list of salient responsibilities
Related Authority	What can be done independent of prior consultation
Reporting Relationship	Who does the incumbent report to?
Required Qualifications	Minimum and preferred qualifications
Other Special Requirements	Other expectations not previously mentioned

Figure 19.1   Basic Components of a Job Description

Caution: In many cases, you will find that job descriptions you are looking for are either in need of refinement, or are nonexistent. In either case, your task must be to leave the situation in a better state than you found it.

19.4 Hiring

Hiring is a very important management function. No manager can function without people (human resources). The software engineer may be called upon to participate in the process of hiring suitably qualified individuals to be part of the project team.

Below are some important considerations for the hiring process:

1. Clearly define the position to be filled. If it is a new one, then approval from a senior level of management may be required. In any case, a clearly defined job description should be in place.
2. Advertise for applicants to fill the position.
3. Convene an interview panel.
4. Arrange for interviews of the applicants. Depending on the organization, the interview schedule may vary. Typically, job interviews are done in three stages: Firstly, a technical interview scrutinizes the technical and professional preparedness of the applicant. Next, a human resource interview looks at the overall individual and tries to determine whether they would be suitable for the job. A final interview is usually done to make an offer to the selected person. Please note that in some instances, the stages may be merged. For instance, technical and human resource interviews may be merged as one.
5. Select the most suitable individual.

Before conducting an interview, the software engineer must make the required preparation:

1. Prepare a set of criteria to be met by the incumbent.
2. Review the applicant's curriculum vitae.
3. Prepare questions that are consistent with the defined criteria.

It is standard practice to have in each organization, an interview evaluation form. This may vary with different departments, as well as with the positions being considered. Obviously, for SE/IT jobs, the interview evaluation forms must be prepared by the SE/IT executive in charge, and then given to members of the interview panel. After each interview, the panel members complete their respective evaluations of the applicant. Figure 19.2 illustrates a sample interview evaluation form.

LMX Software Inc.		
Interview Evaluation Form, Web Technology Department		

A. Basic Information

Interviewee:	
Position Considered:	
Date of Interview:	
Interviewer:	

B. Performance on Specific Benchmarks

Criterion	Comment	Evaluation Score
Required Qualification		[ ] of [20]
Professional Preparation		[ ] of [20]
Familiarity with Required Technology		[ ] of [20]
Human Relations Skills		[ ] of [20]
Problem-Solving Skills		[ ] of [20]
Motivation and Drive		[ ] of [20]
Overall Performance		**[ ] of [120]**

C. Recommendation

[ ] Hire	
[ ] Do not Hire	
Overall Comment:	

Figure 19.2   Sample Interview Evaluation Form

In conducting the interview, below is a checklist of some of the things you should probe for each candidate:

- **Required Qualification:** This may require asking questions about the institutions attended by the candidate, and the courses pursued.
- **Professional Preparation:** This may require asking questions relating to prior working experiences of the candidate.
- **Familiarity with Technology:** This may require asking technical questions relating to the technologies and methodologies relating to the job.
- **Human Relations Skills:** This may require asking questions relating to the candidate's handling of human relations challenges in the past, as well as simply observing how the candidate handles pressure.
- **Motivation and Drive:** Here you try to assess the candidate's enthusiasm for the job.
- **Problem-Solving Skills:** Is the candidate adept at solving various problems on short notice? Here, problem scenarios are posed to the interviewee, and he/she is asked to describe how they would address the various problems.

19.5 Maintaining the Desired Environment

Having a team in place is good. But how do you create and maintain the working environment that you desire? And how do you get the team to cooperate and support you? This is one of the challenges of management. There are no straight answers to these questions, but there are guidelines. If, as the team leader, you are coming into this position due to a promotion, your challenge may be different from the situation where you are coming in as a new addition to a team. The spectrum of possible human reactions ranges from jealousy and vindictiveness on one extreme, to complete commendation and adulation on the other.

Whether you got a promotion or were hired into the position, you are likely to face initial challenges. Following are some uncomfortable situations that you may face:

- There may be animosity in the camp as to whether you were most deserving of the position. This possibility is increased if members of the team were considered for the position and then bypassed.
- There may be some resistance to changes that you want to put in place, in your new capacity.
- Individuals may try to challenge your mettle during the early period of your administration.

If you are in a position of leading software engineer, you will want to create an environment where negative factors that could potentially undermine the success of your team are discouraged, and positive factors are encouraged and reinforced. With this focus in mind, it is a good idea to schedule an initiation meeting, shortly after assuming your responsibilities. At this meeting the following activities should occur:

1. Meet the team members (if the team is very large, then meet the key players, for instance, people who report to you, along with the supervisors).
2. Clearly outline what your expectations are, and perhaps the mode of operation that you will pursue.
3. Find out what the team members expect from you, and determine whether you can deliver on those expectations.
4. If you are new to the organization, try to get an appreciation of each (major) team member's job (prior to the meeting, you should familiarize yourself with the company norms and policies).

In order to achieve and maintain the desired environment, the following strategies will be also useful:

- Be a good motivator.
- Develop good conflict resolution skills.
- Be an effective communicator.
- Be generous on rewarding outstanding achievement, and consistent in treating errant actions.

Motivation was discussed in the previous chapter; that discussion is applicable here. Suffice it to state that in order to keep a software engineering team optimally engaged, knowledge and application of appropriate motivation strategies is imperative. We will address the other issues here.

19.5.1 Effective Communication

Successful managers are usually effective communicators, both orally and in written form. This is also true for software engineers. Having prepared the requirements specification and design specification for your software engineering project, it will be imperative that team members are sold on the project. If the team members were involved during the preparatory work, then this should not be difficult.

Below are a few experiential tips on effective communication:

- Have the mission statement of the organization and that of the division or department attractively framed and strategically positioned in each office.
- Consider putting beautifiers (e.g. potted plants) in various offices. Plants are known to provide therapeutic value and natural beauty to corporate offices.
- Consider strategically placing memory gems to remind employees of their responsibilities. Below are two examples of motivational quotes:
  o Excellence Begins Here
  o Excellence is our Standard Requirement
- Avoid being perceived as antagonistic or confrontational.
- Involve team members in the determination of development strategies so that they can experience ownership of the plans.

19.5.2 Conflict Resolution

If you are successful at being an effective communicator, this will pave the way for minimizing conflicts and resolving them. Conflict, by definition, is not necessarily a bad thing, in fact, it can serve to provide different perspectives to a problem, and this leads to a more comprehensive solution. What you want to discourage is employees become too personal over their disagreements.

The following approaches (Figure 19.3) may be employed in resolution of conflicts. Although they offer no guarantees, you should often find them useful.

Strategy	When to Use
Ignore	Conflict is between two members and it does not threaten the team's success
Confront	Conflict threatens the team's success
Assign someone else to resolve	Conflict is between two members and does not threaten team's success

Figure 19.3   Conflict Resolution Strategies

In confronting a conflict, consideration should be given to the nature of the conflict:

- If the conflict is about personal feelings of team members and yourself, be prepared to apply some empathy, or to be conciliatory.
- If the conflict is about work (e.g. how to approach a problem) be objective and impartial.
- If the solution sought does not seriously threaten the success of the project, be prepared to make or accept compromises.

Reputed author of organization theory, Gareth Morgan makes a number of insightful observations about the source and usefulness of conflict (see [Morgan 2006]). He notes that people develop interests based on their backgrounds, values, desires, expectations, and orientations. People, through their own selfishness will act in pursuit or defense of their interests. This confluence of self-interests creates conflicts and tensions which are resolved through power and politics.

There are four sets of interest that employees try to gain balance for (the first three have been identified by Morgan):

- Task interests—relating to their job responsibilities
- Career interests—relating to their progress in their careers
- Personal (extramural) interests—relating to their private lives as well as personal preferences in the workplace
- Group interest—relating to a group of employees that they are invested in

Morgan goes on to observe that conflict may be interpersonal, personal, or between rival groups within the organization. The conflict may be structural (arising due to structural composition of the organizational), or a result of competition for scarce resources; it may be clandestine or overt. When managed well, conflicts can be the source of organizational growth.

In attempting to manage conflict, the following are some suggested resolution strategies:

- Understand that people are entitled to their own opinions. The solution should be evidence-based, not opinion-based.
- Get all involved parties to discuss the problem related to the conflict.
- Start the discussion on a positive note.
- Obtain an accurate representation of each party's viewpoint; ask for clarification if necessary.
- Focus on the problem; not the individual(s).
- Communicate assertively and clearly, but not contemptuously.
- Listen without interrupting the speaker(s).
- Avoid prejudging one's thoughts or actions.
- Avoid assigning motives to others.
- Avoid the cheap psychology of projecting your own flaws and/or fears on others.
- Identify possible solutions to the problem.
- Where possible, explore win-win scenarios that allow the participants to feel satisfied that their interests have been addressed.
- If a resolution cannot be immediately found, invite the participant(s) to table the matter for a subsequent discussion. Alternately, you may "agree to disagree."
- Thank the participant(s).

Please note, application of these strategies greatly increases the likelihood of successful conflict resolution. However, there is no guarantee that success will always be achieved for every crisis. The key is to understand the source and nature of conflict, knowing that it can be an opportunity to enhance the success of your effort.

19.5.3 Treating Outstanding Achievements and Errant Actions

It is a healthy practice to have in place a system that recognizes outstanding achievement, and discourages errant actions. This must be communicated to all the team members so that there are no surprises. Better yet, if the employees participated in determining the rules and consequences, they will not feel cheated or victimized when they are applied.

Reward for outstanding achievement may come in different forms, for example adding the employees name to a prestigious list of achievers, providing a special gift, etc. Discipline for errant actions is usually in the form of a letter of reprimand, suspension, or termination of employment.

19.6 Preserving Accountability

It is the responsibility of every manager to put in place a system of accountability which ensures that team members perform according to expectations. This will ultimately ensure that important deadlines are met and goals are accomplished. The software engineer is by no means excluded from this responsibility.

In preserving accountability, the software engineer may pursue a number of strategies:

- Design and assign the work schedules of team members.
- Have a system of evaluating performance against assigned work.
- Have a system of reward/recognition for outstanding performance (as discussed in the previous section).

19.6.1 Designing and Assigning Work

As software engineer in charge of a project, you must be able to define work, provide clarification, and motivate team members to perform the desired activities. *Management by objectives* (MBO) is a term that was often used in management literature to describe the situation where employees are set clear objectives over an evaluation period, and are then evaluated on those objectives.

The work assigned must align with the corporate objectives as well as established strategies of your division or department. The following guidelines should be useful:

- Each employee must be assigned work that is commensurate with his/her job, level of expertise and professional capability.
- Deadlines must be clearly established.
- Deadlines must be realistic.
- The employee must agree to both the assigned work and the deadline, in a cordial discussion. If there are differences, they must be resolved.
- The employee must be made to clearly appreciate how their work fits into the big picture of the department, division, and by extension, the organization.
- If required, there must be clearly established checkpoints to different activities on the employee's work schedule.

Figure 19.4 illustrates a sample work schedule of a team member on a software engineering project. Note that the assignment clearly identifies the project, the specific activities, and target date(s).

LMX Software		
Software Development & Research Department		
Software Engineering Assignment		
Name: Jacob Lambert		
Project: College/University Administrative Information System (CUAIS)		
Activities:		
1. Gather information for Finance, Payroll & Insurance.		
2. Prepare Requirements Specification for areas above.		
3. Prepare Requirements Specification Documents for all areas		
4. Prepare Design Proposal for Finance.		
Due Date: February 27, 2006		
Status on Due Date:		
Decision Taken:		
Supervisor:		

Figure 19.4   Sample Work Schedule

19.6.2 Evaluating Performance

Performance evaluation is typically done at the end of the determined evaluation period. Typically, companies have monthly, quarterly, and annual evaluations.

If the organization is on an MBO program, then the annual evaluation is usually significant, since it might mean a large or small bonus, depending on how favorable of the evaluation is. Here are a few important points about performance evaluation:

- The evaluation must be based on the established objectives and assigned activities that were agreed upon with the employee at the start of the evaluation period.
- The evaluation must be objective, and with supportive evidence.
- The evaluation must be signed by both appraised (i.e. the employee) and appraiser.
- Where the employee might have missed established targets, or performed below expectations, measures must be put in place to help the individual to improve on the next appraisal.

Figure 19.5 illustrates an employee appraisal. Note that the appraisal clearly identifies the evaluation criteria, the assessment period, comments, and signature by the appraiser as well as the appraised.

LMX Software		
Software Development & Research Department		
Employee Performance Evaluation		

Name & Title: Jacob Lambert, Software Engineer
Supervisor & Title: Elvis Foster, Software Development Manager
Appraisal Period: January – March 2006

Appraisal Criteria	Comments	Ratings
Availability		
Punctuality	Seldom late	4/5
Availability for Emergencies	Usually available	5/5
Attendance	Needs improvement	4/5
Performance on the Job		
Ability to work on own initiative	Manages on his own	5/5
Productivity level	Needs to improve throughput	4/5
Capacity to Follow Instructions	Excellent	5/5
Capacity to see Jobs to Completion	Excellent	5/5
Completion of Task on Time	Needs improvement	4/5
Dependability	Very dependable	5/5
Professional Development		
Attitude to New Concepts/Challenges	Excellent	5/5
Capacity for Critical Thinking	Excellent	5/5
Willingness to Learn	Excellent	5/5
Attitude to Authority	Excellent	5/5
Human Relations		
Relationship with Peers	Gets along with everyone	5/5
Relationship with Team Members	Excellent	5/5
Relationship with Superiors	Excellent	5/5
Communication Skills	Could be improved	4/5
Dress and Deportment	Excellent	5/5
Overall Evaluation		85/90 = 94.4%

Authentication

Appraiser's Comment: Excellent employee who needs to improve in areas identified.
Appraisee's Comments: I think the evaluation is accurate. I will work on the weaknesses identified.

Employee's Signature: _____   Date: _____

Supervisor's Signature: _____   Date: _____

Figure 19.5   Sample Employee Performance Appraisal

19.7 Grooming and Succession Planning

Succession planning is a very important function of any conscientious manager; it is particularly critical in the case of information management, since any failure of the organizations information systems could prove disastrous. A responsible software engineer will therefore have a small cadre of (one to three) individuals (depending on the size of the division or department) who are specially trained and prepared to assume overall responsibility, should he/she temporarily or permanently depart the organization. Although selfishness often inhibits this being done, prudence and professionalism demand it.

A few points to note:

- The super leader and path-goal leadership styles both facilitate the principles of succession planning, without any additional effort on the part of the manager.
- The individuals being groomed need not know that they are in fact being specially prepared for possible takeover; this knowledge sometimes creates tensions between (or among) the contenders.
- Implementing a succession plan is not usually easy. In many instances, there are fallouts that might result in some individual(s) leaving the organization. Case in point: When General Electric's CEO and chairman stepped down (in 2000) and named his successor, the other two contenders immediately started making arrangements to leave the company.

Despite possible negative repercussions from implementing a succession plan, it remains a good idea as the benefits far outweigh the possible drawbacks. Among the benefits to be gained are the following:

- Continuity on mission-critical projects, even in the face of sudden unavailability of the project manager
- High level of motivation among team members

19.8 Summary and Concluding Remarks

It is now time to summarize what has been covered in this chapter:

- The functions of a manager may be summarized as planning, organizing, coordinating, leading, and hiring. Software engineers with management responsibilities are often called upon to carry out these functions.
- Among the leadership styles available to the software engineering project manager are the following: autocratic style; democratic style; laissez-faire style; goal-path leadership style; transformational leadership style; super leadership style; task-oriented leadership style; relations-oriented leadership style; contingency leadership style.
- The software engineer should be comfortable updating or creating job descriptions for himself/herself or other members of the software engineering team.
- The software engineer may be called upon to engage in the process of hiring new members of the software engineering team. Knowing the standard approach for this function is therefore important.
- The software engineer should know how to create and maintain an environment that is conducive to the success of the software engineering projects. This includes effective communication, excellent conflict resolution, recognizing outstanding achievements, and discouraging errant behavior.

- The software engineer should know how to preserve accountability in the workplace. This includes designing and assigning work, as well as evaluating performance of team members.
- A responsible and smart software engineer always has a grooming and succession plan.

Another issue that needs to be addressed during development and refined over the effective life of the software system is the matter of software economics. This will be discussed in the next chapter.

19.9 Review Questions

1. What are the main responsibilities and functions of a manager? Why are they relevant to software engineering?

2. What are the different management styles that are available to the software engineer? For each style, identify the advantages and disadvantages; also identify a scenario that would warrant the application of this approach?

3. Assume that you are a software engineer with the responsibility of managing a software engineering project. Develop a job description for a programmer on your team.

4. What criteria should guide your assessment of a prospective team member?

5. If you were a software engineer in charge of a software engineering project, what principles would guide your effort to create and maintain a productive and congenial working environment?

6. If you were a software engineer in charge of a software engineering project, what principles would guide your effort to create and maintain accountability in the workplace?

7. What are the pros and cons of implementing a succession plan?

References and Recommended Reading

[Lewis 1993] Lewis, Phillip V. 1993. *Managing Human Relations*. Boston, MA: Kent Publishing.

[Morgan 2006] Morgan, Gareth. 2006. *Images of Organization*, updated ed. Thousand Oaks, CA: Sage Publications. See chapter 6

[Sayless 1989] Sayles, Leonard R. 1989. *Leadership: Managing in Real Organizations*, 2nd ed. New York, NY: McGraw-Hill.

[Reece 1999] Reece Barry, L. and Rhonda Brandt. 1999. *Effective Human Relations in Organizations*, 7th d. Boston, MA: Houghton Mifflin Company.

[Sommerville 2016] Sommerville, Ian. 2016. *Software Engineering*, 10th ed. Boston, MA: Pearson. See chapter 24.

Chapter 20

Software Economics

Software economics was first introduced in Chapter 3 (Section 3.7) though not by that term. At that time, we were discussing the feasibility of the software engineering project. After reading Chapter 3, one may get the impression that software cost is equal to development cost. In this chapter, you will see that the two are often different; that development cost is just one component of software cost; and that there are other factors. You will also see that determination of software cost, price, and value are difficult issues that continue to be the subject of research and further refinement. The chapter proceeds under the following captions:

- Software Cost versus Software Price
- Software Value
- Assessing Software Productivity
- Estimation Techniques for Engineering Cost
- Summary and Concluding Remarks

20.1 Software Cost versus Software Price

There are few cases where software cost is the same as software price; in most cases, they are different. Let us examine both matters.

20.1.1 Software Cost

As mentioned in Chapter 3 (Section 3.7), there are many components that go into the cost of developing a software system. These cost components are summarized in Figure 20.1.

Software engineering is a business as well as an applied science. The business approach that many organizations employ is simply to put a price tag on each of these components based on experience as well as business policies: The equipment cost, facilities cost, and operational cost will continue to remain purely business issues. The standard business approach for determining the engineering cost is to

DOI: 10.1201/9780367746025-24

multiply the organization's prescribed hourly rate by the estimated number of hours required for the project. However, as you will see later, there have been efforts to apply more scientific and objective techniques to the evaluating engineering cost through the exploration of deterministic models.

Cost Component	Comments
Equipment Cost	Made up of hardware cost, software cost, cost of transporting equipment, depreciation over the acquisition and usage period
Facilities Cost	Includes computer installation cost, cost of electricity and cooling, security cost
Engineering Cost	Includes investigation, analysis and design cost; development cost; implementation cost; training cost
Operational Cost	Consists of cost of staffing; cost for data capture and preparation; supplies and stationery cost; maintenance cost

Figure 20.1   Software Engineering Cost Components

20.1.2 Software Price

Determining the software cost is important but it does not complete the software evaluation process. We must also determine the software (selling) price. Naturally, the biggest contributor to software price is software cost. Three scenarios are worth noting:

- If your organization is a software engineering firm and/or the product is to be marketed, then determining a selling price that consumers will pay for the product is critical. In this case, the price may be significantly discounted since software cost can be recovered from multiple sales.
- If your organization is not a software engineering firm and/or the product will not be marketed to the consuming public, then the software price is likely to be high, since its cost cannot be recovered from sale. In this case, the minimum price is the software cost.
- If your organization is a software engineering firm contracted to build a product exclusively for another organization, then the software price is likely to be high, since the software engineering firm must recover its costs and make a profit.

Figure 20.2 summarizes some of the main factors affecting software pricing. Please note that there may be other factors that affect the pricing of the software (for instance, other quality factors). The intent here is to emphasize that pricing a software system is not a straightforward or trivial matter.

Factor	Clarification
Market Opportunity	The selling price may be significantly discounted in order to break into a new market. The software engineering firm may anticipate huge benefits in other areas of the market, and therefore make huge pricing concessions on the product. You will recall the recent rivalry between Microsoft and Netscape, and how it ended: Microsoft bundled its Internet Explorer software for free with its Windows operating system. Netscape protested. After a contemptuous court battle, Netscape won the battle but Microsoft won the war. Today, Netscape is no longer around, while Microsoft remains the world's largest software engineering firm.
Uniqueness	The more unique the product, the higher the price is likely to be.
Usefulness and Quality	The larger the anticipated use of the product, the lower the price is likely to be. However, this may be offset by packing more features into the product. The price of the software is also a function of its quality (and whether this quality is known by potential users). A good case in point is IBM products. In many cases, the consuming public is asked to pay more for a product from IBM than they would for the same product from another company. This has not significantly affected IBM sales, because generally speaking (and irrespective of your like or dislike of the company), when you purchase an IBM product, you are getting a product of high quality.
Contracting Terms	If the software engineering firm was contracted to build the product for a particular customer, then if the customer retains ownership of the source code, its price significantly increases. On the other hand, if the software engineering firm can retain the source code and use it for other purposes, the price is significantly reduced.
Requirements Volatility	If the requirements are likely to change over the short or medium term, a software engineering firm may lower its bidding price in order to win a contract, and include a clause that treats changes in the requirements as a separate price item.
Financial Health	Generally speaking, software engineering firms that are financially strong are more reluctant to lower their prices than those that are not. Of course, there are exceptions to this observation. In recent years, the consuming public has benefited from huge concessionary overtures from leading software engineering firms such as Oracle, Microsoft, and IBM.
Software Engineering Cost	As clarified in figure 20-1.

Figure 20.2   Some Factors Affecting Software Pricing

20.2 Software Value

The value of a software system may be different from the price. Whether your organization intends to market the software or not, it is often useful to place a value on the product, for the following reasons:

- If the product is being marketed, placing a value on it can provide a competitive advantage to the host organization (i.e. the organization that owns the product) if the value is significantly more than the price. The host organization can then use this as marketing advantage to appear generous to its prospective consumers.
- If the product is being kept by the host organization, then having a value that is significantly higher than its price/cost makes the acquisition more justifiable.
- Whatever the situation, it makes sense to place a value on the product from an accounting point of view.

The big question is, how do we place a value on a software product? There is no set formula or method for answering this question; therefore, we resort to estimation techniques. A common-sense approach is to assume that software value is a function of any or each of the following:

- Software cost
- Productivity brought to the organization (and how this translates to increased profit)
- Cost savings brought to the organization

In the end, determining a value for the software system is a management function that is informed by software engineering. For this reason, we will examine some of the estimation techniques that are available (Section 20.4). Before doing so, we will take a closer look at evaluating software productivity.

20.3 Evaluating Software Productivity

There are two approaches to assessing software productivity. One is to concentrate on the effort of the software engineer(s). The other is to concentrate on the added value to the enterprise due to the software product. Most of the proposed models tend to concentrate on the former approach. In keeping with the theme of the text, and in the interest of comprehensive coverage, we will examine both approaches.

A common method of assessment of software productivity is to treat it as a function of the productivity contribution of the software product, and the collective engineering effort. Two common types of metrics have been discussed in the software engineering literature. They are *size-related metrics* and *function-related metrics*. A third approach is a *value-added metrics*; it attempts to determine and associated value added by a software system to an organization. A brief discussion of each approach follows.

20.3.1 Size-related Metrics

Size-related methodologies rely on the software size as the main determinant of productivity evaluation. Common units of measure are the number of lines of source code, the number of object code instructions, and the number of pages of system documentation. To illustrate, a size-related metric may compute a software productivity index for a project, based on the formulas in Figure 20.3.

PI{Code} = (LOC)/(DT)

Key:
1. PI{Code} is the productivity index based on code. It is expressed in lines of code per programmer-month i.e. LOC/pm.
2. LOC is the number of lines of code.
3. DT is the development time, expressed in programmer-months (pm). A programmer-month is the period of one month (assuming 8-hour day) for one programmer.

PI{Doc} = (POD)/(DT)

Key:
PI{Doc} is the productivity index based on documentation. It is expressed in pages of documentation per programmer-month i.e. POD/pm.
POD is the number of pages of documentation.
DT is the documentation time, expressed in programmer-months (pm). A programmer-month is the period of one month (assuming 8-hour day) for one programmer.

Figure 20.3 Formulas for Size-related Metrics

Associated with this approach are the following caveats:

- If different programming languages are used on the same project, then making a single calculation for PI{Code} would be incorrect, since the level of productivity varies from one software development environment to another. A more prudent approach would be to calculate the PI{Code} for each language and take a weighted average.
- Most of the documented size-related metrics concentrate on lines of code, with scant or no regard for documentation. This plays squarely into the fallacy that software engineering is equivalent to programming. The lower half of Figure 20.3 has been added to provide balance and proper perspective to the analysis. If as proposed by this course, more effort ought to be placed on software design than on software development, and that the latter should be an exciting and enjoyable follow-up of the former, then any evaluation of software cost or productivity should reflect that perspective.

The main problem with this model is that it does not address the important matter of software quality. What if the software product and documentation are both voluminous due to faulty design? The model does not address this concern.

20.3.2 Function-related Metrics

Function-related methodologies rely on the software functionality as the main determinant of productivity evaluation. The common unit of measure for these methodologies is the *function-points* (FP) per programmer-month (a programmer-month is the time taken by one programmer for one month, assuming a normal work week of 40 hours). The number of function points for each program is estimated based on the following:

- External inputs and outputs
- User interactions
- External inquiries
- Files used

Additionally, a complexity weighting factor (originally in the range of 3 to 15) is applied to each program. Next, an *unadjusted function-point count* (UFC) is calculated by cumulating the initial function-points count times the weight, for each component:

UFC = $\sum[(FC) * (W)]$ where FC is the function-point count for a program component and W is the complexity weight for that program component.

Next, complexity factors are assigned for the project based on other factors such as code reuse, distributed processing, etc. The UFC is then multiplied to this/these other complexity factor(s) to yield an *adjusted function-point count* (AFC). Finally, the productivity index is calculated.

Figure 20.4 summarizes the essence of the approach. Associated with this approach are the following caveats:

- The approach is language independent.
- It is arguable as to how effective this approach is for event-driven systems and systems developed in the OO paradigm. For this reason, *object-points* have been proposed to replace function points for OO systems. We will discuss object-points later (Section 20.4.3).
- The approach is heavily biased towards software development rather than the entire software engineering life cycle.
- The approach is highly subjective. The function-points, weights, and complexity factors are all subjectively assigned by the estimator.

---

1. Determine the FC and W for each component
2. UFC = $\sum[(FC) * (W)]$
3. Identify other complexity factors for the project ... C1, C2, ...Cn
4. AFC = UFC * C1 * C2 * .... * Cn
5. PI = (AFC)/(DT)

Key:
1. FC is the function-point count for a program component and W is the weight for that component.
2. UFC is the unadjusted function-point count.
3. AFC is the adjusted function-point count.
4. PI is the productivity index expressed in function-point counts per programmer-month i.e. FC/pm.
5. DT is the development time, expressed in programmer-months (pm).

Figure 20.4   Calculations for Function-Related Metrics

The matter of software quality still remains a concern, though to some extent, it has been addressed in the software's functionality. Indeed, it can be argued that to some extent, a software system's functionality is determined by the quality of the software design.

20.3.3 *Assessment Based on Value Added*

There is much work in the area of value-added assessment in the field of education as well as other more traditional engineering fields. Unfortunately, the software engineering industry is somewhat lacking in this area. In value-added software assessment, we ask and attempt to obtain the answer to the question, what value has been added to the organization by introduction of a software system or a set of software systems? In pursuing an accurate answer to this question, there are two alternatives that are available to the software engineer:

- Evaluate the additional revenue that the software system facilitated.
- Evaluate the reduced expenditure that the software system has contributed to.

These alternatives are by no means mutually exclusive; in fact, in many instances they both apply. One way to conduct the analysis is to estimate the useful economic life of the software system, and compute the above-mentioned values over that period. In hindsight, this may be challenging, but by no means insurmountable. However, the reality is, in most cases, it is desirable to conduct this analysis prior to the end of the economic life of the system. Moreover, in many cases, this analysis is required up front, as part of the feasibility study for a software engineering venture (review Section 3.7).

Figure 20.5 provides two formulae that may be employed in estimating the value added by a software system. The first facilitates a crude estimate, assuming that interest rates remain constant over the period of analysis (of course, this is not practical, which is why it's described as a crude estimate). Since this approach involves taking the difference between the value added and the acquisition cost, for convenience, let's call this approach the *difference method*. The second formula computes *the net present value* (NPV), with due consideration to interest rates; it is considered a more realistic estimate. The simple adjustment to be made here is to ensure that the *cash flow per annum* includes additional revenue due to the system as well as reduced expenditure due to the system.

$SV = [(EVA) - (AC)]/(EL)$

Key:
1. SV is the software value in dollars.
2. EVA is the estimated value added (which is increased revenue + reduction in expenditure).
3. AC is the acquisition cost of the software.
4. EL is the estimated economic life of the software in years.

$NPV = -A_0 + A_1 / (1+r_1) + A_2 / (1+r_2)^2 + \ldots + A_n / (1+r_n)^n$

Key:
1. NPV is the net present value. It allows you to pull forward into the present, the future value of the software system after **n** years.
2. A_0 is the initial investment.
3. A_i is the annual cash flow (which is increased revenue + reduction in expenditure).
4. r_i is the rate of interest.

Figure 20.5   Estimating Software Value-added

20.4 Estimation Techniques for Engineering Cost

In the foregoing discussion, the importance of the engineering cost and the challenges to accurately determining it were emphasized. In Section 20.1, it was mentioned that the standard business approach to estimating engineering cost is to multiply the estimated project duration (in hours) by the organization's prescribed hourly engineering rate. In this section, we will examine the engineering cost a bit closer, and look at other models for cost estimation.

Our discussion commences with the work of Barry Boehm [Boehm 2002]. According to the Boehm model for cost estimation, there are five approaches to software cost estimation (more precisely, the software engineering cost estimation) as summarized below:

- **Algorithmic Cost Modeling:** The cost is determined as a function of the effort.

- **Expert Judgment:** A group of experts assess the project and estimate its cost.
- **Analogy:** The project is compared to some other completed project in the same application domain, and its cost is estimated.
- **Parkinson's Law:** The project cost is based on convenience factors such as available resources, and time horizon.
- **Pricing Based on Consumer:** The project is assigned a cost based on the consumer's financial resources, and not necessarily on the software requirements.

Obviously, the latter four proposals are highly subjective and error-prone; they will not be explored any further. The algorithmic approach has generated much interest and subsequent proposals over the past twenty-five years, some of which have been listed for recommended readings.

20.4.1 *Algorithmic Cost Models*

Algorithmic cost models assume that project cost is a function of other project factors such as size, number of software engineers, and possibly others. As such, each cost model uses a mathematical formula to compute an index for the software engineering effort (E). The cost is then determined based on the evaluation of the effort. Figure 20.6 outlines the typical cost model.

$E = A + B * P^C$

Key:
1. E denotes the effort in programmer-months.
2. A, B and C are empirically derived constants.
3. P is the primary input in LOC or FP.

Figure 20.6   Typical Cost Model

By way of observation, the exponent C typically lies in the range {0.8 .. 1.5}. The constants A, B, and C are called adjustment parameters, and they are determined by project characteristics (such as complexity, experience of the project team members, the development environment, etc.). Pressman [Pressman 2015] lists a number of specific cases-in-point of this cost model (in each case the acronym KLOC means thousand lines of code):

• $E = 5.2 * (KLOC)^{0.91}$	...	the Watson-Felix model
• $E = 5.5 + 0.73 * (KLOC)^{1.16}$	...	the Bailey-Basili model
• $E = 3.2 * (KLOC)^{1.05}$	...	the COCOMO Basic model
• $E = 5.288 * (KLOC)^{1.047}$	...	the Doty model for KLOC > 9
• $E = -91.4 + 0.355 * (FP)$	...	the Albrecht & Gaffney model
• $E = -37 + 0.96 * (FP)$	...	the Kemerer model
• $E = -12.88 + 0.405 * (FP)$	...	the small project regression model

20.4.2 The COCOMO Model

Boehm first proposed the *Constructive Cost Model* (COCOMO) in 1981, and since then it has matured to the status of international fame. The basic model was of the form

$E = B * P^C * (EAF)$ where EAF denotes the effort adjustment factor (equal to 1 in the basic model)

Boehm used a three-mode approach, as follows (Figure 20.7 provides the formula used for each model):

- **Organic Mode**—for relatively simple projects.
- **Semi-detached Mode**—for intermediate-level projects with team members of limited experience.
- **Embedded Mode**—for complex projects with rigid constraints.

Organic:	$E = 2.4 * (KLOC)^{1.05}$
Semi-detached:	$E = 3.0 * (KLOC)^{1.12}$
Embedded:	$E = 3.6 * (KLOC)^{1.20}$

Figure 20.7   COCOMO Formulas

The original COCOMO model was designed based on traditional procedural programming in languages such as C, Fortran, etc. Additionally, most (if not all) software engineering projects were pursued based on the waterfall model. The next subsection discusses Boehm's revision of this basic model.

20.4.3 The COCOMO II Model

Software engineering has changed significantly since the basic COCOMO model was first introduced. At the time of introduction, OOM was just an emerging idea, and most software engineering projects followed the waterfall model. In contrast, today, most software engineering projects are pursued in the OO paradigm, and the waterfall model has given way to more flexible, responsive approaches (review Chapter 1). In 1997, Boehm and his colleagues introduced a revised COCOMO II model to facilitate the changes in the software engineering paradigm.

The COCOMO II model is more inclusive, and receptive to OO software development tools. It facilitates assessment based on the following:

- Number of lines of source code
- Number of object-points or function-points
- Number of lines of code reused or generated
- Number of application points

The changes relate to application points and code reuse/generation—features of contemporary OO software development tools. We will address both in what is called the *application composition model*. Two other sub-models of COCOMO II are the *early design model* and the *post-architecture model*.

20.4.3.1 Application Composition Model

In the application composition model, we are interested in *object points* (OP) as opposed to function points (FP). The model can be explained in seven steps as summarized below:

1. The number of object points in a software system is the weighted estimate of the following:
   - The number of separate screens displayed
   - The number of reports produced
   - The number of components that must be developed to supplement the application
2. Each object instance is classified into one of three complexity levels—simple, medium or difficult—according to the schedule in Figure 20.8.

No. of Views/Sections Contained in a Screen or Report	Number and Source of Data Tables Required		
	Total < 4	Total 5 – 7	Total >= 8
< 3	Simple	Simple	Medium
3 – 7	Simple	Medium	Difficult
>= 8	Medium	Difficult	Difficult

Figure 20.8   Object Instance Classification Guide

3. The number of screens, reports, and components are weighted according to the schedule in Figure 20.9. The weights actually represent the relative effort required to implement an instance of that complexity level.

Object Classification	Simple	Medium	Difficult
Screen	1	2	3
Report	2	5	8
Component			10

Figure 20.9   Complexity Weights for Object Classifications

4. Determine the *object point count* (OPC) by multiplying the original number of object instances by the weighting factor, and summing to obtain the total OPC. Figure 20.10a clarifies the calculation and Figure 20.10b provides an illustration.

```
OPC = ∑[(NI) * (W)]

Key:
1.  NI is number of instances of a particular object classification (screen, report, or component)
    and W is the weight for that object classification.
2.  OPC is the object point count.
```

Figure 20.10a Calculating the OPC

Object Classification	Occurrence	Complexity	Weight	OPC
Screen	10	Simple	1	10
Screen	10	Medium	2	20
Screen	6	Difficult	3	18
Component	6	Difficult	10	60
Total OPC				108

Figure 20.10b Illustrating Calculation of the OPC

5. Calculate the *number of object points* (NOP) by adjusting the OPC based on the level of code reuse in the application:

$$NOP = (OPC) * \left[(100 - \%Reuse) / 100 \right]$$

6. Determine a *productivity rate* (PR) for the project, based on the schedule of Figure 20.11. Note that the schedule that it is in the project's best interest to have a project team of very talented and experienced software engineers, and to use the best software development that is available.

Team Quality	Very Low	Low	Nominal	High	Very High
Development Tool Quality	Very Low	Low	Nominal	High	Very High
Productivity Rate	4	7	13	25	50

Note:
1. Team Quality includes the capabilities and talent of the team members.
2. Development tool quality includes the maturity and capabilities of the software development tool employed.

Figure 20.11 Productivity Rate Schedule

7. Compute the effort (E) via the formula

$$E = (NOP) * (PR)$$

Figure 20.12 summarizes the steps in the application composition model. With practice on real projects, you will become more comfortable with this cost estimation technique. The important thing to remember about the model is that the software cost is construed to be a function of the engineering effort, and the complexity of the software.

1. Identify screens, reports and components in the software system.
2. Classify each object instance as simple, medium or difficult, based on the schedule of figure 17-8.
3. Assign a complexity weight to each project screen, report, or component according to the schedule of figure 17-9.
4. Calculate the OPC based on the formula OPC = ∑[(NI) * (W)]
5. Calculate the NOP based on the formula NOP = (OPC) * [(100 - %Reuse)/100]
6. Determine the PR for the project based on the schedule of figure 17-11.
7. Calculate the effort based on the formula E = (NOP) * (PR)

Figure 20.12   Summary of the Application Composition Model

20.4.3.2 *Early Design Model*

The early design model is recommended for the early stages the software engineering project (after the requirements specification has been prepared). A full discussion of the approach will not be conducted here, but a summary follows.

The formula used for calculating effort is

$$E = B * P^C * (EAF) \text{ where EAF denotes the effort adjustment factor}$$

As clarified earlier, B and C are constants, and P denotes the primary input (estimated LOC or FP). Based on Boehm's experiments, A is recommended to be 2.94, and C may vary between 1.1 and 1.24. The EAF is a multiplier that is determined based on the following seven factors (called *cost drivers*). In the interest of clarity, the originally proposed acronyms have been changed:

- Product Reliability and Complexity (PRC)
- Required Reuse (RR)
- Platform Difficulty (PD)
- Personnel Capability (PC)
- Personnel Experience (PE)
- Facilities (F)
- Schedule (S)

$$EAF = (PRC) * (RR) * (PD) * (PC) * (PE) * (F) * S$$

20.4.3.3 *Post-Architecture Model*

The post-architecture model is the most detailed of the COCOMO II sub-models. It is recommended for use during actual development, and subsequent maintenance of the software system (Chapter 18 discusses software maintenance).

The formula used for calculating effort is of identical form as for the early design model:

$$E = B * P^C * (EAF) \text{ where EAF denotes the effort adjustment factor}$$

In this case, B is empirically recommended to be 2.55 and C is calculated via the formula

$$C = 1.01 + 0.01 \sum (CW) \text{ where CW represents capability weights}$$
calibrated based on characteristics of the project team
and the host organization.

The criteria (also called *scale factors*) used to assign capability weights (CW) about the project and the project team are as follows:

- **Precedence:** Does the organization have prior experience working on a similar project?
- **Development Flexibility:** The degree of flexibility in the software development process.
- **Architecture/Risk Resolution:** How much risk analysis has been done, and what steps have been taken to lessen of eliminate these risks?
- **Team Cohesion:** How cohesive is the team?
- **Process Maturity:** What is the process maturity of the organization? How capable is it in successfully pursuing this project?

Figure 20.13 provides the recommended schedule for determining the capability weights for these criteria. As can be seen from the figure, the weights range from 0 (extra high) to 6 (very low).

Capability Factor	Very Low	Low	Nominal	High	Very High	Extra High
Precedence	4.05	3.24	2.42	1.62	0.81	0.00
Development/Flexibility	6.07	4.86	3.64	2.43	1.21	0.00
Architecture/Risk Resolution	4.22	3.38	2.53	1.69	0.84	0.00
Team Cohesion	4.94	3.95	2.97	1.98	0.99	0.00
Process Maturity	4.54	3.64	2.73	1.82	0.91	0.00

Figure 20.13   Capability Weights Schedule

The EAF is determined from a much wider range of factors than the early design model. Here, there are seventeen factors (cost drivers). Figure 20.14 lists these cost drivers along with their assigned ratings. In the interest of clarity, the original acronyms have been changed. The EAF is the product of these cost multipliers.

You have no doubt observed that this is quite a complex cost model. It must be emphasized that in order to be of any use, the model must be calibrated to suit the local situation, based on local history. Moreover, there is considerable scope for uncertainty in estimation of values for the various factors. The model must therefore be used as a guideline, not a law cast in stone.

Capability Factor	Very Low	Low	Nominal	High	Very High	Extra High
Product:						
Required Software Reliability (RSR)	0.75	0.88	1.00	1.15	1.39	-
Database Size (DBS)	-	0.93	1.00	1.09	1.19	-
Product Complexity (Cplx)	0.70	0.88	1.00	1.15	1.30	1.66
Required Reusability (RR)		0.91	1.00	1.14	1.29	1.49
Documentation (Doc)		0.95	1.00	1.06	1.13	
Platform:						
Execution Time Constraint (ETC)	-	-	1.00	1.11	1.31	1.67
Main Storage Constraint (MSC)	-	-	1.00	1.06	1.21	1.57
Platform Volatility (PV)	-	0.87	1.00	1.15	1.30	-
Personnel:						
Analyst Capability (AC)	1.50	1.22	1.00	0.83	0.67	-
Programmer Capability (ProgC)	1.37	1.16	1.00	0.87	0.74	-
Personnel Continuity (PrsnlC)	1.24	1.10	1.00	0.92	0.84	-
Applications Experience (AE)	1.22	1.10	1.00	0.89	0.81	-
Platform Experience (PE)	1.25	1.12	1.00	0.88	0.81	-
Language and Tool Experience (LTE)	1.22	1.10	1.00	0.91	0.84	
Project:						
Software Tools (ST)	1.24	1.12	1.00	0.86	0.72	-
Multi-site Development MD)	1.25	1.10	1.00	0.92	0.84	0.78
Development Schedule (DS)	1.29	1.10	1.00	1.00	1.00	-

Figure 20.14   Post-Architecture Cost Drivers Schedule

20.5 Summary and Concluding Remarks

Let us summarize what we have discussed in this chapter:

- Software cost is comprised of equipment cost, facilities cost, engineering cost, and operational cost. Equipment cost, facilities cost, and operational cost are determined by following standard business procedures. Engineering cost may be determined using standard business procedures also, but in software engineering, we are also interested in fine-tuning the estimation of this cost component by exploring more deterministic models.
- Software price is influenced by factors including (but not confined to) software cost, market opportunity, uniqueness, usefulness and quality, contracting terms, requirements volatility, and financial health of the organization that owns the product.
- Software value is influenced by software cost, productivity brought the organization, and cost savings brought to the organization.
- Software productivity may be evaluated based on the software engineering effort employed in planning and constructing the product, or based on the value added to the organization. Two metrics used for evaluating the engineering effort are the size-based metrics, and the function-based metrics. Two methods for evaluating value added are the difference method and the NPV analysis method.
- The size-based metrics estimate productivity based on the number of lines of code and pages of documentation of the software.
- Function-based metrics estimate productivity based on the number of function-points of the software.

- Value-added metrics attempt to estimate the value added to the organization by the software.
- Techniques for estimating engineering costs include algorithmic cost modeling, expert judgment, analogy, Parkinson's Law, and pricing based on consumer. Algorithmic cost modeling presents much research interest in software engineering.
- The typical formula for evaluating engineering effort in algorithmic cost modeling is $E = A + B * P^C$
- The COCOMO model uses three derivations of the basic algorithmic cost modeling formula. The model relates to traditional systems developed in the FO paradigm.
- The COCOMO II model is a revision of the basic COCOMO model, to facilitate more contemporary software systems developed in the OO paradigm. It includes an application component model, an early design model, and a post-architecture model.
- The application composition model outlines a seven-step approach for obtaining an evaluation of the engineering effort of the software system. This is summarized in Figure 20.13.
- The early design model uses an adjusted formula for evaluating the engineering effort. The formula is $E = B * P^C * (EAF)$, and certain precautions must be followed when using it.
- The post-architecture model uses the same formula $E = B * P^C * (EAF)$; however, the precautions to be followed are much more elaborate.

We have covered a lot of ground towards building software systems of a high quality. The deliverable that comes out of the software development phase is the actual product! It is therefore time to discuss implementation and maintenance. The next three chapters will do that.

20.6 Review Questions

1. What is the difference between software cost and software price?

2. What are the components of software cost? What are the challenges to determining software cost?

3. State and briefly discuss the main factors that influence software price?

4. How is software value determined? What are the challenges to determining software value?

5. State three approaches to evaluating software productivity. For each approach, outline a methodology, and briefly highlight its limitations.

6. Identify Boehm's five approaches to software cost estimation. Which approach provides the most interest for software engineers?

7. Describe the basic algorithmic cost model that many software costing techniques employ.

8. Describe the COCOMO II model.

9. Choose a software engineering project that you are familiar with.
   a. Using the COCOMO II Application Composition Model, determine an evaluation of the engineering effort of the project.
   b. Using the COCOMO II Early Design Model, determine an evaluation of the engineering effort of the project.
   c. Using the COCOMO II Post-Architecture Model, determine an evaluation of the engineering effort of the project.
   d. Compare the results obtained in each case.

References and Recommended Reading

[Albrecht 1979] Albrecht, A. J. 1979. *"Measuring Application-Development Productivity."* *AHARE/GUIDE IBM Application Development Symposium*. See chapter 26.

[Boehm 1981] Boehm, Barry W. 1981. *Software Engineering Economics*. Englewood Cliffs, NJ: Prentice Hall.

[Boehm et. al. 1997] Boehm, Barry W., C. Abts, B. Clark, and S. Devnani-Chulani. 1997. *COCOMO II Model Definition Manual*. Los Angeles, CA: University of Southern California.

[Boehm 2002] Boehm, Barry W. et al. 2002. COCOMO II. Accessed July 2017. http://sunset.usc.edu/research/COCOMOII.

[Fenton 1997] Fenton, Norman E. and Shari L. Pfleeger. 1997. *Software Metrics: A Rigorous and Practical Approach*. Boston, MA: PWS Publishing.

[Humphrey 1990] Humphrey, Watts S. 1990. *Managing the Software Process*. Reading, MA: Addison-Wesley Publishing.

[Jones 1997] Jones, Capers. 1997. *Applied Software Measurement: Assuring Productivity and Quality*, 2nd ed. New York, NY: McGraw-Hill.

[MacDonell 1994] MacDonell, Stephen G. 1994."Comparative Review of Functional Complexity Assessment Methods for Effort Estimation." *BCS/IEE Software Engineering Journal* 9(3), pp. 107–116.

[Pressman 2015] Pressman, Roger. 2015. *Software Engineering: A Practitioner's Approach*, 8th ed. New York, NY: McGraw-Hill. See chapter 30.

[Putnam 1992] Putnam, Lawrence H. and Ware Myers. 1992. *Measures for Excellence: Reliable Software on time, within Budget*. Englewood Cliffs, NJ: Yourdon Press.

[Schach 2011] Schach, Stephen R. 2011. *Object-Oriented & Classical Software Engineering*, 8th ed. New York: McGraw-Hill. See chapter 9.

SOFTWARE IMPLEMENTATION AND MANAGEMENT

This division addresses the implementation and management of computer software. The objectives of this division are as follows:

- To emphasize the importance of having a wise software implementation plan
- To underscore the importance of good software management as an essential aspect of good software engineering
- To discuss the alternatives for organizing software engineering enterprises and/ or ventures

The division consists of the following chapters:

- Chapter 21: Software Implementation Issues
- Chapter 22: Software Management
- Chapter 23: Organizing for Effective Management

DOI: 10.1201/9780367746025-25

Chapter 21

Software Implementation Issues

Having successfully engineered the software product, it must be implemented in an environment where end users will find it useful. Unless this is done, the whole effort involved in investigation, analysis, design, and development of the product would have been pointless. This chapter discusses important software implementation issues under the following subheadings:

- Introduction
- Operating Environment
- Installation of the System
- Code Conversion
- Change Over
- Training
- Marketing of the Software
- Summary and Concluding Remarks

21.1 Introduction

Irrespective of how the software is acquired (review Chapter 1), it must eventually be implemented. Planning and preparation for this system implementation should start long before the completion of the acquisition. If the implementation is not carefully planned and all factors considered, the exercise can be very frustrating and misrepresenting of the system and professionals responsible for its introduction.

Poor implementation can cause the failure and rejection of a well-designed software system that actually meets the needs of its intended users. This underscores that software engineering does not end after product development. Users must be trained to use the product. To this end, the system must be installed, configured, and monitored. If your organization is in the business of marketing computer

DOI: 10.1201/9780367746025-26

software, then part of the implementation would be the development and pursuit of a marketing plan for the product. These and other related matters will be addressed in the upcoming sections.

21.2 Operating Environment

Consideration about the operating environment addresses the following questions:

- Will the system be centrally operated, or will a distributed computing environment be in place?
- What precautions will be necessary, given the environmental constraints?

21.2.1 Central System

The centralized approach is the traditional approach, where the software system runs on a particular machine and is inaccessible except through that machine. Traditionally, data processing (DP) departments used this approach to manage centralized information systems that provided information services for the other departments in the organization. All data entry, maintenance, and processing were done centrally and reports were sent to various departments. This approach was (and still is) particularly useful in a batch-processing environment.

Advantages of this approach are as follows:

- There is a central locus of control, which allows for easy tracking of system problems.
- There is little or no ambiguity about accountability.
- The approach is ideal for embedded systems that do not need to interact with multiple end users.

The approach suffers from the following disadvantages:

- For information systems, the likelihood of interdepartmental delays regarding the transfer of information is very high. Information may therefore not arrive at the computing center on time. This would trigger a late entry of data into the system, which in turn would cause the late generation of reports. This lateness factor could ripple through the entire organization, causing untold problems.
- The organization's information system would not very responsive.
- The approach is very restrictive and inflexible.

21.2.2 Distributed System

The distributed approach is the contemporary approach to software system implementation. All users have access to the software system (via work-stations). Data enters the system at various points of origin—via workstations operated by users in different departments, or via electronic transfers (that may be transparent to end users). The software system is accessed by various end users as required without any intervention from the software engineering team.

Two broad approaches can be identified:

1. Distributed workstations are connected to a central network server in a local area network (LAN), a wide area network (WAN), or a metropolitan area network (MAN).
2. A distributed network (LAN, WAN, or MAN), consisting of a conglomeration of servers and workstations, and encompassing different departments and/or branches, is constructed and managed.

There may be various configurations (topologies) of each approach—fully-connected (also called mesh), ring, star, bus, and tree—but this is a subject for another course (in either *electronic communication systems* or *computer networks).*

Advantages of the distributed approach include the following:

- The approach is ideal for multi-user environments.
- For information systems, immediate access to the system (especially mission critical information) is provided.
- For information systems, accountability on data accuracy is shifted from the software engineering team to user departments.
- The software engineering team can concentrate on ensuring that the system provides users with appropriate interfaces, data validity, system performance, and other technical issues.
- Through user training and interaction, a clearer understanding between end users and software engineering team is enhanced.

Disadvantages of the approach include the following:

- User training can be challenging. For information systems, even after comprehensive training, the odd user may key in inaccurate data and try to blame "the system" or the "IS/IT Department."
- In the absence of adequately trained personnel, this approach could be a prescription for chaos.

21.2.3 Other Environmental Issues

Other environmental issues to be addressed in software implementation include (review Chapter 3)

- The availability of adequate power supply
- Cooling (or heating) requirements
- Physical security and accessibility requirements

21.3 Installation of the System

In preparation for system installation, a fundamental question to address is whether the installation will involve new machines or just software.

If new machines are involved, an installation diagram (plan) showing where certain equipment will go, is required. Consideration must be given to the various installation alternatives (for example, in the case of a computer network various topologies apply) and the most appropriate one chosen.

If installation involves software only, then consideration should be given to

- The effect on the existing system
- Precautions involved
- Sites of installation (in case of a distributed system)

If installation involves both hardware and software, then obviously, considerations must be given to both areas as outlined above.

21.4 Code Conversion

Code conversion is applicable where software system replaces an existing one, and the coding system used to identify data in one system is different in the other. In such a circumstance, the software engineer must do the following:

- Analyze both systems and clearly identify points of differences in the coding systems.
- Design and test a methodology of linking the differing coding systems.
- Design and test a methodology for converting data from the old format to the new format.

In many cases, some amount of interface coding is necessary. Figure 21.1 represents the main components of a code conversion system, while Figure 21.2 provides an example. The interface program(s) provide(s) conversion and communication between old and new codes.

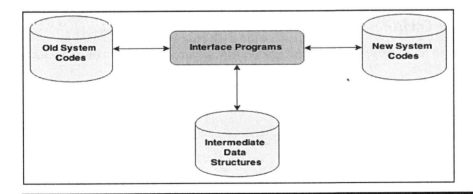

Figure 21.1   Components of a Code-Conversion System

Suppose that an existing information system that used an alphanumeric code (A5, say) for account number is being replaced by a new information system that uses a numeric code (N9, say) for account numbers. Suppose that there is no logical relationship between old codes and new codes. However, it is desired to link new accounts of the new system with their old counterparts in the old system (the old system might contain a lot of historical but needed data which will eventually be converted).

The intermediate database table (**OldNewLink**, say) could be designed to store both old and new codes as follows:

Old Code (A5)	New Code (N9)
AB001	000450000
...	...
XY999	099590001

Each **old code** value is assigned a **new code** value. Intermediate ADD and UPDATE operations would allow for maintenance of the table. Another intermediate RETRIEVAL operation would retrieve a **new code** value, given an **old code** value and vice versa.

Figure 21.2   Code Conversion Example

21.5 Change Over

The previous section provided a clue to the current section: change over from an old software system to a new one. Four strategies can be identified:

- Direct change over
- Parallel conversion
- Phased conversion
- Distributed conversion

21.5.1 Direct Change Over

The essence of direct changeover approach is that on a given date, the old system is dropped and the new system is put into use. After changeover, users cannot use the old system.

This approach can only be pursued if the new system (with all interfaces) is thoroughly tested and shown to be completely acceptable. At best, a minimal delay is expected.

21.5.2 Parallel Conversion

In the parallel conversion approach, the new and old systems are simultaneously run until users are satisfied with the new system. Then and only then is the old system discarded. This is traditionally the strategy used in converting from manual to computerized systems, but it may also be applied to change over from an old computerized system to a newer one.

The approach gives users a sense of security, but is costly in terms of effort. This is so because in some cases, two sets of workers have to be employed to work on the parallel systems; in others cases, employees may be called upon to work overtime.

21.5.3 Phased Conversion

In phased conversion, change from old to new is gradual over a period of time. Aspects of the new system are gradually introduced. The approach is consistent with the phased prototype model of Chapter 1, but also applies to situations where the system was acquired by other methods apart from direct development.

The main drawback in phased conversion can be protracted over an extended period of time. This could potentially create anxiety between the project team and end users.

One advantage is that each component is thoroughly tested before use. Also, users are familiar with each component before it becomes operational. The main drawback is that of interfacing among components.

21.5.4 Distributed Conversion

In the distributed conversion approach, many installations of the same system are contemplated. Based on the evaluations of these installations a decision is made with respect to global implementation.

The approach is very common in large financial institutions with several branch offices (e.g. introducing ATM machines at various sites, a bank take-over, etc.).

21.6 Training

Training of end users is an integral part of software engineering. The following issues often determine the training strategies:

* Who is being trained?
* Who is conducting the training?
* What resources are available for the training?
* Where the training will be conducted?

Training may be conducted by software engineers, vendors, or in-house users of the software. The following table (Figure 21.3) provides a rough guideline as to who may conduct training, given certain circumstances.

Scenario	Recommended Trainer
Principal users being trained for the first time	Software Engineer or qualified Vendor
Secondary users who are power players	Software Engineer
In-house (review) training of principal users	Specially trained in-house users or Vendor
New marketing opportunity	Software Engineer or specially trained Marketer

Figure 21.3   Who Should Train?

Training may be conducted at a specific (and specially prepared) training center, or on-site (at the request of the organization or department receiving the training), the latter being more expensive.

Training objectives (and performance criteria) must be clearly communicated to trainees as this helps them to appreciate what is expected of them, and to adequately prepare for the exercise. Who is being trained will affect the training objectives. For example, clerical staff for data entry requires a different approach from managers of an organization.

Methods used in training should involve visual, hearing, and practical (hands-on training) as is appropriate. Training materials must be well prepared. Training materials are usually in the form of user guides, summary sheets, and workbooks.

If trainee evaluation is required, it must be objective and based on the established performance criteria set. In some instances, the instructor may also be evaluated by the trainees, as a means of quality assurance by the company (or department) offering the training.

21.7 Marketing of the Software

If the product was developed by a software engineering company, to be placed on the market as a consumer product, then the earlier mentioned issues would be part of a much larger marketing strategy.

Issues such as operating environment, installation, and code conversion, would be addressed in the product documentation. Changeover would be left up to the purchasing consumer. Training might be handled by the company or other vendors who might be marketing the product.

Contemporary marketing strategies are typically based on the following five principles:

- **Product:** This relates to the product line(s) the software will be marketed with/as.
- **Pricing:** This relates to how the software will be priced (review Chapter 16).
- **Promotion:** This includes both *positioning* and *packaging* of the software. Positioning relates to how the product is introduced and marketed in the market place. Packaging relates to what other products that are marketed with the software.
- **Placement (or Physical Distribution):** This relates to the avenues through which the product eventually gets to the consumer.
- **People:** This relates to the customer services (including training and support) that will be provided for the product.

The marketing strategy is typically guided by a market research, which is planned and conducted by suitably qualified individuals. The findings of the research are then used to drive the marketing strategy.

21.8 Summary and Concluding Remarks

Here is a summary of what we have discussed in this chapter:

- Planning the software implementation is critical to the user acceptance of the software system.

- The operating environment for the software product may be a central system or a distributed system.
- System installation must be carefully planned.
- If the software system is replacing a pre-existing system, then the matters of code conversion and system changeover become very important.
- System changeover may be direct, parallel, phased, or distributed.
- Implementation often requires training of the end users. This must be carefully planned.
- If the software system is to be marketed to the consuming public, then a marketing plan for the product must be developed and followed.

The following deliverables should be available at the end of a software engineering project:

- Initial System Requirements
- Feasibility Analysis Report
- Requirements Specification
- Design Specification
- System Manual(s)
- User's Guide(s)
- Help System
- The operational Software Product

The requirement specification, design specification, system manual(s), user guide(s), and help system constitute the software documentation. These items must be maintained along with the operational product. The next chapter will discuss software maintenance.

21.9 Review Questions

1. What are the critical issues to be managed during the software implementation?

2. If you were in charge of outlaying your college or university with a strategic information system for its core functions:
   - What operating environment would you consider? Why?
   - Describe a plan for the training of the faculty and staff.
   - What strategy would you use for transitioning from an archaic system on which the institution relied heavily, to your new system? What precautions would you take?

3. Discuss the importance and relevance of code conversion during software implementation.

4. What are the possible approaches to system changeover? When would you use each approach?

5. What guidelines would you follow in conducting user training for a new software system?

Recommended Readings

[Kendall 2014] Kendall, Kenneth E., and Julia E. Kendall. 2014. *Systems Analysis and Design*, 9th ed. Boston, MA: Pearson. See chapter 16.

[Pfleeger 2006] Pfleeger, Shari Lawrence. 2006. *Software Engineering Theory and Practice*, 3rd ed. Upper Saddle River, NJ: Prentice Hall. See chapters 10 and 11.

[Schach 2011] Schach, Stephen R. 2011. *Object-Oriented & Classical Software Engineering*, 8th ed. New York: McGraw-Hill. See chapter 15.

Chapter 22

Software Management

This chapter discusses software management as a vital part of the software engineering experience. As you are aware, it is the final phase of the SDLC. However, the truth is, unless you are very lucky, it is the most likely entry point into a career as a software engineer: In the marketplace, you will most likely have to start off by carrying out management responsibilities on existing software before you get a chance to develop a complete system from scratch. Being part of a project team that develops a major system from scratch is an achievement that every software engineer should aspire to experience at least once in his/her career. However, it is probably worth remembering that many practicing software engineers never experience this.

Software management involves product maintenance, product integration, and product reengineering. Against this background, the chapter proceeds under the following captions:

- Introduction
- Software Maintenance
- Legacy Systems
- Software Integration
- Software Re-engineering
- Summary and Concluding Remarks

22.1 Introduction

Once the software has been implemented, it then enters a usage phase. During this time, the requirements of the software may change. Software maintenance ensures that the software remains relevant for the duration of its life cycle. Failure to maintain the software can reduce its life cycle, thus making it irrelevant and/or unwanted.

If, due to design and/or development flaws, or obsolescence, it has been determined that the cost of maintaining the software (nearly) outweighs the cost of replacing it, then at such point, a decision needs to be taken about the future role and usefulness of the software as well as its possible replacement. However, in the absence of such critical circumstance, the software must be maintained.

As you proceed through this chapter, bear in mind that various factors will affect the management decisions taken about a software product. Three such factors are mentioned in Figure 22.1.

Factor	Clarification
Size and Complexity	As the size and complexity of the software product increases, the likelihood of having a proliferation of such type (category) of software degreases, and the likelihood that companies involved in such projects will seek to maintain their existing products increases.
Technology	If a product is based on obsolete technology, its parent company will eventually be faced the prospects of discontinuing marketing of the product, replacing the product, or reengineering the product.
Support	The maintenance provided is dependent on the availability technical, financial, technological, and human resources.

Figure 22.1 Factors Affecting Software Management Decisions

22.2 Software Maintenance

In order to ensure that the software product remains relevant, despite the changing requirements of the user environment(s), software maintenance is necessary. We shall examine software maintenance under the following subheadings:

- Software Modifications
- Software Upgrades and Patches
- Maintenance Cost

22.2.1 Software Modifications

Software modifications may be put into three broad classifications:

- Enhancements
- Adaptive changes
- Corrective changes

Enhancements are modifications that add to the usefulness of the software. Enhancements may add/improve features, add/improve functionality, or broaden the scope of the software. Most software changes tend to be enhancements.

Adaptive changes are peripheral changes, caused mainly due to changes in the operational environment of the software.

Corrective changes address flaws (bugs) in the software; a well-designed software product will have few bugs. Poorly designed software systems tend to have more bugs than well-designed ones.

Software modifications are not to be made in a flippant manner, but in a manner that ensures that established standards are conformed with. Typically, the software manager establishes a formal organizational procedure for effecting software change. It may include activities such as

- Request by user on a standard change-request form (as illustrated in Figure 22.2)
- Evaluation of the request by a user liaison officer and/or the software manager
- Evaluation of the request by a software engineer
- Decision taken on the request
- Action taken in response to the request

Request Date:	
Requesting Department:	
Department Representative:	
Existing Software System:	

1. State the observed system limitation:

2. State your required system modification:

3. Explain how this change will affect your work:

Figure 22.2   System Modification Request

22.2.2 Software Upgrades and Patches

A software *upgrade* is a modification to a software product to improve its functionality or performance in specific areas. The upgrade may be in any combination of the following possibilities:

- Addition of new features (and possibly program files) that seamlessly integrate into the existing software
- Improvement of existing features and/or functionalities by replacement of existing program files with revised program files
- Removal of known bugs by replacement of existing program files with revised program files

Software engineering companies that market products to the consuming public usually issue software upgrades in the form of releases. Traditionally, software releases are typically numbered in the format that the sections of this text are numbered. However, a more recent trend is to number the versions based on the year of release and to add descriptive qualities to the version names. Following are three examples in three different product lines:

- Early versions of Borland Delphi were of the form V1Rx (or V1.x) up to V8Rx (or V8.x). This was followed by releases of Delphi 2006 Professional, Delphi 2006 Architect, and Delphi 2006 Enterprise; this trend continued into 2010 versions of the product line (except that it is now marketed by Embarcadero, and not Borland). Since then, the version names have been Delphi XE, Delphi XE2, Delphi XE3, and Delphi XE4.
- Microsoft product lines have been consistently named based on the year of introduction.
- Companies such as Oracle and IBM tend to favor the more traditional approach along with descriptive names, but Oracle applies a slight deviation, using version names such as Oracle 9I, Oracle 10G, Oracle 11G, Oracle 12C, and Oracle 19C.

If you are working for an organization that does not market its software systems, this meticulous version naming convention may not be necessary, but would nonetheless be advantageous.

Software upgrades from these established software engineering companies are usually made available via common portable storage media as well as via downloads from the World Wide Web (WWW). Again, if your organization is not in the business of marketing its software products, then this approach may or may not be necessary, depending on the prevailing circumstances.

A software *patch* is a modification to a software product to improve its functionality or performance in specific areas. The patch is more narrowly focused than an upgrade; it may be in any combination of the following possibilities:

- Improvement of existing features and/or functionalities by replacement of existing program files with revised program files

- Removal of known bugs by replacement of existing program files with revised program files

Software engineering companies that market products to the consuming public usually issue software patches in the form of service patches (SP). They are typically numbered sequentially (example SP1, SP2, etc.), with clear indications to consumers as to what versions (releases) of the software product, the patches relate to. Patches are commonly made available via the WWW, but may also be placed on common portable storage media.

22.2.3 Maintenance Cost

The more poorly designed the software, the more maintenance will be required and the higher will the maintenance cost be. Conversely, the better the design, the lower the maintenance cost is likely to be.

Maintenance cost is influenced by the following factors:

- **Lifetime of the software:** In many (but not all) cases, longer lifetime means more maintenance.
- **Quality of the software design:** Generally speaking, poor design leads to high maintenance cost.
- **Amount of changes required:** The greater the number of changes required, the higher the maintenance cost is likely to be.
- **Quality and stability of the support staff:** A very skilled and stable support staff is likely to reduce the maintenance cost.
- **Complexity of software design and configuration:** The more complex the design, the higher the maintenance cost is likely to be (there may be exceptions to this).
- **The development software used** (programming language, DBMS, etc.): Maintenance cost may be constrained by the cost of the software development tools employed.
- **Software size** (in terms of components and source code): Generally speaking, bigger the software system, the higher the maintenance cost (there may be exceptions to this).

As software engineering becomes more standardized, the cost of software maintenance should progressively lessen. If the maintenance cost is going in the opposite direction as time increases, this is a certain alert signal to the organization to be prepared to replace the product at some point in the future.

22.3 Legacy Systems

A *legacy system* is a system that was constructed using relatively old technology, but its useful life continues within the organization. Companies have legacy systems due to a number of compelling reasons:

1. A company, having made huge investments in a system, is not prepared to throw it out, since this would mean significant financial loss.
2. The system might be very large and/or complex. Replacing it would therefore be very costly.
3. The system is mission critical: the company's life revolves around it.
4. The complete system specification may not be known or documented, so that replacing it could be risky.
5. Business rules may be embedded (hidden) in the software and may not be documented anywhere else. Attempting to replace the software would therefore be risky.
6. Any attempt to replace the system must also include the conduct of a data conversion process that offers (infinitesimally close to) 100% guaranteed success. Companies prefer to delay this ultimate activity until they have no alternative, and/or they are guaranteed success.

There are three alternatives for treating legacy systems:

- Continue maintaining the system. If the system is stable and there are not many user requests for changes, this is the most prudent alternative.
- Scrap the old system and replace it with a new one. This alternative should be pursued when the following conditions hold:
  o The old system is showing signs of being a liability rather than an asset, that is, it has (almost) reached its scope of usefulness.
  o It is determined that further maintenance would be an effort in futility.
  o The company is confident that it has gathered enough information to successfully redesign the system.
- Transform the system to improve its maintainability. This process is referred to as software re-engineering, and will be discussed shortly.

22.4 Software Integration

In many cases, and for a number of reasons, a software product may be excellent in the provision of certain desirable services but deficient in the provision of others. In other cases, it might be difficult to find a single software product that provides all the services that an organization is looking for. Should such organizations abandon the investments made in these individual products, in search of a single product? Not necessarily. Rather, the organization could explore the path of software integration.

In software integration, a number of component software products are made to peacefully coexist—i.e. they are integrated. This is consistent with the CBSE model of earlier chapters (review Chapters 1 and 9). In order to provide them with the set of services that they require, large organizations such as banks, insurance companies, and universities often use a conglomeration of software products, merged together in an integrated *information infrastructure*. The term *information infrastructure* is used to mean the complete environment (hardware and software) of the organization, and would be more appropriately discussed in a course in *strategic information management*.

Software integration has become a very important aspect of software engineering. In recognition of this, large software engineering companies (for example Microsoft, IBM, Oracle, Borland, etc.) typically have the services of integration experts who are specially trained to provide technical advice and expertise to organizations that need to integrate component software products. Evidently, these experts are trained to promote the bias of the companies they represent. However, with some research (and by asking pointed questions), you can obtain a good perspective of what your integration options are. In many cases, the component products may be using different software standards. If this is the case, then the integration team must be prepared to write some interface coding.

Software integration has become a commonplace in contemporary software engineering. As more software products are written to ubiquitous standards, we can anticipate further proliferation of integration options. This is good for the industry, as it will force higher quality software products.

22.5 Software Re-engineering

In a situation where the system quality continues to deteriorate, and the level of user request for changes, as well as the cost of maintenance continues to be high, software re-engineering is a viable treatment for legacy systems.

Re-engineering may include the following:

- Revising the underlying database of the system (a huge undertaking)
- Using a more modern or sophisticated development software (CASE tool, DBMS suite, or software development kit)
- Refining and/or revising business processes in the organization
- Superimposing a new user interface (wrapper) on top of an existing database

The main activities of software re-engineering are as follows:

- Source code translation or replacement
- Database transformation or replacement (a huge undertaking)
- Reverse engineering (i.e. the existing software is analyzed and the information extracted is used to document its organizational and functional requirements)
- Modernization (related components are grouped together, and redundancies removed; this may involve architectural transformations)
- Data conversion (from old format(s) to new format(s))

Software re-engineering is quite an involved and complex exercise and must be carried out by qualified and experienced software engineers.

22.6 Summary and Concluding Remarks

It is time once again to summarize what we have covered in this chapter:

- During the usage phase of a software system, it must be managed. Factors affecting effective management of the software include size and complexity, technology, and support.
- Software maintenance involves modifications, upgrade, and the management of maintenance cost. Software modifications may be in the form of enhancements, adaptive changes, or corrective changes. Software upgrades come in the form of releases and service patches. The maintenance cost varies inversely to the software quality.
- A legacy system is a system that was constructed using relatively old technology, but its useful life continues within the organization. The software engineer must be familiar with the different ways of treating legacy systems.
- Software integration is the act of merging different software components so that they peacefully coexist. Component-based software engineering (CBSE) has become very popular in recent years.
- Software re-engineering is useful in situations where the system quality continues to deteriorate, and the level of user request for changes, as well as the cost of maintenance continues to be high.

Effective management of a software system can lengthen its useful life, and thereby increase the organization's return on its investment.

22.7 Review Questions

1. Describe how factors such as complexity, technology, and support affect decisions taken about the maintenance of a software product.

2. Consider four categories of software. For each category, discuss how the above-mentioned factors (complexity, technology, and support) have an influence on the maintenance programs for software products in each respective category.

3. Describe three types of changes that are likely to be made to a software product during its useful life.

4. What is the difference between a software upgrade and a software patch? Using appropriate examples, explain how engineering companies typically handle software upgrades and patches.

5. State four factors that affect the cost of software maintenance.

6. What are legacy systems? Why do they abound and will likely continue to do so for the foreseeable future? Describe three approaches for dealing with legacy systems.

7. When should software integration be considered? What benefits are likely to accrue from software integration?

8. When should software reengineering be considered? Describe the main activities that are involved in the reengineering process.

References and/or Recommended Readings

[Pfleeger 2006] Pfleeger, Shari Lawrence. 2006. *Software Engineering Theory and Practice*, 3rd ed. Upper Saddle River, NJ: Prentice Hall. See chapter 11.

[Pressman 2015] Pressman, Roger. 2015. *Software Engineering: A Practitioner's Approach*, 8th ed. New York, NY: McGraw-Hill. See chapter 29.

[Schach 2011] Schach, Stephen R. 2011. *Object-Oriented & Classical Software Engineering*, 8th ed. New York, NY: McGraw-Hill. See chapter 16.

[Sommerville 2016] Sommerville, Ian. 2016. *Software Engineering*, 10th ed. Boston, MA: Pearson. See chapters 25 and 26.

Chapter 23

Organizing for Effective Management

We have discussed various aspects of software engineering. At this point it should be clear to you that software engineering is predominantly a team-based discipline. But how can software engineering teams be organized for effective work? The issue was first introduced in Chapter 1, and revisited in Chapters 2 and 3. This chapter revisits the matter once more, this time focusing on the organizational structure that must be in place in order to support and facilitate good software engineering.

Let us for the moment concentrate on non-software engineering organizations: Some of these organizations have Information Services (IS) departments/divisions; others have Information Technology (IT) departments/divisions; others have Software Engineering (SE) departments/divisions. Generally speaking, IT is regarded as the broader term, and when used, often includes IS or SE in its scope. However, in many circumstances, IS/SE is used loosely to include IT functions as well. Whether a division or a department is in place is to a large extent, a function of the size of the organization. Large organizations tend to favor IT, SE, or IS divisions that are in turn made up of two or more departments; each department may consist of two or more units or sections. Smaller organizations tend to have IT, SE, or IS departments that may consist of smaller units or sections. Whether a division or a department is in place, there is usually a top IT/SE/IS professional who is ultimately in charge of all the IT-related operations. This individual typically operates under the job title of Director or Chief Information Officer (CIO). It is imperative that the appropriate authority and scope of control be accorded to the CIO. In many cases, this translates to the incumbent reporting to the President or the Chief Executive Officer (CEO). The scenario where the CIO reports to the Chief Financial Officer (CFO), though prevalent in many smaller, more traditional organizations, is hardly tenable.

In software engineering firms, the approach is somewhat different. Not only do these organizations need IT support, they are in business to provide IT/SE services to the public.

DOI: 10.1201/9780367746025-28

This chapter examines both scenarios. The chapter proceeds under the following captions:

- Introduction
- Functional Organization
- Parallel Organization
- Hybrid Organization
- Organization of Software Engineering Firms
- Summary and Concluding Remarks

23.1 Introduction

Chapter 3 alluded to the alternatives to organizing a software engineering team, without much discussion. We shall revisit this issue here, and add more clarity. How the IT, SE, or IS division/department is organized will vary from one organization to the other, depending on the following factors:

- The role that IT plays in the organization
- The size of the organization
- The complexity and scope of the organization's information infrastructure
- The nature of the business
- The preference of the CIO

Three approaches to organizing the IT, SE, or IS division/department have been observed:

- The functional approach
- The parallel (project-oriented) approach
- The hybrid approach

Large software engineering companies tend to be more creative in how they apply these models. We will therefore examine each model and then take a look at how they are applied in large software engineering companies.

23.2 Functional Organization

The functional approach is the most stable and widely used approach. Sections (meaning departments of a division or units of a department) are defined to reflect a specialization (division) of labor. Figure 23.1 illustrates this structure.

Advantages of this approach are as follows:

- There are well-defined areas of specialization; employees in these areas will be very proficient at their work.
- The reduced span of control promotes effective communication within the functional sections.

- The approach forces good product documentation.
- The software product created is likely to be one of high quality.
- It provides an orderly approach to achievement of long-term goals.
- The approach provides a sense of stability to team members; for instance, people like the idea of having clearly defined job titles, offices, and roles.

Disadvantages of this approach are as follows:

- Inter-group communication may be strained.
- Because a single project is spread over several sections, it might be more difficult to meet targets.
- The approach promotes a lack of overall perspective of a given project, or the broader information infrastructure among employees of any section. Having job rotations can significantly minimize this drawback.

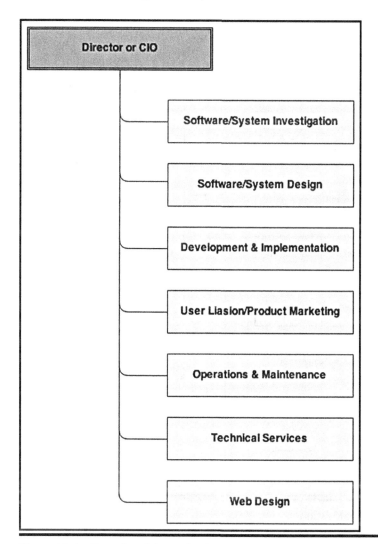

Figure 23.1 Functional Organization

This approach is ideal for a medium/small organization with a limited (manageable) number of projects in a given time horizon. However, it is widely used in small, medium, and large organizations due to its stability.

23.3 Parallel Organization

In the parallel organization (also called project-oriented) approach, the division/department is split based on specific projects. The idea is to preserve the coherence of software engineering projects. Figure 23.2 illustrates this structure.

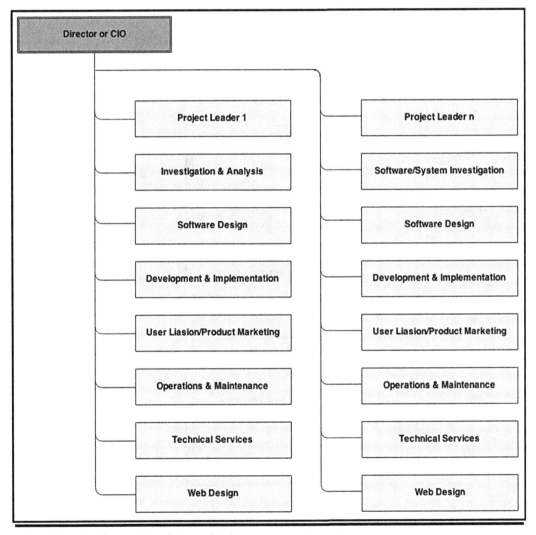

Figure 23.2   Project-oriented Organization

Advantages of this approach are

- There could be parallel development of several projects.
- There is likely to be good communication among team members.
- This approach could enhance motivation of team members.
- The likelihood of short-term goals being realized is enhanced.
- Members of a section have a broader perspective of their assigned projects.

Disadvantages of this approach are

- The project leader could be easily overloaded.
- The possibility of non-uniformity of standards between teams is increased. This could be mitigated by first establishing global standards for all projects.
- Members of a particular project team could be totally oblivious to other projects and by extension, the global information infrastructure. This problem can be minimized by having project rotations, where employees get the chance to work on different projects.

This approach is ideal for a large or medium-sized organization with a large number of projects within a given time horizon.

23.4 Hybrid (Matrix) Organization

The hybrid approach seeks to maximize the advantage of the other two approaches while avoiding the disadvantages. This idea is to have people with default job descriptions, who can be pulled and assigned to various projects. Figure 23.3 illustrates this structure.

This approach provides the following advantages:

- It facilitates assignment of teams depending on the need.
- It promotes lateral and vertical communication.
- It can be both long term and short term.

The following disadvantages may result from the approach:

- The approach could result in conflict of project priority among team members.
- There could also be conflict on reporting authority.
- Conflict on resolving project progress versus personal performance could arise.

This approach is ideal for the very large, competitive, sophisticated organization, where project teams are dynamically formed. These companies are likely to specialize in several large projects within a limited time horizon. It is not as widely used as the functional approach or the project-oriented approach.

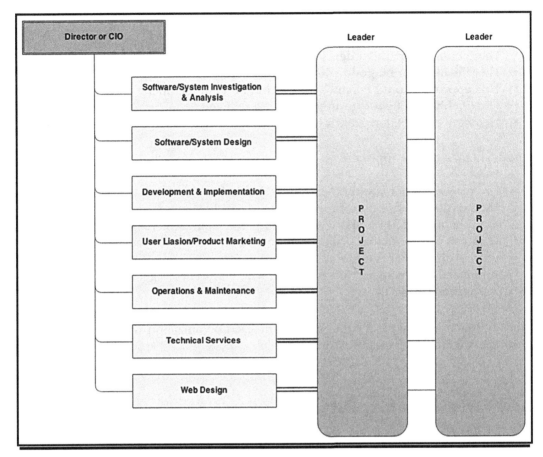

Figure 23.3   Hybrid Approach

23.5 Organization of Software Engineering Firms

Software engineering firms are in the business of software construction and marketing. As you can now attest, software engineering is a very wide field with opportunities in various areas of focus. Typically, a software engineering firm focuses its attention on a finite number of product lines that are consistent with its areas of focus. This products-set will expand as the company grows and widens its scope of interest. Figure 23.4 provides a summary of the software product lines for three of the leading software engineering firms in the industry. For more information on these companies, visit their respective websites as provided in the References and/or Recommended Readings for the chapter.

Due to the wide range of interests that these leading companies have, software engineers are necessarily placed into various teams as required. For these companies, the parallel organization and the hybrid organizations (particularly the latter) are therefore the more suited organizational approaches. Note also that in addition to writing software as one of its primary functions, these companies also need IT/IS/SE divisions/departments for their internal operations.

You may be wondering, what about companies like Alphabet Inc. (parent company of Google) and Amazon? Quite simply, these companies are too massive to be

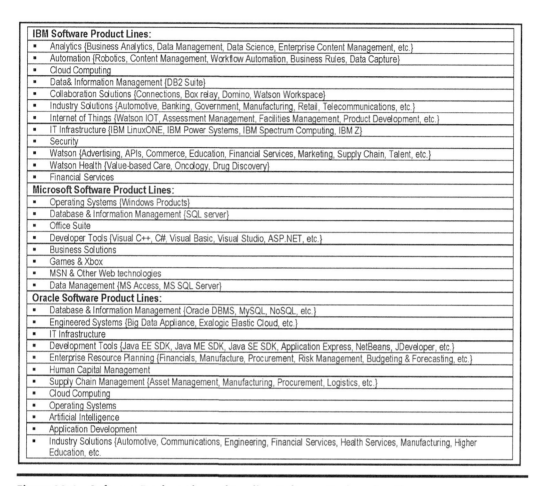

IBM Software Product Lines:
- Analytics {Business Analytics, Data Management, Data Science, Enterprise Content Management, etc.}
- Automation {Robotics, Content Management, Workflow Automation, Business Rules, Data Capture}
- Cloud Computing
- Data& Information Management {DB2 Suite}
- Collaboration Solutions {Connections, Box relay, Domino, Watson Workspace}
- Industry Solutions {Automotive, Banking, Government, Manufacturing, Retail, Telecommunications, etc.}
- Internet of Things {Watson IOT, Assessment Management, Facilities Management, Product Development, etc.}
- IT Infrastructure {IBM LinuxONE, IBM Power Systems, IBM Spectrum Computing, IBM Z}
- Security
- Watson {Advertising, APIs, Commerce, Education, Financial Services, Marketing, Supply Chain, Talent, etc.}
- Watson Health {Value-based Care, Oncology, Drug Discovery}
- Financial Services

Microsoft Software Product Lines:
- Operating Systems {Windows Products}
- Database & Information Management {SQL server}
- Office Suite
- Developer Tools {Visual C++, C#, Visual Basic, Visual Studio, ASP.NET, etc.}
- Business Solutions
- Games & Xbox
- MSN & Other Web technologies
- Data Management {MS Access, MS SQL Server}

Oracle Software Product Lines:
- Database & Information Management {Oracle DBMS, MySQL, NoSQL, etc.}
- Engineered Systems {Big Data Appliance, Exalogic Elastic Cloud, etc.}
- IT Infrastructure
- Development Tools {Java EE SDK, Java ME SDK, Java SE SDK, Application Express, NetBeans, JDeveloper, etc.}
- Enterprise Resource Planning {Financials, Manufacture, Procurement, Risk Management, Budgeting & Forecasting, etc.}
- Human Capital Management
- Supply Chain Management {Asset Management, Manufacturing, Procurement, Logistics, etc.}
- Cloud Computing
- Operating Systems
- Artificial Intelligence
- Application Development
- Industry Solutions {Automotive, Communications, Engineering, Financial Services, Health Services, Manufacturing, Higher Education, etc.

Figure 23.4   Software Product Lines of Leading Software Engineering Firms

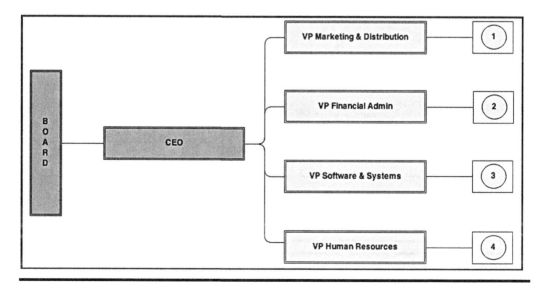

Figure 23.5   Sample Organization Chart for a Software Engineering Firm

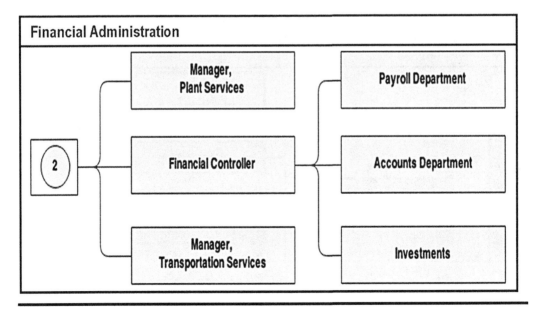

Figure 23.5   (Continued)

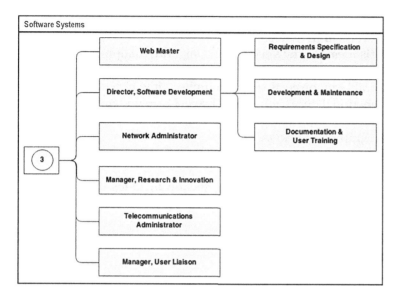

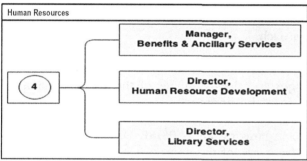

Figure 23.5 **(Continued)**

reasonably covered in this brief discourse. To gain more insight into these organizations, an appropriate start would be to visit their respective websites.

Figure 23.5 provides an illustrative organization chart that could serve a small or medium-sized software engineering firm. In the figure, the Software & Systems Division could use the structure shown as the base from which individuals are drawn for various projects.

23.6 Summary and Concluding Remarks

Let us summarize what has been covered in this chapter:

- An IT, SE, or IS division/department may be organized using the functional approach, the project-oriented approach, or the hybrid approach.
- The functional approach is ideal for a medium/small organization with a limited (manageable) number of projects in a given time horizon. However, it is widely used in small, medium, and large organizations due to its stability.

- The parallel approach is ideal for a large or medium-sized organization with a large number of projects within a given time horizon.
- The hybrid approach is ideal for the very large, competitive, sophisticated organization as project teams are dynamically formed. It is ideal for companies that will specialize in several large projects within a limited time horizon. It is not as widely used as the functional approach or the project-oriented approach.
- Large software engineering companies tend to organize themselves based on product lines and geographic locations.
- Small software engineering firms tend to be functional in their organization.

This completes part E of the course. The upcoming section includes three chapters on advanced software engineering concepts.

23.7 Review Questions

1. Describe the functional approach to organizing an IT/IS/SE Division/Department. Identify the advantages and disadvantages of the approach. State a scenario that would warrant the use of this approach.

2. Describe the project-oriented approach to organizing an IT/IS/SE Division/Department. Identify the advantages and disadvantages of the approach. State a scenario that would warrant the use of this approach.

3. Describe the hybrid approach to organizing an IT/IS/SE Division/Department. Identify the advantages and disadvantages of the approach. State a scenario that would warrant the use of this approach.

4. Which organizational structure is more suited for a medium-sized or large software engineering firm? Defend your answer.

References and/or Recommended Readings

[IBM 2017] IBM. 2017. Accessed July 2017. http://www.ibm.com.
[Microsoft 2017] Microsoft. 2017. Accessed July 2017. http://www.Microsoft.com.
[Oracle 2017] Oracle. 2017. Accessed July 2017. http://www.oracle.com.

ADVANCED SOFTWARE ENGINEERING CONCEPTS

This division of the text contains three topics addressing advanced software engineering concepts and ideas. The objectives of the division are as follows:

- To help you gain an informed understanding of management support systems (MSSs) and their relevance to contemporary software engineering;
- To introduce you to software engineering frameworks;
- To briefly discuss some new frontiers in software engineering, with a look to the future.

The chapters are as follows:

- Chapter 24: Using Database Systems to Anchor Management Support Systems
- Chapter 25: Software Architectures, Frameworks, and Patterns
- Chapter 26: New Frontiers in Software Engineering

DOI: 10.1201/9780367746025-29

Chapter 24

Using Database Systems to Anchor Management Support Systems

In Chapter 13 (Section 13.1), the relevance and importance of database systems to software engineering was mentioned. At this point in the course, this nexus between the two related disciplines of computer science should be very clear to you. This chapter (drawn from [Foster 2016]) reinforces that understanding even more by introducing the concept of *management support systems* (MSSs). Next, the chapter revisits a concept that was raised in Chapter 16—the idea that for many software systems requiring stringent security mechanisms, incorporating such requirements into the underlying database design is a viable alternative. This is followed by discussion of a case study, and some insights into selected MSS project ideas. The chapter advances through the following captions:

- Overview of Management Support Systems
- Building System Security through Database Design
- Case Study: Dynamic Menu Interface Designer
- Selected MSS Project Ideas
- Summary and Concluding Remarks

24.1 Overview of Management Support Systems

As pointed out at the outset of the course, databases typically exist to provide storage requirements for software systems that use them. Put another way, there are many software systems that rely on underlying databases for information support. Additionally, for many software systems, it can be observed that the user interface and system security are intricately linked. One family of software systems that exhibits these two features is the group referred to as *management support systems* (MSSs).

DOI: 10.1201/9780367746025-30

Drawing from a previous research recorded in [Foster 2015], MSSs refer to a family of software systems that are geared towards the promotion and facilitation of efficient and effective management and decision-making in the organization. Included among MSSs are the following categories: *strategic information systems* (SISs), *decision support systems* (DSSs), *executive information systems* (EISs), *expert systems* (ESs), *knowledge management systems* (KMSs), *business intelligence systems* (BISs), and *enterprise resource planning systems* (ERPSs). Following is a brief clarification on each member of the MSS family.

- A SIS is defined in [Wiseman 1988] as a software system that is designed to be aligned with the corporate and strategic vision of an organization or group of related organizations, thus giving strategic and competitive advantages to the host organization(s).
- A DSS has been defined by [Keen & Scott Morton 1978] as a software system that provides information that enables managers and executives to make informed decisions.
- An EIS is a special DSS that focuses exclusively on information reaching the business executive [Rockart & DeLong 1988].
- An ES is a software system that emulates a human expert in a particular problem domain. The classic text, *Introduction to Expert Systems*, by Peter Jackson, represents a significant work in this area (see [Jackson 1999]).
- A recent addition to the MSS family is KMS. This was recognized by Thomas Clark and colleagues in 2007, when they defined MSS to include DSS, EIS, BIS, and KMS (see [Clark, Jones, & Armstrong 2007]). These systems have emerged out of the need for organizations to have access much larger volumes of (often unstructured) information than at any point in the past.
- A BIS is a software system that incorporates a set of technologies that allow a business to operate on relevant information that is made available to its decision makers (see [Mulcahy 2007] and [Oracle 2016]).
- An ERPS is a comprehensive software system that facilitates strategic management in all principal functional areas of operation of a business enterprise. The ERPS typically includes several interrelated sub-systems each of which may qualify as a software system in its own right (see [APTEAN 2015]; [Foster 2015]).

In addition to their intricate connection to the decision-making process in the organization, MSSs are characterized by their data-centeredness, i.e., they rely on data pulled from underlying database(s) to drive the information they provide for decision support. Additionally, MSSs typically require stringent security mechanisms, allowing only authorized access the data that they manage. These two observations will become clearer in the upcoming discussions; hopefully, you will also develop a more profound understanding of the importance of carefully designing the underlying database for each of these software systems.

24.2 Building System Security through Database Design

Contemporary software engineering is typically influenced by quality factors including (but not limited to) development speed, precision, interoperability, user-friendliness, usefulness, and reusability. Software consumers have become quite impatient, and reluctant about persisting with products that do not meet their expectations. Moreover, software developers are expected to deliver projects on schedule, or face the wrath of disgruntled consumer(s).

As you are aware from earlier discussions (review Chapter 14]), the user interface is an integral part of a software system. Additionally, planning, constructing, and testing the user interface for a software system is a time-consuming and labor-intensive effort that often takes relatively more time than other aspects of the software design [Kivisto 2000]. Despite this, system flaws due to lack of thorough testing, are quite common. Moreover, there appears to be a preponderance poorly designed user interfaces for software systems that are heavily used in corporate organizations. Given the importance of the user interface, taking meticulous steps to get it right is of paramount importance.

In foregoing discussions (review Section 16.6), the importance of system security has been emphasized. It turns out that in many cases, user interface and system security are intricately related. This is so because end users typically access system resources via the user interface. It therefore stands to reason that if we could build into the software design the capacity for integrating these two areas, such an achievement would significantly reduce the time spent on software design as well as software development. One way to achieve this end is through the development and application of a software application for that purpose; let's call it a *dynamic menu interface designer* (DMID).

The idea of a DMID is creation of a generic software system that facilitates the specification of the operational components and constraints of other software systems being constructed, as well as the user access matrices for those software systems. This information is then used to dynamically generate the user interface for each end user who attempts to access the target software system(s). The generated user interface contains only resources to which the user has been given the authority to access. Moreover, multiple software systems can be facilitated, and end-user access to these systems is facilitated only through the DMID. In the upcoming section, the basic architecture of the DMID is described.

The DMID as proposed, offers a number of advantages to the software engineering experience, particularly in the area of information systems (IS) development. Included among the proposed advantages are the following:

- User interface specification is significantly simplified by transforming the problem to mere data entry. By providing the facilities for menus to be defined (via data entry) and loaded dynamically, based on the user's access rights, the software engineer is spared the responsibility of major user interface planning and design. The time gained here could be used in other aspects of the project. This, in practice, should significantly shorten the SDLC.

- The shortening of the SDLC could result in noticeable improvement in software engineering productivity, particularly for large projects. The hours gained in not having to program a user interface could be spent on other aspects of the project.

- The DMID not only addresses menu design, but user accessibility also. It is constructed in such a way as to enable the following two constraints: Firstly, only users who are defined to the DMID can gain access to the software system(s) employing it. Secondly, through logical views, each user gets a picture of the system that is based the user's authorization log. Only those resources, to which the user is authorized, will show up on the user's display. So apart from not being able to access other resources, the user is given the illusion that they do not exist. Hence, user's perspective of the system is as narrow or broad as his/her span of authorization.

- The DMID provides support for future changes in the operational environment of the software system, without forcing a corresponding change in the underlying code. This will become clear as we proceed.

24.3 Case Study: Dynamic Menu Interface Designer

This section provides excerpts from other resources previously developed for the DMID project (see [Foster 2011] and [Foster 2012]). We will examine the database requirements followed by the user interface requirements.

24.3.1 Database Requirements of the DMID

The basic architecture of the DMID calls for the use of a few relational tables which are described in Figure 24.1. Notice that to aid subsequent referencing, each table is assigned a reference code (indicated in parentheses); attributes of each table are also assigned reference codes; all reference codes are indicated in square brackets; foreign key references are indicated in curly braces.

The relationship details of Figure 24.1 are clarified in the ERD of Figure 24.2. As the figure conveys, there are 10 entities and 12 relationships connecting the entities. Also observe that entities **E6 Subsystem (Menu) Constituents, E7 User Authorization to Operations, E8 User Authorization to Menu, E9 User Authorization to Systems,** and **E10 Organization—System Mapping** are associative entities (i.e. intersecting relations) that resolve M:M relationship between the respective connected entities.

E01. System Definitions: for storage of internal identifications of all information systems that use the DMID. Each system is assigned a unique identifier. Essential attributes include:
- System Code [SysCode]
- System Name [SysName]
- System Abbreviation [SysAbbr]
- Home Path [SysHome]

The primary key: {SysCode}

E02. Participating Organizations: for storage of internal identification(s) of the organization(s) that have access to the software system(s) of the host organization. Essential attributes include:
- Organization Code [OrgCode]
- Organization Name [OrgName]
- Organization Abbreviation [OrgAbbr]

The primary key: {OrgCode}

E03. System Users: for identification of all legitimate users of the system. Each user must belong to an organization that is recognized by the system. Essential attributes include:
- User Identification Code [UsrCode]
- User Login Name [UsrName]
- User First Name [UsrFName]
- User Last Name [UsrLName]
- User' Organization [UsrOrgCode] **{Refers to E02}**
- User Classification [UsrClass]
- User Password [UsrPssWrd]
- User Password Change Ceiling in days [UsrPssCeil]
- Date of Last Password Change [UsrPssChgD]

The primary key: {UsrCode}

E04. System Operations: for definition of all user operations (options) used. Essential attributes include:
- Operation Code [OpCode]
- Operation Implementation Name [OplName]
- Operation Descriptive Name [OpDName]
- Operation Description [OpDscr]
- Operation Home Path [OprHome]

The primary key: {OpCode}

Figure 24.1   Normalized Relations for the DMID

E05. Subsystem (Menu) Definitions: for definition of all major menus and sub-menus used in the (possibly different) software system(s). Each menu is assigned a unique identifier, and is attached to an information system. Essential attributes include:
- Menu Code [MnuCode]
- Menu Implementation Name [MnulName]
- Menu Descriptive Name [MnuDName]
- Menu Description [MnuDscr]
- Menu's System Code [MnuSysCode]   {**Refers to E01**}
- Menu's Home Path [MnuHome]

The primary key: {MnuCode}

E06. Subsystem (Menu) Constituents: the implementation of a M:M relationship between System Menu Definitions (E05) and System Operations (E04). Essential attributes include:
- Menu Code [MC_MnuCode] {**Refers to E05**}
- Menu Sequence Number [MC_MnuSeqN]
- Constituent Operation Code [MC_OpCode] {**Refers to E04**}

Candidate keys: {MC_MnuCode, MC_MnuSeqN}; {MC_MnuCode, MC_OpCode}

E07. User Authorization to Operations: the implementation of a M:M relationship between System Users (E03) and System Operations (E04). Essential attributes include:
- User Identification Code [AO_UsrCode] {**References E03**}
- Authorized Operation Code [AO_OpCode] {**References E04**}

The primary key: {AO_UsrCode, AO_OpCode}

E08. User Authorization to Menus: the implementation of a M:M relationship between System Users (E03) and System Menu Definitions (E05). Essential attributes include:
- User Identification Code [UM_UsrCode] {**Refers to E03**}
- Authorized Menu Code [UM_MnuCode]   {**Refers to E05**}
- User Menu Sequence Number [UM_MnuSeqN]

Candidate keys: {UM_UsrCode, UM_MnuCode}; {UM_UsrCode, UM_MnuSeqN}

E09. User Authorization to Systems: the implementation of a M:M relationship between System Users (E03) and System Definitions (E01). Essential attributes include:
- User Identification Code [US_UsrCode] {**Refers to E03**}
- Authorized System Code [US_SysCode] {**Refers to E01**}
- User System Sequence Number [US_SysSeqN]

Candidate keys: {US_UsrCode, US_SysCode}; {US_UsrCode, US_SysSeqN}

E10. Organization – System Mapping: the implementation of a M:M relationship between Participating Organizations (E02) and System Definitions (E01). Essential attributes include:
- Organization Code [OS_OrgCode]   {**Refers to E02**}
- System Code [OS_SysCode] {**Refers to E01**}
- System Sequence Number [OS_SysSeqN]

Candidate keys: {OS_OrgCode, OS_SysCode}; {OS_OrgCode, OS_SysSeqN}

Figure 24.1   (Continued)

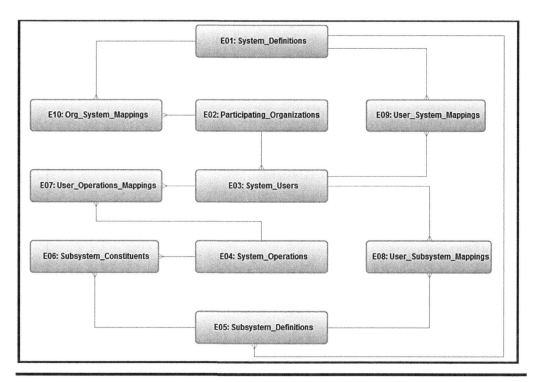

Figure 24.2   ERD for the DMID

Through these relational tables, one can accurately and comprehensively define the constituents, structure, and security constraints of the user interface for any software system that requires the use of user menu(s). Specifically, here is a summary of information that could be specified for each software system:

- Definitional details for the software system(s) employing the DMID
- Basic information about the participating organization(s)
- Basic information about users who will access their respective software system(s) via the DMID
- Definitional details for the operations that comprise each participating software system
- Definitional details for the menus and submenus for each participating software system
- The structure and operational constituents of each menu/submenu comprising each participating software system
- User authorization matrix for subsystems comprising each participating software system
- User authorization matrix for operations comprising each participating software system
- User authorization matrix for participating software systems
- Mapping of participating system(s) for each participating organization (particularly relevant if there are multiple participating organizations and multiple systems)

To facilitate the users having different perspectives of the software system(s), based on their authorization schedule, a number of logical views are required; the salient ones are described in Figure 24.3.

User's System Overview [DM_UsrSysA_LV1]: This is the logical join of User-System Authorizations (E09) with System Definitions (E01), and System Users (E03). Attributes will be read-only:
- User Identification Code [US_UsrCode]
- User Login Name [UsrName]
- User First Name [UsrFName]
- User Last Name [UsrLName]
- System Code [US-SysCode]
- User System Sequence Number [US_SysSeqN]
- System Name [SysName]
- System Abbreviation [SysAbbr]

User Menus Summary [DM_UsrMnuA_LV1]: This is the logical join of User-Menu Authorizations (E08) with Subsystem (Menu) Definitions (E05) and System Users (E03), and Subsystem (Menu) Definitions (E05) with System Definitions (E01). Attributes will be read-only:
- User Identification Code (UM_UsrCode)
- User Login Name [UsrName]
- User First Name [UsrFName]
- User Last Name [UsrLName]
- Menu Code (UM_MnuCode)
- Menu Implementation Name [MnuIName]
- Menu Descriptive Name [MnuDName]
- Menu's System Code [MnuSysCode]
- System Name [SysName]
- System Abbreviation [SysAbbr]
- User Menu Sequence Number (UM_MnuSeqN)

Organization-System Mapping [DM_OrgSysM_LV1]: This is the logical join of Organization-System Mapping (E10) with Participating Organizations (E02), and System Definitions (E01). Attributes will be read-only:
- Organization Code [OS_OrgCode]
- Organization Name [OrgName]
- Organization Abbreviation [OrgAbbr]
- System Sequence Number [OS_SysSeqN]
- System Code [OS_SysCode]
- System Name [SysName]
- System Abbreviation [SysAbbr]

System Users [DM_User_LV1]: This is the logical join of System Users (E03) with Participating Organizations (E02). Attributes will be read-only:
- User Identification Code [UsrCode]
- User Login Name [UsrName]
- User First Name [UsrFName]
- User Last Name [UsrLName]
- User' Organization [UsrOrgCode]
- Organization Name [OrgName]
- Organization Abbreviation [OrgAbbr]
- User Classification [UsrClass]
- User Password [UsrPssWrd]
- User Password Change Ceiling in days [UsrPssCeil]
- Date of Last Password Change [UsrPssChgD]

Figure 24.3   Important Logical Views for the DMID

User Operations Summary [DM_UsrOprA_LV1]: This is the logical join of User-Operation Authorization (E07) with Subsystem (Menu) Constituents (E06), System Operations (E04), System Users (E03), and Menu Definitions (E05). Attributes will be read-only:

- User Identification Code [UO_UsrCode]
- User Login Name [UsrName]
- User First Name [UsrFName]
- User Last Name [UsrLName]
- Authorized Operation Code [UO_OpCode]
- Operation Implementation Name [OpIName]
- Operation Descriptive Name [OpDName]
- Menu Code [MC_MnuCode]
- Menu Sequence Number [MC_MnuSeqN]
- Constituent Operation Code [MC_OpCode]
- Menu Implementation Name [MnuIName]
- Menu Descriptive Name [MnuDName]

System Menu Definitions [DM_MenuD_LV1]: This is the logical join of Subsystem (Menu) Definitions (E05) and System Definitions (E01). Attributes will be read-only:

- Menu Code [MnuCode]
- Menu Implementation Name [MnuIName]
- Menu Descriptive Name [MnuDName]
- Menu Description [MnuDscr]
- Menu's System Code [MnuSysCode]
- System Name [SysName]
- System Abbreviation [SysAbbr]

System Menu Constituents [DM_MenuC_LV1]: This is the logical join of Subsystem (Menu) Constituents (E06), Subsystem (Menu) Definitions (E05), System Definitions (E01), and System Operations (E04). Attributes will be read-only:

- Menu Code [MC_MnuCode]
- Menu Implementation Name [MnuIName]
- Menu Descriptive Name [MnuDName
- Menu's System Code [MnuSysCode]
- System Name [SysName]
- System Abbreviation [SysAbbr]
- Menu Sequence Number [MC_MnuSeqN]
- Constituent Operation Code [MC_OpCode]
- Operation Implementation Name [OpIName]
- Operation Descriptive Name [OpDName]

Key:
- Attributes in this blue-green color are taken from the reverencing relation
- Attributes in this rust-brown color are taken from the referenced relation(s)

Figure 24.3   (Continued)

With these logical views in place, the next step is to superimpose on the database, a user interface that provides the user with the operational objectives mentioned in Section 24.2, but reiterated here for reinforcement:

- Specification of operational components of software system(s) being constructed;
- Specification of user access matrices for end users of the software system(s);
- User access to related software system(s) via the DMID;
- Automatic generation of user menu(s) for each user based on his/her access matrix.

24.3.2 *Overview of the DMID's User Interface Requirements*

Having described the database requirements, let us now examine the user interface requirements for the DMID. This user interface will be superimposed on top of the relational database. Through the DMID's user interface, it must be possible to define and configure the user interface requirements for other software system (s). In its first incarnation, the DMID's user interface itself is subject to future enhancements; however, it is fully operational.

The operational objectives of the DMID were stated in the previous subsection. From an end-user perspective, the DMID should also meet the following end-user objectives:

- **Usefulness:** Software engineers must be able to use the DMID as a means of shortening the development time on software systems that they have under construction.
- **Interoperability:** The DMID must be applicable across various software systems and operating systems platforms.
- **User-friendliness:** The system must be user friendly and easy to use. The interface must be intuitive so that there is a very short learning curve.
- **Reusability:** It must be possible to use the DMID on various independent software systems, as well as a conglomeration of software systems operating as part of a larger integrated system.
- **Flexibility:** The DMID must provide users with the flexibility of specifying the relative order of menu options comprising the software system being constructed.

The user's first interaction with the DMID is via logging on. In logging on, the user specifies the following: user identification code or name; organization the user represents; and password (not displayed in the interest of security). This information is checked against the internal representations stored in the database. If a match is found, the user is taken to the next stage; otherwise, the user is given an appropriate error message, and allowed to try logging on again (the number of allowable attempts may be appropriately restricted). The current incarnation does not encrypt the password, but this is an enhancement that could be pursued.

In its preliminary configuration, two classes of users are facilitated—end users and system administrators (user classification is specified when the user account is defined). System administrators have access to the Administrative Specification Management Subsystem (ASMS). This subsystem provides SUDI privileges to all the data contained in the DMID's internal database. This means the administrator can carry out functions of defining the operational requirements and constraints of software systems, subsystems, and users (as described in the upcoming subsection). End users have access to the End-user Access Control Subsystem (EACS), which will be elaborated on shortly.

24.3.3 *Management of System Constraints via the DMID*

The ASMS provides facilities for defining, reviewing, and modifying the operational scope of participating software systems using the DMID, as well as the operational constraints for end users of the participating software system(s). Among the related activities for this feature are the following:

- Addition of new system(s) and subsystems
- Deletion of (obsolete) system component(s)
- Addition of new operations and/or menu options
- Deletion of menu options
- Configuring/restructuring of system menus
- Setting user authorization schedules in respect of access to systems, subsystems, and operations
- Change of user authorization schedules in respect of access to systems, subsystems, and operations
- Addition of new users
- Deletion of users
- Setting the Organization–System Mapping
- Changing the Organization–System Mapping
- Reviewing (querying) system constraints information

Figure 24.4 shows a screen capture from the ASMS. In this illustration, a test user called Lambert is working with system definitions. At first entry into this option, an initial list of all software systems being managed through the DMID is provided. As these system definitions are modified, or if items are removed during the session, the list is updated. New entries can also be made. Finally, notice that there is a **Search** button to the bottom right of the panel. If this is clicked, a related operation is invoked as illustrated in Figure 24.5. This will allow the user (in this case Lambert) to peruse through the system definitions using any combination of the search criteria provided.

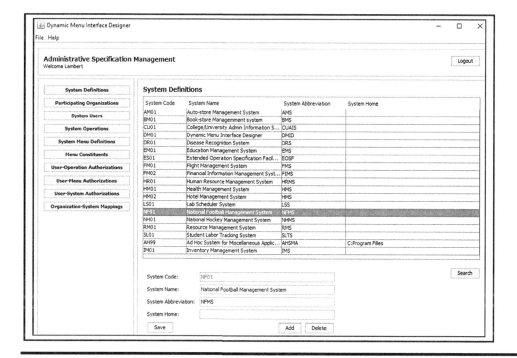

Figure 24.4   Screen-shot from the DMID's ASMS—Managing System Definitions

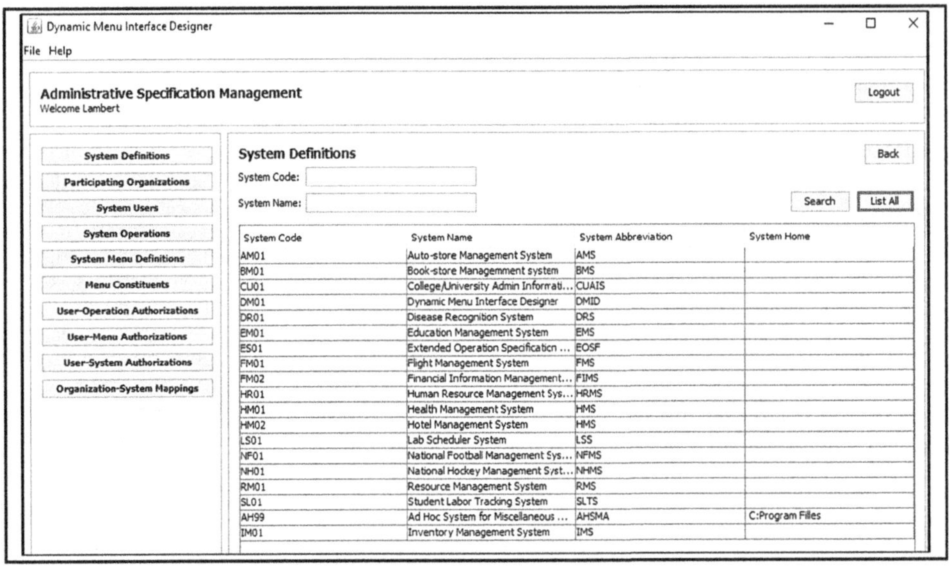

Figure 24.5 Screen-shot from the DMID's ASMS—Searching on System Definitions

It should be noted that the illustrations of Figures 24.4 and 24.5 represent the most basic operational features of the ASMS. There are other more sophisticated operations involving access of the logical views of subsection 24.2 (revisit Figure 24.3). Figure 24.6 provides an example. Here, user-menu authorizations are being managed; specifically, user Lambert is perusing a logical view that joins multiple tables (review the spec for **User Menus Summary [DM_UsrMnuA_LV1]** in Figure 24.3). He has the option of listing all user-menu combinations stored, or narrowing the search by specifying data for any of the six search criteria shown. Please note that each operation follows a similar design concept.

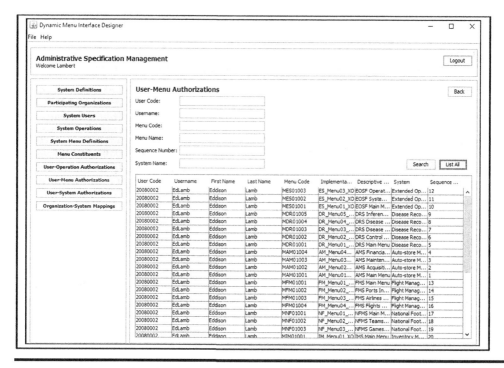

Figure 24.6   Screen-shot from the DMID's ASMS—Searching User-Menu Authorizations

Only individuals who are duly authorized to carry out the functions of software system configuration and management will have access to the ASMS; they must have the *administrator* classification. Database administrators (DBAs) and software engineers would typically fall in this category.

24.3.4 Access to System Resources

Let us now revisit the EACS and provide some additional clarifications. Through this subsystem, an end user can only perform functions defined by his/her authorization schedule; in fact, only these capabilities will appear on that user's menu. End users will be oblivious to system resources that they are not authorized to access; and if by stroke of luck or unauthorized means they attempt to access such resources, such attempted access will be denied.

Assuming successful logon, the user gets a dynamic system-generated menu, depending on his/her authorization schedule that is stored in the underlying DMID database. Three mutually exclusive scenarios are likely:

a. Being an end user, the user gets a menu with the software system(s) to which he/she has access rights.
b. The end user is provided with a blank menu, representing zero access to resources.
c. The user, being a system administrator, gains access to the ASMS menu. Administrative users have access to all resources.

From here on, the user's display panel varies according to what system resources he/she is authorized to access; only resources to which the user is authorized are shown on his/her display panel. Figures 24.7 and 24.8 illustrate two extremes: one in which a user has access to all the resources managed through the DMID, and another in which the user has negligible access to system resources.

In Figure 24.7a, the main menu for a user called Edison is shown. Notice that Edison has access to all the software systems managed via the current instance of the DMID; you can tell this is so by comparing the screen-shot portrayed in Figure 24.7a with the one displayed in Figure 24.5. Also observe from the Figure 24.7a that user Edison has selected the *Auto-store Management System* for the session depicted. In Figure 24.7b, the user (in this case Edison) is provided with a submenu of subsystems of the previously selected system, to which he/she has access privileges. Then, in Figure 24.7c, after selecting the *Acquisitions and Sales* subsystem from the previous screen, another submenu of related operations to which the user is authorized is provided. At this point, the user may select any desired operation for execution. At each level, the user's display panel is filled with options so that he/she is able to scroll and select the option of choice (by highlighting and clicking the **Select** push button). The end user is never shown a resource that he/she is not authorized to access.

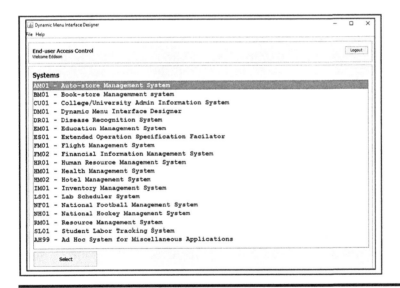

Figure 24.7a   Screen-shot from the DMID's EACS—Main Menu

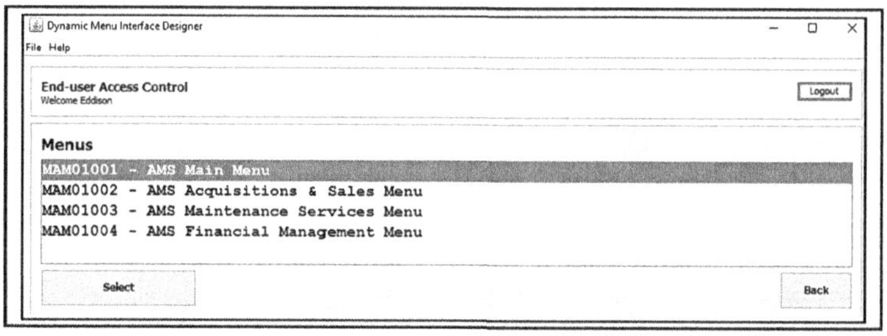

Figure 24.7b   Screen-shot from the DMID's EACS—Subsystems Menu for a Specific Software System

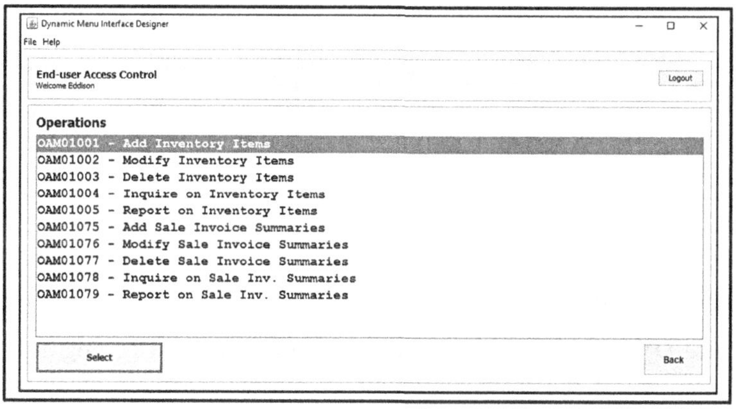

Figure 24.7c   Screen-shot from the DMID's EACS—Operations Menu for a Specific Subsystem

In Figures 24.8a through 24.8c, the user (in this case Ann Marie) has more limited access to systems, subsystems, and operations managed by the DMID. In the most extreme scenario, a user may have no access to any resource (system, subsystem, or operation), in which case a blank screen would appear, and the user would not be able to do anything.

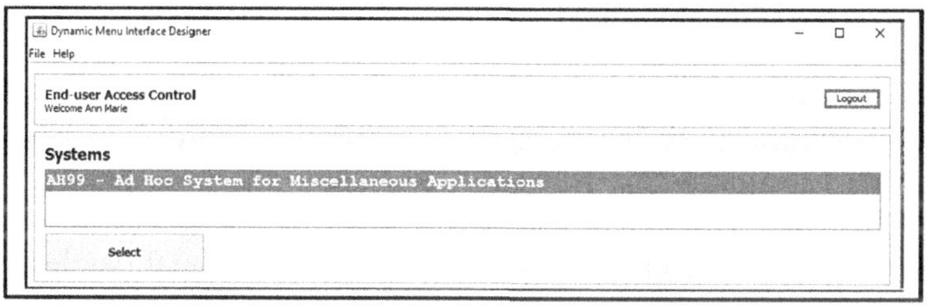

Figure 24.8a   Screen-shot from the DMID's EACS—Main Menu with only one Option

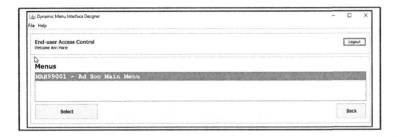

Figure 24.8b   Screen-shot from the DMID's EACS—Subsystems Menu with only one Option

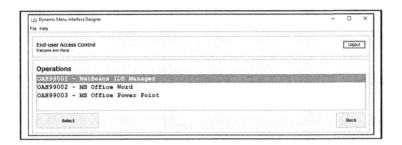

Figure 24.8c   Screen-shot from the DMID's EACS—Operations Menu with Limited Options

24.3.5 Usefulness and Applicability of the DMID

The design concepts for the DMID, as described and illustrated in the forgoing sub-sections, may be replicated for any set of one or more complex software systems. Moreover, the DMID prototype that has been described above can be easily adopted and used for any software engineering endeavor requiring an elaborate user interface and stringent security arrangements for various end users. If used in this context, the DMID could significantly reduce the timeframe for the software development life cycle (SDLC) on specific software engineering projects by transforming previously arduous activities of and time-consuming user interface design and development to mere data entry. When one considers the push towards more ubiquitous software components (for instance, see [Niemelä & Latvakoski 2004]), this design approach offers some promise.

24.4 Selected MSS Project Ideas

MSSs appear in various aspects of life, including (but not confined to) engineering, management, education, health care, etc. Following are selected examples of MSSs and the kinds of information they manage; these examples are drawn from the author's ongoing software engineering research projects with undergraduate students of computer science. For each case presented, it would be necessary to carefully design the identified information entities to conform to relational database norms as discussed in [Foster 2016] and summarized in Chapter 13. Also remember that for each project, only main entities are mentioned; it is anticipated that other related entities would emerge as part of the database design process. Finally, for each of the projects mentioned, the concepts of a DMID (as described and illustrated earlier) would be applicable.

24.4.1 Electoral Management System

An *Electoral Management System* (EMS) is an example of a SIS (as defined in Section 24.1) that could be used to support local as well as national elections. Among the information entities that would be tracked are the following:

- Country employing the EMS (stored to avoid hard-coding)
- Provinces/Territories in the Country
- Electoral Districts in each Province
- Cities/Towns in each Electoral District
- Voter Information for each eligible voter
- Candidate Information for each candidate
- Election Definition for each election managed by the EMS
- Election Schedule for each election managed by the EMS
- Election Referendum Issues being tested
- Election Result by Candidate
- Election Result by Referendum Issue
- Election Summarized Result—a summarized READ-ONLY derived view that anonymizes each voter in the final result

Figure 24.9a shows an information topology chart (ITC) for the project, while Figure 24.9b depicts an object flow diagram (OFD). While the list of information entities for this EMS appears very small, the following must be noted:

1. Data volume on some of these entities would be very large, depending on the population participating in the election in question.
2. The system could be configured to facilitate multiple local as well as national elections.
3. Stringent data security mechanisms would be required to ensure the integrity of the voting and vote-counting.
4. Depending on the country and/or territory, use of special voting technology may be required. However, it is possible to develop a generic solution that uses readily available technologies and methodologies.

Electoral Management System (EMS)	
Logistics Management Subsystem (LMS)	E01: Country Definitions
	E02: Province Definitions
	E03: Electoral District Definitions
	E04: City/Township Definitions
	E05: Voter Definitions
Ballot Management Subsystem (BMS)	E06: Election Definitions
	E07: Candidate Definitions
	E08: Election-Country Mappings
	E09: Election-Province Mappings
	E10: Election-Electoral-District Mappings
	E11: Election-Township Mappings
	E12: Political Parties
	E13: Election Referendum Issues
	E14: Positions (Public Offices) Sought
	E15: Election Vote Finality Stamps
	E16: Election Results Country-wide
	E17: Election Results by Province
	E18: Election Results by Electoral District
	E19: Election Results by Township/City
	E20: Referendum Results Country-wide
	E21: Referendum Results by Province
	E22: Referendum Results by Electoral District
	E23: Referendum Results by Township/City
	E24: Election Detailed Votes
	E25: Referendum Detailed Votes
	E26: Electoral Custodians
Voter Interaction Subsystem (VIS)	E05: Voter Definitions
	E24: Election Detailed Votes

Figure 24.9a   Information Topology Chart for the EMS Project

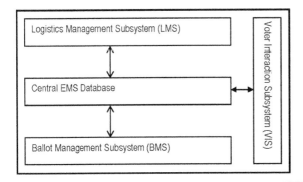

Figure 24.9b   Object Flow Diagram for the EMS Project

24.4.2 Health Information Management System

Another example of an SIS is a *Health Information Management System* (HIMS). This HIMS could be used to facilitate management of the health information for a conglomerate of health organizations (hospitals and clinics), a local town, a county, a province/state, or a whole nation. Following are some of the main information entities that would be managed in such a system:

- Country employing the HIMS (stored to avoid hard-coding)
- Provinces/Territories in the Country
- Patron Information (for individuals requiring health care)
- Medical Professional Information (for medical professionals)
- Health Institutions
- Medical Examinations of Patrons
- Medical Prescriptions for Patrons
- Medical Records
- Health Insurance Providers
- Health Insurance Programs
- Common Medical Cases
- Answers to Frequently-asked Questions

The overview diagrams—ITC and OFD—are shown in Figures 24.10a and 24.10b, respectively. At first glance, this list of information entities for the HIMS appears very small; however, bear in mind the following:

1. Similar to the previous case, data volume on some of these entities would be very large, depending on the population being served.
2. Medical records often involve the use of high-precision graphical information (for example X-rays and other similar records); the system must be designed to efficiently and effectively handle such data.
3. Stringent data security mechanisms would be required to ensure the integrity of all information stored and accessed, with a zero-tolerance for unauthorized access. This security mechanism should include a strategy for encrypting the data.
4. The system would be tailored for the application of various technologies used in modern medicine.

Health Information Management System (HIMS)	
Control Information Subsystem (CIS)	E01: Country Definitions
	E02: Province Definitions
	E03: City/Township Definitions
	E04: Health Institutions
	E05: Medical Professionals
Health Management Subsystem (HMS)	E06: Medical Patrons
	E07: Medical Examinations
	E08: Medical Diagnoses
	E09: Medical Prescriptions
	E10: Medical Records
	E11: Health Insurance Providers
	E12: Health Insurance Programs
	E13: Medical Cases
	E14: Frequently Asked Questions
	E15: Known Diseases

Figure 24.10a   Information Topology Chart for the HIMS Project

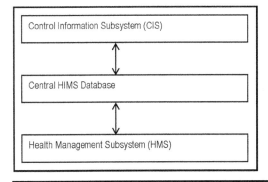

Figure 24.10b   Object Flow Diagram for the HIMS Project

24.4.3 *Strategic Education Management System*

Most governments have a vested interest in its education system at the local, provincial, and national levels. A *Strategic Education Management System* (SEMS) is an SIS that would be geared towards facilitation of the management of an education system at various levels all the way up to the national level, and for various categories of educational institutions. The core information entities that would be managed in a SEMS are the following:

- Country implementing the SEMS (stored to avoid hard-coding)
- Provinces/Territories in the country
- Educational Districts in each province
- Cities/Towns in each province
- Categories of Educational Institutions (elementary school, high school, college, university, vocational)
- Educational Institutions in various categories

- Performance Criteria for Institutions in various categories
- Performance Criteria for Students in various categories
- Evaluation of Institutional Performance at various levels against established institutional performance criteria
- Evaluation of the Student Performance at various levels against established student performance criteria
- Tracking of Outputs (i.e. graduates) from the various Institutions of Learning
- Forecasting Future Outputs and Demands based on historical trends

By way of illustration, Figure 24.11a shows the project's ITC while Figure 24.11b depicts its OFD. As in the previous two examples, the list of information entities for this HIMS may appear to be small; however, the following must be noted:

1. Similar to the previous case, data volume on some of these entities would be very large, depending on the population being served.
2. Educational records often involve the use of high-precision graphical information (for example X-rays and other similar records); the system must be designed to efficiently and effectively handle such data.
3. Stringent data security mechanisms would be required to ensure the integrity of all information stored and accessed, with a zero-tolerance for unauthorized access. This security mechanism should include a strategy for encrypting the data.
4. The system would be tailored for the application of various technologies (for library management, financial management, learning management, etc.) used in modern education.

Strategic Education Management System (SEMS)		
Logistics Management Subsystem (LMS)		E01: Country Definitions
		E02: Province Definitions
		E03: School District Definitions
		E04: City/Township Definitions
		E05: User Category Definitions
		E06 Users
		E07 Institution Category Definitions
		E08 Institutions of Learning
Institutional Performance Subsystem (IPS)		E09: Institutional Performance Criteria
		E10: Institutional Performance Logs
Student Performance Subsystem (SPS)		E11: Student Performance Criteria
		E12: Students
		E13: Student Performance Logs

Figure 24.11a   Information Topology Chart for the SEMS Project

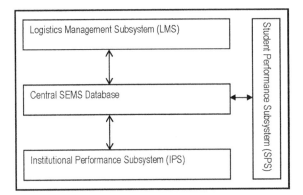

Figure 24.11b   Object Flow Diagram for the SEMS Project

24.4.4 Flight Information Management System

The definition for a DSS is provided in Section 24.1. A *Flight Information Management System* (FLIMS) qualifies as DSS that facilitates effective management of flights at an airport. Moreover, the FLIMS could be designed to be customizable for a group of airports for an entire country. Following are some of the main information entities that would be managed in a FLIMS:

- Countries of Interest
- Ports of Interest in various countries
- Airlines with which the port does business
- Aircraft information on each aircraft for each airline that uses the airport
- Information on each runway (including Terminals and Gates)
- Information on all pilots and flight attendants
- Standard Flight Schedule of all flights to and from the airport
- Log of Actual Arrivals/Departures for flights to and from the airport

The overview diagrams—ITC and OFD—are shown in Figures 24.12a and 24.12b, respectively. Depending on airport where implementation takes place, FLIMS could be a data-intensive system. For instance, at a busy airport such as Logan International, there are over 20,000 flights per month on the average. At a less busy airport, this average may be down to a few hundred flights per month. Additionally, security and data integrity would be extremely important objectives, with zero-tolerance for unauthorized access or incorrect data.

Flight Information Management System (FLIMS)	
Port Management Subsystem (LMS)	E01: Country Definitions
	E02: Province Definitions
	E03: Port Definitions
	E04: Runway Definitions
	E05: Terminal Definitions
	E06: Gate Definitions
Airlines and Aircrafts Subsystem (IPS)	E07: Airlines
	E08: Aircraft Types
	E09: Aircrafts
	E10: Flight Officials
Flights Tracking Subsystem (SPS)	E11: Flight Schedules
	E12: Flight Logs
	E13: Student Performance Logs

Figure 24.12a   Information Topology Chart for the FLIMS Project

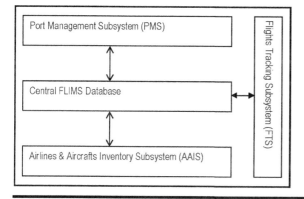

Figure 24.12b   Object Flow Diagram for the FLIMS Project

24.4.5 *Financial Information Management System*

Though this example has become an almost trivial solution in many organizations due to the abundance of existing solutions; it still remains a rather challenging problem for undergraduate students of computer science to appreciate, hence its inclusion: A *Financial Information Management System* (FIMS) is an example of an EIS (as defined in Section 24.1). The FIMS qualifies as an EIS because it provides useful decision support for the financial manager in any organization. Among the core information entities managed are the following:

- Country in which FIMS is being implemented (stored to avoid hard-coding)
- Provinces/Territories in the country
- Organization in which the FIMS is being implemented (stored to avoid hard-coding)
- Chart of Accounts—a list containing all valid account numbers and clarifications about them

- Account Balances
- Financial Transactions
- Transaction Classifications
- Purchase Payment Plans for payments to suppliers
- Payments Made to suppliers
- Financial Institutions
- Sales Payment Plans for payments from customers
- Payments Received from customers
- Investments Logs
- Departments of Interest within the organization
- Customers (representing sources of revenue)
- Suppliers (representing sources for expenditure)
- Inventory/resource Items (representing items used to conduct business)
- Purchase Orders generated and sent out to suppliers
- Purchase Invoices received from suppliers
- Purchase Returns made to suppliers
- Sale Orders generated and sent to customers
- Sale Invoices generated and sent to customers
- Sale Returns received from customers

In Figures 24.13a and 24.13b, the project's ITC and OFD are, respectively, shown. This list of entities gets even longer after normalization (for more insight, see Appendix C). Here are some additional insights:

1. Depending on the size, scope, and complexity of the organization in which implementation takes place, data volume on some of these entities would be very large.
2. As in the previous cases, stringent data security mechanisms would be required to ensure the integrity of all information stored and accessed, with a zero-tolerance for unauthorized access.
3. The system would be tailored for the application of various technologies used in financial management.
4. Depending on the context of implementation, entities from Departments of Interest to the end of the list may already have internal representations in other existing software systems within the organization. Should that be the case, integration (or at minimum, collaboration) with such system(s) would be necessary.

Financial Information Management System (FIMS)		
Controls Management Subsystem (CMS)	E01: Country Definitions	
	E02: Province Definitions	
	E03: City/Township Definitions	
	E04: Organizations	
	E05: Departments	
	E06: Employees	
	E07: Customers	
	E08: Suppliers	
	E09: Inventory Categories	
	E10: Inventory Items	
	E11: Financial Institutions	
	E12: Chart of Accounts	
Acquisitions & Expenses Management Subsystem (AEMS)	E12: Purchase Orders	
	E14: Purchase Invoices	
	E16: Purchase Returns	
	E18: Financial Transactions (Payments Made)	
	E19: Payment Plans	
	E20: Payroll Logs	
	E21: Account Balances	
Sales & Investments Management Subsystem (SIMS)	E22: Sale Orders	
	E24: Sale Invoices	
	E26: Sale Returns	
	E28: Financial Transactions (Payments Received or Investments)	
	E29: Payment Plans	
	E30: Investment Logs	
	E21: Account Balances	

Figure 24.13a Information Topology Chart for the FIMS Project

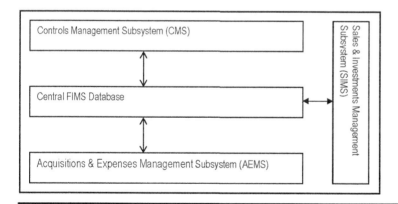

Figure 24.13b Object Flow Diagram for the FIMS Project

24.4.6 Disease Recognition System

ESs (as defined in Section 24.1) appear frequently in various aspects of twenty-first century lifestyle. Among the areas that such systems frequent are engineering, business, manufacturing, health care, etc. One example of an ES is a *Disease Recognition System* (DRS), tailored to facilitate preliminary diagnosis of certain diseases in specified problem domains. In this regard, the DRS fulfills the role of a medical expert. The required database to support this DRS would include the following core entities:

- Disease Categories
- Disease Definitions
- Disease Symptoms—listing various symptoms for the diseases of interest
- Disease-Symptoms Matrix—for mapping each disease to its various symptoms
- Countries of Interest—for addressing purposes
- Provinces/Territories within each country of interest—for addressing purposes
- Vising Patrons—for tracking individuals that use the DRS
- Standard Prescriptions—stored for various diseases
- Diagnosis Summaries—for summarizing each diagnosis
- Diagnosis Symptoms—for noting the observed symptoms of the diagnosed disease
- Diagnosis Prescriptions—for storing information on prescriptions issued
- Medical Professionals—for storing information about the participating physicians

By way of illustration, Figure 24.14a shows the project's ITC while Figure 24.14b depicts its OFD. Since the number of known diseases is extremely large, it is impractical to think that such a system covers all possibilities. More realistically, the system would be configured to cover a specific set of diseases. Following are some additional insights:

1. Depending on the set of diseases studied as well as the targeted number of participants, data volume on some of these entities could be very large or relatively small.
2. As in the previous cases, stringent data security mechanisms would be required to ensure the integrity of all information stored and accessed, with a zero-tolerance for unauthorized access. This security mechanism should include a strategy for encrypting the data.
3. For the system to work, it will be necessary to observe a number of software engineering procedures: constructing an information-gathering instrument for specific diseases and their symptoms; codification of that information and entry into the database; construction of an inference engine; testing and refinement. The intent here is to highlight the role of the underlying database for such a system.
4. In recent years, IBM's supercomputer Watson, which started out with emulation of the game Jeopardy, is taking on the role of a DRS as described above (for more information on this, see [IBM 2015]). This DRS initiative started long before Watson, and no doubt, there will be other initiatives after Watson. Nonetheless, the Watson initiative offers huge potential benefits to the fields of expert systems as well as health.

Disease Recognition System (DRS)	
Disease Information Subsystem (DIS)	E01: Disease Category Definitions
	E02: Disease Definitions
	E03: Disease Symptom Definitions
	E04: Disease-Symptoms Matrix
Control Information Subsystem (CIS)	E05: Country Definitions
	E06: Province/State Definitions
	E07: Town/City Definitions
	E08: Medical Institutions
	E09: Physician Definitions
	E16: System Messages
	E17: Sessions Log
Disease Diagnosis Subsystem (DDS)	E10: Visiting Patron Details
	E11: Patron Visitations
	E12: General Prescription Specifications
	E13: Diagnosis Summary Entries
	E14: Diagnosis Detailed Symptoms
	E15: Diagnosis Detailed Prescriptions
Patron's Access Subsystem (PAS)	E10: Visiting Patron Details
	E11: Patron Visitations
	E14: Diagnosis Detailed Symptoms
	E15: Diagnosis Prescriptions

Figure 24.14a   Information Topology Chart for the DRS Project

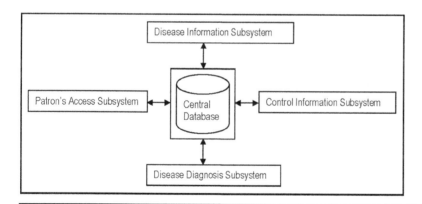

Figure 24.14b   Object Flow Diagram for the DRS Project

24.4.7 Cognitive Leadership Analysis System

Another ES project worthy of mention is a *Cognitive Leadership Analysis System* (CLAS). The CLAS project involves the design, construction, and testing of an ES to identify and encourage the development of leadership styles and qualities in individuals and organizations. The idea is to identify certain leadership styles and qualities, allow respondents to fill out a questionnaire to measure disposition to those leadership styles and qualities, and then generate a statistical profile of the respondent's leadership dispositions based on the responses made. Among the Core information entities included in the underlying database are the following:

- Leadership Styles of interest
- Leadership Qualities of interest
- Leadership Style Questions
- Leadership Quality Questions
- Leadership Analysis Rules
- Individual Evaluation Instruments
- Participating Individuals
- Individual Responses for Style
- Individual Responses for Quality
- Individual Response Analysis for Style
- Individual Response Analysis for Quality
- Participating Organizations
- Organizational Evaluation Instruments
- Organizational Responses for Style
- Organizational Responses for Quality
- Organizational Response Analysis for Style
- Organizational Response Analysis for Quality

The overview diagrams—ITC and OFD—are shown in Figures 24.15a and 24.15b, respectively. Observe that the various leadership styles and qualities are not hard-coded, but treated as data entry. The same is true for the questions geared toward measuring these styles and qualities. The questions are connected to their related styles and/or qualities through leadership analysis rules. These strategies make the system very flexible and customizable for different scenarios of interest. This flexibility is enabled and facilitated by thoughtful database design. Following are some additional insights on the project:

1. The proposed system can be used by individuals or by an organization. Depending on the context of usage as well as the number of participants, data volume on some of these entities could be very large or relatively small.
2. For the system to work, the following software engineering procedures will be necessary: constructing an information-gathering instrument for specific leadership styles and qualities, as well as the criteria (questions) for measuring them; codification of that information and entry into the database; codification of the leadership analysis rules; construction of an inference engine to evaluate the leadership capacity of the respondents; testing and refinement. The intent here is to highlight the role of the underlying database for such a system.

Cognitive Leadership Analysis System (CLAS)	
Leadership Parameters Subsystem (LPS)	E01: Leadership Style Definitions
	E02: Leadership Quality Definitions
	E03: Leadership Style Questions
	E04: Leadership Quality Questions
	E05: Country Definitions
	E06: Province/State Definitions
	E07: Township/City Definitions
	E08: Organizations
	E09: Departments
	E10: Individuals
Individual Analysis Subsystem (IAS)	E11: Individual Evaluation Instruments
	E13: Individual Instrument – Leadership Style Mappings
	E14: Individual Instrument – Leadership Quality Mappings
	E15: Individual Instrument – Leadership Style Questions
	E16: Individual Instrument – Leadership Quality Questions
	E17: Individual Analysis Rules
	E18: Individual Responses on Leadership Style
	E19: Individual Responses on Leadership Quality
	E20: Individual Response Analysis on Style
	E21: Individual Response Analysis on Quality
	E31: Individual Survey Participation Events
Organizational Analysis Subsystem (OAS)	E12: Organizational Evaluation Instruments
	E22: Organizational Instrument – Leadership Style Mappings
	E23: Organizational Instrument – Leadership Quality Mappings
	E24: Organizational Instrument – Leadership Style Questions
	E25: Organizational Instrument – Leadership Quality Questions
	E26: Organizational Analysis Rules
	E27: Organizational Responses on Leadership Style
	E28: Organizational Responses on Leadership Quality
	E29: Organizational Response Analysis on Style
	E30: Organizational Response Analysis on Quality
	E32: Organizational Survey Participation Events

Figure 24.15a   Information Topology Chart for the CLAS Project

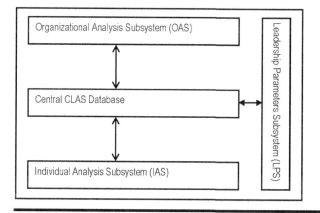

Figure 24.15b   Object Flow Diagram for the CLAS Project

24.4.8 *Financial Status Assessment System*

Here is another example that also qualifies as an ES: A *Financial Status Assessment System* (FSAS) to evaluate the financial health of an individual or organization based on known information about the financial dealings and obligations of the past for that person or organization. This is similar to what banks and other financial institutions do in order to determine one's credit worthiness. Indeed, the current credit score system that is administered by the three leading credit evaluation agencies (Experian, Equifax, and TransUnion), fulfill this role for the American public. The idea is to identify financial performance criteria, evaluate individual and/or organizational performances based on those criteria, and then generate a statistical profile of the respondent's financial profile. However, rather than exacting real penalties and/or rewards on its users, the purpose of such a software system would be educational—working with what-if scenarios and test data, to teach the consequences of financial decisions. Among the core information entities that may be included in the underlying database for this FSAS are the following:

• Country implementing the FSAS (stored to avoid hard-coding)
• Provinces/Territories in the country
• Individual Identity Information
• Sources of Periodic Income/Expense—for identifying all sources of income/expense
• Inflow/Outflow Categories—for analysing inflow/outflow of funds
• Financial Institutions—for storing data on various financial institutions
• Financial Obligations in Loans
• Financial Obligations in regular bills excluding loans
• Cash Inflows—for storing periodic inflow of funds
• Cash Outflows—for storing periodic outflow of funds
• Financial Investments—for storing data on investments made by the individual
• Investment Categories—for analysing investments
• Benchmarks & Evaluation Rules—for storing formulas and/or performance benchmarks
• Personal Evaluation Summaries—for storing the summaries of financial evaluations
• Personal Evaluation Summaries—for storing the details of financial evaluations

By way of illustration, Figure 24.16a shows the project's ITC while Figure 24.16b depicts its OFD. While the specific formulas used by credit agencies are not known (to this course), a software system like this enhances a more profound understanding of what these agencies do, and what financiers look for in customers that they sponsor. Here are some additional insights on the project:

1. As in the case of the previous case, the proposed FSAS project can be used by individuals or by an organization. Depending on the context of usage as well as the number of participants, data volume on some of these entities could be very large or relatively small.
2. As in the previous cases, stringent data security mechanisms would be required to ensure the integrity of all information stored and accessed, with a zero-tolerance for unauthorized access.

3. For the system to work, the following software engineering procedures will be necessary: constructing an information-gathering instrument for specific the financial dealings of individuals/organizations, as well as the criteria for measuring them; codification of that information and entry into the database; construction of an inference engine to evaluate the financial status of the respondents; testing and refinement. Again, the intent here is to highlight the role of the underlying database for such a system.

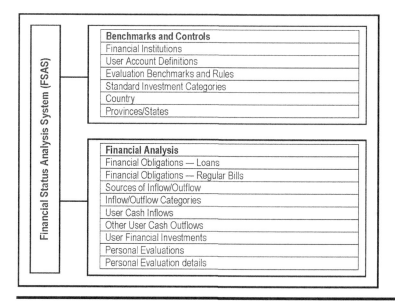

Figure 24.16a   Information Topology Chart for the FSAS Project

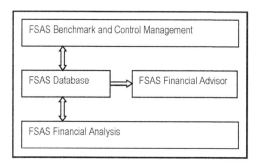

Figure 24.16b   Object Flow Diagram for the FSAS Project

24.4.9 College/University Administrative Information System

Recall from Section 24.1 that an ERPS is a software system that addresses all the principal functional areas of an organization. This final example qualifies as an ERPS: a College/University Administrative Information System (CUAIS). This is a huge software system that facilitates effective management of a college environment. The project encompasses over 100 information entities; full discussion is beyond the scope of this course.

For the purpose of initial discussion, Figure 24.17a shows a partial ITC for the CUAIS project. The ITC presents the main information entities for a software engineering project, organized in the logical areas (sub-systems or modules) where they will likely be managed. Also notice from the diagram that eight subsystems are included— Infrastructure & Resource Management System (IRMS), Public Relations & Alumni Management System (PRAMS), Curriculum & Academic Management System (CAMS), Human Resource Management System (HRMS), Student Affairs Management System (SAMS), Financial Information Management System (FIMS), Library Management System (LMS), and Cafeteria Services Management System (CSMS).

Continuing, Figure 24.17b shows an OFD for the CUAIS project. In this illustration, you will notice the eight aforementioned subsystems and a ninth one for the central database that is accessed by each subsystem. The bidirectional arrows convey that each subsystem has read-write access to the database.

Referring again to Figure 24.17a, note that for each subsystem, only a partial list of data entities is provided. After applying database design principles of Chapter 3 and 5, the number of entities would significantly increase. Here are some additional insights:

1. Depending on the size of the institution and the time period over which the underlying database is used, the data volume on some of these entities would be expected to get very large.
2. As in the previous cases, stringent data security mechanisms would be required to ensure the integrity of all information stored and accessed, with a zero-tolerance for unauthorized access.
3. The system would be tailored for the application of various technologies used in higher education.
4. Similar to the discussion of Section 24.4.3, educational records often involve the use of high-precision graphical information; the system design would need provisions to efficiently and effectively handle such data.

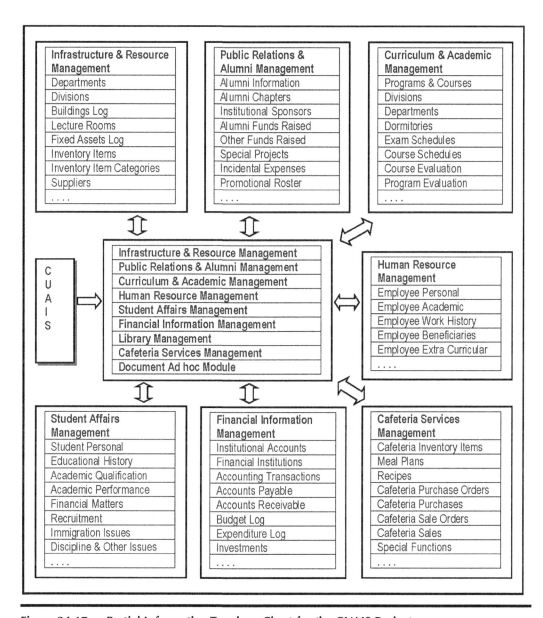

Figure 24.17a Partial Information Topology Chart for the CUAIS Project

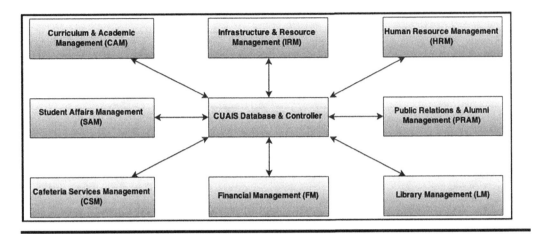

Figure 24.17b   Object Flow Diagram for the CUAIS Project

24.5 Summary and Concluding Remarks

This chapter highlights the dominant role that a carefully designed database can play in the design of a software system. Here is a summary of the salient points covered in the chapter:

MSSs refer to a family of software systems that are geared towards the promotion and facilitation of efficient and effective management and decision-making in the organization. Included among MSSs are the following categories: SISs, DSSs, EISs, ESs, KMSs, BISs, and ERPSs.

In addition to having intricate connections to the decision-making process in the organization, MSSs are characterized by their data-centeredness, i.e., they rely on data pulled from underlying database(s) to drive the information they provide for decision support. Additionally, MSSs typically require stringent security mechanisms, allowing only authorized access the data that they manage. For such systems, the idea of a DMID is relevant.

The concept of a DMID calls for a software component that facilitates the specification of the operational components and constraints of another software system being constructed, as well as the user access matrices for that software system. This information is then used to dynamically generate the user interface for each end user who attempts to access the target software system.

The DMID discussed in this chapter, embodies a database that includes 10 normalized relations and 12 relationships connecting them. Through logical views, these relations are combined to create a virtual environment that facilitates easy enforcement of desirable security constraints. Superimposed on top of this database design is a user interface that facilitates specification of aforementioned constraints, while achieving objectives of usefulness, interoperability, user-friendliness, reusability, and flexibility.

The DMID proposed in this chapter, has two main subsystems—the Administrative Specification Management Subsystem (ASMS) and the EACS. The ASMS is accessed only by system administrators; it provides SUDI privileges to all the data contained in the DMID's internal database. This means the administrator can carry out functions of defining the operational requirements and constraints of software systems, subsystems, and users (as described in the upcoming subsection). End users have access to the EACS, which dynamically generates their menus at login-time. The EACS also controls user access by including on the user's menu only system resources to which they have be granted access privileges.

The design concepts for the DMID may be replicated for any set of one or more complex software systems. Moreover, the DMID prototype that has been described above can be easily adopted and used for any software engineering endeavor requiring an elaborate user interface and stringent security arrangements for various end users. If used in this context, the DMID could significantly reduce the timeframe for the software development life cycle (SDLC) on specific software engineering projects by transforming previously arduous and time-consuming activities of user interface design and development to mere data entry.

MSSs appear in various aspects of life, including (but not confined to) engineering, management, education, health care, etc. Among the examples of MSSs discussed are the following:

- Electoral Management System (EMS)—for facilitating local as well as national elections
- Health Information Management System (HIMS)—for facilitating management of the health information for a conglomerate of health organizations (hospitals and clinics), a local town, a county, a province/state, or a whole nation
- Strategic Education Management System (SEMS)—for enhancing the management of an education system at various levels all the way up to the national level, and for various categories of educational institutions
- Flight Information Management System (FLIMS)—for empowering effective management of flights at an airport
- Financial Information Management System (FIMS)—for providing useful decision support for the financial manager in any organization
- Disease Recognition System (DRS)—for facilitating preliminary diagnosis of certain diseases in specified problem domains
- Cognitive Leadership Analysis System (CLAS)—for identifying and encouraging the development of leadership styles and qualities in individuals and organizations
- Financial Status Assessment System (FSAS)—for evaluating the financial health of an individual or organization based on known information about the financial dealings and obligations of the past for that person or organization
- College/University Administrative Information System (CUAIS)—for effective management of a college environment

As you begin to practice database systems and/or software engineering, you will learn that at the heart of most problematic legacy systems is a poorly designed database system; you will also learn and understand that replacing a poorly designed database after it has been placed in production is often a very daunting task. It therefore behooves every practicing software engineer and/or database expert to invest much time in getting the database properly designed at the outset; in so doing, you will save yourself and/or others much grief later on.

For these more complex software systems, use and application of building blocks is of paramount importance. The next chapter discusses another family of building blocks — software engineering frameworks and patterns.

24.6 Review Questions

1. What two aspects of many software systems are usually intricately linked?

2. What do you understand by the acronym DMID? Briefly explain how a DMID as described in this chapter, allows you to seamlessly tie aforementioned two aspects of a software system together.

3. Describe the essential database components of a DMID.

4. Describe the essential user interface components of a DMID.

5. What does the acronym MSS mean? What is an MSS? Identify the different categories of software systems that comprise the MSS family.

6. Identify two dominant characteristics of MSSs.

7. Identify four examples of MSSs. For each, briefly outline the database requirements.

References and/or Recommended Readings

[APTEAN 2015] APTEAN. 2015. ERP Solutions. Accessed June 1, 2016. http://www.aptean.com/solutions/application/erp-solutions

[Clark, Jones, & Armstrong 2007] Clark, J. Thomas, M. C. Jones, & C. P. Armstrong. 2007. "The Dynamic Structure of Management Support Systems: Theory Development, Research Focus, and Direction." *MIS Quarterly.* 31(3), 579–615. Accessed February 1, 2013. http://ezproxy.library.capella.edu/login?url=http://search.ebscohost.com.library.capella.edu/login.aspx?direct=true&db=bth&AN=25980839&site=ehost-live&scope=site

[Foster 2011] Foster, Elvis C. 2011. *Design Specification for the Dynamic Menu Interface Designer.* Accessed January 2016. www.elcfos.com

[Foster 2012] Foster, Elvis C. 2012. *"Dynamic Menu Interface Designer."* In *Enterprise Management Information Systems. Selected Papers from the 7th and 8th International Conferences on Information Technology and Computer Science*, June 13–16, 2011, May 21–24, 2012. Chapter 23.

[Foster 2015] Foster, Elvis C. 2015. *"Towards Measuring the Impact of Management Support Systems on Contemporary Management."* ATINER 11th Annual International Conference on Information Technology & Computer Science, Athens, Greece, May 18-21, 2015.

[Foster 2016] Foster, Elvis C. with Shripad Godbole. 2016. *Database Systems: A Pragmatic Approach*, 2nd ed. New York, NY: Apress.

[IBM 2015] IBM. 2015. IBM Watson Health. Accessed January 2015. http://www.ibm.com/smarterplanet/us/en/ibmwatson/health/

[Jackson 1999] Jackson, P. 1999. *Introduction to Expert Systems*, 3rd ed. New York, NY: Addison-Wesley.

[Keen & Scott Morton 1978] Keen, P. G. W. & M. S. Scott Morton. 1978. *Decision Support Systems: an Organizational Perspective*. Reading, MA: Addison-Wesley.

[Kivisto 2000] Kivisto, Kari. 2000. *A Third Generation Object-Oriented Process Model*. Department of Information Processing Science, University of Oulu, 2000. Accessed January 2016. http://herkules.oulu. fi/isbn9514258371/isbn9514258371.pdf [July 2010). See chapter 3.

[Mulcahy 2007] Mulcahy, Ryan. 2007. "Business Intelligence Definition and Solutions." *CIO*. Accessed June 1, 2016. http://www.cio.com/article/2439504/business-intelligence/business-intelligence-definition-and-solutions.html

[Niemelä & Latvakoski 2004] Niemelä, E. & J. Latvakoski. 2004. *"Survey of Requirements and Solutions for Ubiquitous Software."* In *Proceedings of the 3rd International Conference on Mobile and Ubiquitous Multimedia MUM 04 (2004), ACM Press*, 71-78.

[Oracle 2016] Oracle. 2016. Business Intelligence and Data Warehousing. Accessed June 1, 2016. http://www. oracle.com/technetwork/topics/bi/whatsnew/index.html

[Rockart & DeLong 1988] Rockart, J. & D. DeLong. 1988. *Executive Support Systems*. Homewood, IL: Dow Jones-Irwin.

[Schach 2011] Schach, Stephen R. 2011. *Object-Oriented and Classical Software Engineering*, 8th ed. Boston, MA: McGraw-Hill.

[Wiseman 1988] Wiseman, C. 1988. *Strategic Information Systems*. Homewood, IL: Irwin.

Chapter 25

Software Architectures, Frameworks, and Patterns

As the field of software engineering has developed over the years, the complexity has increased even more. The paradigm has been changed from the traditional function-oriented approaches to the object-oriented approach. This change of philosophy and focus has resulted in the emergence of numerous applications and platforms. This transformation has been catalyzed by the widespread use of architectures, frameworks, and patterns. So widespread are their applications, many computer scientists would find difficulty differentiating among them. Nonetheless, these three software engineering catalysts are related. The purpose of this chapter is to clearly differentiate them and provide a cursory introduction to each. The chapter proceeds under the following subheadings:

- Software Architecture Tools
- Software Frameworks
- The Model-View-Controller Framework
- Software Patterns
- Summary and Concluding Remarks

25.1 Software Architecture Tools

Early software systems were limited to accepting input, running calculations, and then providing output. These systems would usually run processes in a batch mode and be used for business or scientific calculations. However, as the field of computer science grew, especially with the advent of object-oriented methodologies, software developers could modularize their code into reusable components and arrange them in different structures to provide solutions for more complex computing needs. These different formations would eventually be described as the software architectures. *Software Architecture* then is the overall structure of all the software components in

DOI: 10.1201/9780367746025-31

an application and how they interrelate with each other. The software architecture's goal is to promote the following:

- **Code Reuse:** Thanks to principles of inheritance and polymorphism, software developers have been arranging their code in order to maximize reuse. Code reuse is one of the 'magic-bullets' in software engineering as it leads to reduced effort and increased efficiency.
- **Easier Maintenance of the Code:** If code is standardized and well-organized, this typically leads to easier debugging and modification. If you consider some of your past software development projects (formal or informal), you can probably think of areas that could be enhanced. The same phenomenon exists in professional software development. If the code in a project is not organized, or structured, then it often digresses into an unusable state. Many software architectures will force developers to adhere to a certain standard, thus making it easier to find errors in the code.
- **Modularity:** Along the same lines as the first two points, software architectures generally promote modularity. A software system exhibits modularity if it appears in smaller components that share common functionalities; as mentioned in Section 9.3, modularity may be fine-grained or large-grained. The class structure itself is a good example of modularity. Software architectures often define different subsystems and/or modules that interact with each other.
- **Scalability of the code:** Although it is not ideal to start a software project with the idea that it is going to change along the way, it often happens. In fact, it should be assumed that eventually, something is going to have to change in your code. Either there will be new functionality or existing functionality will need to be modified forcing your code to be changed. Some software architectures focus on this allowing for easy, or easier, modification of the system as a whole.
- **Robustness of the code:** By now you should be catching on, but each of the above items helps promote the others. Software architectures will often take advantage of code organization to help promote robustness in code, including runtime errors and system malfunctions. Many software architectures can leverage its own structure to provide an intrinsic method for error recovery. Although this is not necessarily an explicit goal of architecture, it often happens naturally due to the organization of the code.

While software architecture refers to the fundamental structures and components of a software system, the term is also used to reference a large family of ubiquitous software components that enable the design and construction of other software systems that embody sound software design principles. These software components are also called software architectures. The term *software architecture* is therefore widely used in software engineering as a descriptive concept as well as a noun. Software architecture as a descriptive concept was discussed in Chapter 12 (Section 12.3). The remainder of this section focuses on software architecture as a noun to reference a family of software architecture tools. Due to the expansive nature of software engineering, it is impossible to list all the known software architectures. Instead, we will list and briefly clarify several generic types of architectures so you can understand how they are used in different situations. Among the generic software architectures are the following categories:

- Structural Framework
- Layered Architecture
- Hierarchical Architecture
- Publish/Subscribe Architecture

Figure 25.1 provides a list of commonly used software architecture products that are readily available. You will likely recognize some of these products; know that there are several others not included in the list. The remainder of this section provides a brief clarification on each of the aforementioned categories of software architecture.

Application	Platform	Architecture Classification	Capability Summary
Visual Studio (via form project)	Microsoft Windows	Structural Architecture	Allows parallel development for desktop applications. Separates visual UI and business code.
NetBeans (via form project)	Cross Platform	Structural Architecture	Allows parallel development for desktop applications. Separates visual UI and business code.
Code Igniter	Cross Platform (requires web-server)	Structural Architecture	Provides a robust format in which to write PHP web applications. This will separate the program into visual, business logic and database modules.
Ruby on Rails	Cross Platform (requires web-server)	Structural Architecture	Provides a robust format in which to write Ruby web applications. This will separate the program into visual, business logic and database modules.
Robotic Operating System	Linux	Publish/Subscribe	Provides a mechanism for inter-process communication. Allows for multiple computers to send messages to and from each other and handles random disconnects.
TCP/IP Sockets	Provided by OS to most programming languages	Layered	This functionality is provided by most operating systems and allows an application to communicate directly to another application on a different computer.

Figure 25.1   Some Widely-used Software Architecture Applications

25.1.1 Structural Frameworks

The structural framework is actually both architecture and a framework described later. Because of the ambiguity of the terms, the two are often interchanged for one another. The structural framework forces the software developer to place specific portions of code in specific places. Consequently, this approach tends to limit some use of the programming language to ensure good coding practice is adhered to. However, this form of architecture is useful in producing code that is maintainable and modularized. This style of architecture also assists in breaking up code implementation into parallel paths. Usually these architectures allow for some degree of scalability but this is more of an implicit benefit of the modularity.

Figure 25.2 provides an example of structured framework that can be found in Microsoft Visual Studio, if you create a form application. Here the code is broken down into two parts, the graphical user interface and the business logic for your program. This is a simple example of a structural framework; we will explore a more advanced form later this chapter.

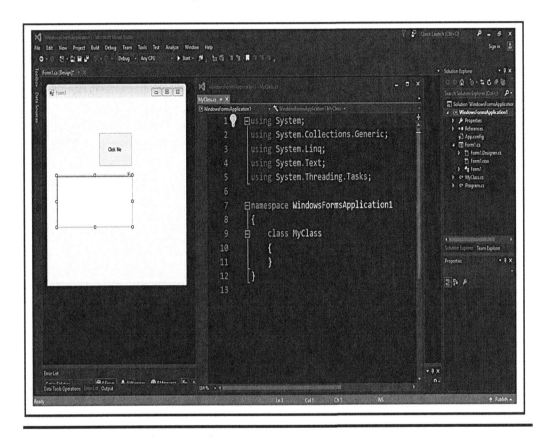

Figure 25.2   A Microsoft Form Project

25.1.2 Layered Architecture

The layered architecture was made famous by the OSI/ISO and TCP/IP network models (Figure 25.3 left). As mentioned in Section 12.3.1, the layered architecture is a common approach to creating complex software systems such as operating systems and communication protocols. Each layer will do some narrowly-defined portion of the overall (bigger) task. The layers will communicate either with the layer above or below itself. The implementation for each layer is independent of the entire system as long as the interface with its neighboring layers does not change. Therefore, should a specific layer need to be updated or replaced entirely, it can be done without disturbing the other layers.

This architecture is also used to modularize complex software problems to more manageable sub-problems. This architecture can also be scaled quite easily by adding additional layers on top, or replacing various layer implementations (Figure 25.3

right). Overall, this architecture will usually result in very robust, manageable systems.

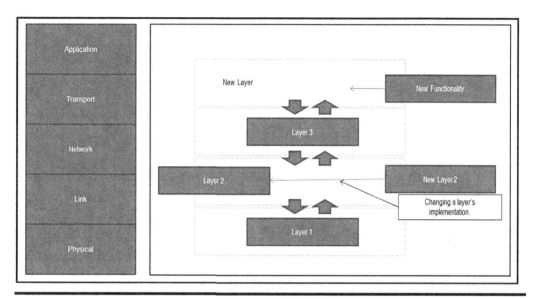

Figure 25.3   The TCP/IP Project (left) and the Generic Layered Architecture (right)

25.1.3 *Hierarchical Architecture*

As the name implies, the hierarchical architecture facilitates design and development via hierarchical structures with classes. The approach benefits from the fact that it is commonly used as the primary organization technique for software systems. The approach also aligns naturally with well-known benefits such as inheritance, code re-use, and polymorphism. This architecture is typically more abstract at the top-level (for most generic functionalities), and the most concrete at the bottom-levels (for most specific functionalities). There is no limit to the number of layers that may appear between the topmost level and the bottommost level; this is normally dictated by the complexity of the software system and design approach.

Another significant spinoff stemming from related principles of inheritance, code reuse, and polymorphism is low maintenance effort. This is particularly important in complex software design scenarios, since the level of duplicate coding is kept at a minimum. If you have well-designed hierarchical inheritance structure in place, this will significantly minimize your effort at fixing bugs. If careful planning is done at the beginning of the project, this architecture can be quite scalable. Software extensions can be achieved by either adding a new leaf class(es) or adding a new code segment(s) in the upper classes.

A common area that takes advantage of aforementioned principles is the video game industry (see Figure 25.4). Here, a vast amount of the code can be implemented in the inherited class(es), thus reducing the programming effort to create a new character or object in the game. Moreover, developers can reuse previously tested components (review Section 13.5 on code reuse).

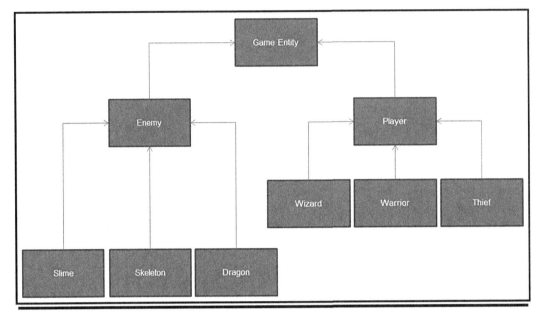

Figure 25.4 **A Typical Hierarchical Structure for a Video Game**

25.1.4 Publish/Subscribe Architecture

The publish/subscribe architecture is less known than the previous three. Nonetheless, there are scenarios for which it is quite relevant. The approach allows different nodes (often representing separate processes) to broadcast or publish information. Other nodes can listen, or subscribe, to this information. A node can both publish and subscribe should it wish to. This publish/subscribe strategy allows software systems to be broken down into smaller components and reduce the inter-dependency between nodes.

Most cases of the publish/subscribe architecture mandate very strict formatting for the messages and type of information that is exchanged among participating nodes. However, there are typically very little other restrictions for the programming of the node itself.

There are several benefits to the publish/subscribe architecture. For one, the approach is distributed, and when coupled with TCP/IP, can encompass multiple computers simultaneously. Secondly, the approach modularizes the problem into smaller, easier tasks. Finally, with appropriate planning, this architecture can be very robust and recover from node failure.

The publish/subscribe architecture is most widely used in the areas of sensor networks and robotics; in fact, this approach tends to dominate these two problem domains. Figure 25.5 shows a graph for a problem scenario requiring this approach. The problem scenario here is the ROS (Remote Operating System) architecture.

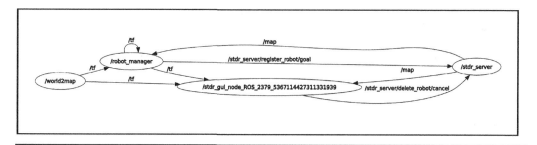

Figure 25.5   Graph Visualizing the ROS Publish/Subscribe Architecture

25.2 Software Frameworks

A software framework is an abstraction that provides general family of functionalities for some purpose but allows the developers to adopt those functionalities and overwrite code behavior whenever necessary. The framework provides resources that can be used in subsequent software development. Among the resources that may be included in a framework (in any combination) are the following: application programming interfaces (APIs), support programs and/or classes, compilers, tool-sets, and code libraries. Essentially, a framework can be anything that extends the functionality of a programming language, meaning everything from a structured framework, to a third-party library. Among the desirable features of software frameworks are the following:

- The framework is designed to be ubiquitous, generic, and reusable.
- The framework contains resources that make software development easier.
- The framework will be implemented, usually specific to a particular language and application domain (for example Web-application, database, graphics, etc.).

The main difference between a framework and an architecture is that a framework does not necessarily have to dictate the code structure. Also, architectures are technically a blueprint for the code and do not have to be implemented. With that said, if someone designed an architecture that provided generic functionality and then implemented it in such a way the user could overwrite components, it would be both an architecture and a framework.

Learning a framework requires an initial investment of time and effort. Moreover, initial use of the framework may result in the code becoming bulkier than development without the framework—a phenomenon called *code-bloat*. However, once that initial learning curve has been negotiated, use of the framework results in increased efficiency and productivity.

There are several software frameworks that are available in the software engineering industry. Among the main categories are the following: structural frameworks; frameworks that augment functionality of an existing language; and frameworks that make existing functionalities easier to use. Figure 25.6 identifies some of the commonly-used frameworks (but be aware that the list provided is by no means exhaustive). The remainder of this section provides a brief overview of each framework category.

Application	Applicable Language(s)	Framework Classification	Capability Summary
Graphics (OpenGL/Glut)	Most C/C++ Compilers	Improve Existing Functionality	Provides an easier interface with OpenGL, allowing for quicker deployment of 3D applications.
Boost C++	C++ Compilers	Augmenting functionality of an existing language	Allows for more versatile data-structures and algorithm native support in C++.
Code Igniter	PHP/Webserver	Structural Framework	Provides a robust format in which to write PHP web applications. This will separate the program into visual, business logic and database modules.
Jenga	Python/Webserver	Structural Framework	Provides a robust format in which to write Python web applications. This will separate the program into visual, business logic and database modules.
Ruby on Rails	Ruby/Webserver	Structural Framework	Provides a robust format in which to write Ruby web applications. This will separate the program into visual, business logic and database modules.
Java API	Java	Improve Existing Functionality	The Java API provides extensive implementation of different data-structures, functionality and class types. This library makes programming on Java easier, since the programmer does not have to re-create the wheel.

Figure 25.6   Some Widely used Frameworks

25.2.1 *Structural Frameworks*

The structural framework is the most defined framework. This particular framework provides added functionality to existing languages by dictating how the user will organize their code (thus fulfilling the role of architecture as well). This framework is very common in different languages where the programming language may be lacking in providing coding standards for software developers. This particular framework will emphasize modularity and software maintenance. The structural framework is commonly used in web-applications such as Ruby on Rails (Ruby), Jenga (Python), and Code-Igniter (PHP).

25.2.2 *Augmenting Functionality of Existing Language*

This type of framework provides additional/missing functionalities to a particular language. Occasionally these frameworks become part of their native language in a later version. These frameworks are more prevalent with older programming languages, where software paradigms shifted and the language had to adapt. A perfect example would be the Boost C++ library. Originally, it was a 3rd party library but it gained so much popularity that it became part of the C++ standard library.

25.2.3 *Making Existing Functionality Easier*

This third type of framework is the broadest and most widely used. This framework makes existing programming functionalities easier to use. Included in this category are API's, libraries, or wrapper functions that abstract particularly difficult function calls. While it is not necessarily introducing anything new to the programming language, they are often more convenient to use versus the built-in facilities. A good example of this would be the GLUT library to OpenGL. OpenGL is a 3D rendering API for C/C++ and technically speaking you do not need any other libraries to render 3D objects to the screen. However, GLUT is a supplemental library that abstracts some of the more difficult elements in OpenGL and makes 3D programming significantly easier.

25.3 The Model-View-Controller Framework

As a case in point, one of the most well-known frameworks, which is also a software architecture, is the Model-View-Controller (MVC) framework. This framework forces the programmer to write their code in one of three places. As the name suggests the three areas are model, view, and controller. The MVC framework is very common on web-applications and multiple versions have been written across different programming languages. The most common languages to employ this framework are python, Ruby, and PHP.

25.3.1 Properties of the MVC Framework

Below are the three areas, or components of the MVC architecture:

- **Model:** This component contains code that will interact with a database or handle data. All reading, writing, and editing of stored data will be implemented in one or more model(s).
- **View:** This component deals with what the user sees. This will contain all code for the user interface and provide a mechanism for the user to interact with the system. These components are often coded a little more loosely than the other two as they often have HTML elements intertwined with code.
- **Controller:** This component acts as the coordinator between the database and the view. The controller will also perform business logic of the program.

The flow of an MVC program generally follows the procedure summarized below and illustrated in Figure 25.7:

1. User request or sends data which is received by the controller.
2. The controller will parse input data, possibly error check, and run any business logic necessary.
3. If necessary, the controller will send or request data from the model and perform any logic that is required.
4. The controller will then create a data-structure, often a dictionary, for the response to the user.
5. The controller will then send this data-structure to the view, which in turn, will generate the user interface for the end user.

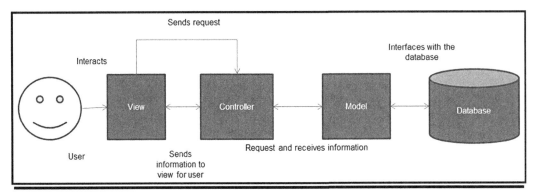

Figure 25.7   An Example of a Typical MVC Framework

There are several different names for the MVC framework, each with its own idiosyncratic features. Among the commonly used names are the following:

- Hierarchical-View-Adapter
- Model-View-Presenter
- Model-View-Viewmodel
- Presentation-Abstraction-Control

25.3.2 Benefits of the MVC Framework

Working with the MVC framework leads to a wide range of benefits. This subsection highlights the main benefits in three categories—functionality, development constraints, and modularity.

Functionality: The MVC framework provides a simple mechanism to allow for a project to scale and increase in functionality. Normally, only a small set of functions are required for the model, (CRUD—Create, Read, Update and Delete functions) and most of the time you will not need to add any functions to the model unless it is for additional convenience(s). Adding new components of the interface usually means creating new view files; this prevents the user interface components from becoming monolithic. Finally, adding new functionalities may require corresponding adjustments for the controller. In the worst-case scenario, the controller may become bloated if the programmer is not careful, but still more manageable than most alternatives. Also, in the event that a project becomes too large or unwieldly, the MVC framework typically allows for multiple controller objects.

Development Constraints: Another benefit of the MVC framework is the fact it restricts how a program can be written. At first glance, this would appear to be a disadvantage. However, as mentioned in Chapter 9 (Section 9.5), having enforced software development standards is a requisite for serious and successful software engineering. As stated earlier, MVC is a common framework for scripting languages, especially the soft-type languages that can be written in line with HTML. With the remarkable advance of Web technology, programmers have come to the realization that not having any restriction on where the code was placed, resulted in the proliferation of bad coding practice and unmanageable projects. This recognition has led to widespread acceptance of the MVC framework.

Modularity: The MVC architecture provides natural modularity to your project. A major component of software engineering is trying to find the balance between modularized code and convenience. You may have seen a project that is monolithic, or one with redundant levels of inheritance serving any real purpose. The MVC takes the guesswork out—you get three major categories of code, and usually, the model is rarely touched after it is initially created. Use of the MVC framework therefore capitalizes on re-use without requiring an unwieldy inheritance tree. Because of this, any application built with an MVC framework can instantly be parallelized between three groups of developers. The database team can work on the model, a second team can work on the controller, and the user interface team can work on the views. The only thing that must be agreed upon is how the data will be interchanged between the three modules.

25.4 Software Patterns

Software patterns (also called software design patterns) are similar to frameworks, providing a generic solution approach to a specific problem domain. However, software patterns are typically not implemented but serve as template algorithms for the programmer. Also, software patterns are usually much smaller in scope than frameworks, focusing on common coding problems in the specific domain of interest.

As an aspiring or practicing software engineer, you may choose to develop your own software pattern. Formally speaking, a software pattern should be accompanied by the following five areas of documentation:

1. A meaningful name to identify the pattern
2. A detailed description of the related problem to which the pattern may be applied
3. Categorical description of the solution, including all of the components and their relationships
4. Description of how the pattern may be used
5. Description of the pros and cons of using the pattern

There are over 20 different known software design patterns. Rather than trying to cover them all, this chapter observes that design patterns may be categorized in the following four ways: creational patterns, structural pattern, behavioral patterns, and concurrency patterns. Figure 25.8 provides examples of popular design patterns. This is followed by a summary of each category of software patterns.

Design Pattern	Pattern Classification	Capability Summary
Factory Pattern	Creation Patterns	Initializes objects optimally. Allows users to create many objects with just one type of class.
Lazy Initialization Pattern	Creation Pattern	Quick object initialization (lazy initialization). Allows an object to be instantiated but does not necessarily have to initialize all components until they are needed.
Adapter Pattern	Structural Pattern	Allows an intermediate class to act as an interface between two other classes.
The Façade Pattern	Structural Pattern	Using a class to simplify function calls of a more complex class to improve ease of use.
Thread-pool Pattern	Concurrency Pattern	Useful for multithreaded applications. Minimizes computational expense by creating the threads up front and assigning them to task when needed.

Figure 25.8   Some Widely Used Software Patterns

25.4.1 Creational Patterns

Creational patterns focus on instantiating objects in the most suitable way for a given set of circumstances. There are several types of creational patterns; following are three members of this family:

- Factory Pattern: In this design pattern, there is an object whose entire purpose is to create new objects. The factory pattern will allow the user to call a function or method which will return different types of objects depending on the parameters specified.
- Abstract Factory Pattern: This is an extension of the factory pattern. Here, an abstraction (typically via an interface) is created, and different factories are spawned from the abstraction. This allows a greater level of flexibility as the programmer can now create all possible objects from the pattern with a single call.
- Lazy Initialization: This pattern is employed when an object can be created but elements of the object are not initialized until they are needed. This can greatly reduce resources for creating objects, especially if there are components that may not be called.

25.4.2 Structural Patterns

Structural patterns attempt to make an object's relationship with other objects easier to design. Remember that when we talk about object relationships we are talking about object-object connection via inheritance, implementation, or aggregation. Three types of structural patterns are described below:

- Adapter Pattern: This software pattern (also called a wrapper or translator pattern) facilitates using the interface of an existing class as another interface, thus providing flexibility to the developer. The effect of the adapter pattern is to allow normally incompatible classes to work together by converting the interface of one class into another form. The ultimate objective is polymorphic behavior.
- Decorator Pattern: This software pattern extends the functionality of a class dynamically at run time. This is achieved by implementing a wrapper class around a concrete class. The wrapper class can then instantiate other sub-classes to forward the new functionality to the intended target object(s).
- Façade Pattern: This pattern creates an interface that masks a more complex implementation of the code. This strategy greatly reduces the complexity of the current code by allowing the developer to simply use the façade pattern instead of trying to design a more complex family of internal classes. The façade pattern may even be interfacing an internal class library.

25.4.3 Concurrency Patterns

The concurrency patterns deal solely with multi-threaded applications. As the names imply, these patterns deal with timing problems of multiple threads.

- Lock Pattern: This design pattern allows a thread to place a lock on a resource, thus preventing other threads from accessing it.
- Read/write Lock Pattern: This pattern is an extension of the lock pattern, but allows multiple concurrent accesses to a resource if, and only if, the threads are only reading information. Should a thread need to write data it will have to have exclusive access to that resource.
- Thread-Pool Pattern: Instantiating many threads throughout the life-cycle of the program can be costly. Many times, threads are quickly finished and the destroyed. This pattern creates a set of threads that will handle incoming requests; the cost of instantiating a thread only occurs when the program is started.

25.4.4 Behavioral Patterns

Behavioral patterns focus on how communication occurs within a set of objects and how to improve communication with that family. Following is a description of three members of this family:

- Publish/Subscribe Pattern: We have already discussed this concept in the section on architecture (review subsection 25.1.4). A publish/subscribe pattern categorizes topics via labels and does not force the sender (publisher) to know who is listening (or subscribing). The ROS architecture would be an implementation of this pattern.
- Chain-of-Responsibility Pattern: This design pattern allows a sender to be capable of sending one command (or message) and not worry about which receiver it is being sent to. Put another way, a client is able to send a request to the chain of objects (represented by the pattern) for appropriate processing. There will be a list (or chain) of receivers and each one will have an opportunity to view the information. If a receiver is incapable of handling the message it simply passes it on to the next receiver. This approach is applicable in scenarios where any combination of the following criteria exist: it is desirous to decouple sender from receiver of a request/message; multiple runtime objects are candidates for handling a request; it is not desirable for handlers to be specified in your code; or the developer wants to issue a request to one of several potential receivers.
- Mediator Pattern: A mediator pattern allows objects to communicate with each other without being directly coupled. A single object will encapsulate all of the necessary communication functionalities; all other objects will communicate directly with it. This approach reduces pressure on the developer to communicate directly with a wider range of different objects in the immediate code.

25.5 Summary and Concluding Remarks

This chapter has provided a brief overview of software architectures, frameworks, and patterns as three families of building blocks that are used in contemporary software design and development. Here is a summary of what has been covered:

Software architecture refers to the fundamental structures and components of a software system. The term is also used to reference a large family of ubiquitous software components that enable the design and construction of other software systems that embody sound software design principles. These software components are also called software architectures. The term [software architecture] is therefore widely used in software engineering as a descriptive concept as well as a noun.

Focusing on software architecture as a noun, four categories of software architecture covered are as follows:

- Structural Framework—forcing the software developer to place specific section of coding in specific areas
- Layered Architecture—requiring the software engineer to design and develop software systems in specified layers
- Hierarchical Architecture—facilitating the design and construction of software systems via hierarchical structures with classes
- Publish/Subscribe Architecture—facilitating the design and construction of software systems based on the transmission of messages among participating nodes

A software framework is an abstraction that provides general family of functionalities for some purpose but allows the developers to adopt those functionalities and overwrite code behavior whenever necessary. The framework provides resources that can be used in subsequent software development. Among the resources that may be included in a framework (in any combination) are the following: application programming interfaces (APIs), support programs and/or classes, compilers, tool-sets, and code libraries.

There are several software frameworks that are available in the software engineering industry. Among the main categories are the following:

- Structural Frameworks—providing additional functionality to a specific programming language by dictating how the user will organize their code
- Frameworks that augment functionality of an existing language
- Frameworks that make existing functionalities easier to use

As a case in point, one of the most well known frameworks, which is also a software architecture, is the Model-View-Controller (MVC) framework. This framework forces the programmer to write their code in one of three areas—model, view, and controller. The MVC framework is very common on web-applications and multiple versions have been written across different programming languages. The most common languages to employ this framework are python, Ruby, and PHP.

Software patterns (also called software design patterns) are similar to frameworks, providing a generic solution approach to a specific problem domain. However, software patterns are typically not implemented but serve as template algorithms for the programmer. Also, software patterns are usually much smaller in scope than frameworks, focusing on common coding problems in the specific domain of interest. Four categories of design patterns are creational patterns, structural pattern, behavioral patterns, and concurrency patterns.

Creational patterns focus on instantiating objects in the most suitable way for a given set of circumstances. There are several types of creational patterns; three members of this family factory patterns, abstract factory patterns, and lazy initialization patterns.

Structural patterns attempt to make an object's relationship with other objects easier to design. Three types of structural patterns are adapter patterns, decorator pattern, and façade pattern.

Behavioral patterns focus on how communication occurs within a set of objects and how to improve communication with that family. Three members of this family are publish/subscribe patterns, chain-of-responsibility patterns, and mediator patterns.

The concurrency patterns deal solely with multi-threaded applications. Three members of this family are lock patterns, read/write lock patterns, and thread-pool patterns.

Figure 25.9 shows that there is overlap among architectures, frameworks, and patterns. Because of this, there is much ambiguity over the terms and their boundaries.

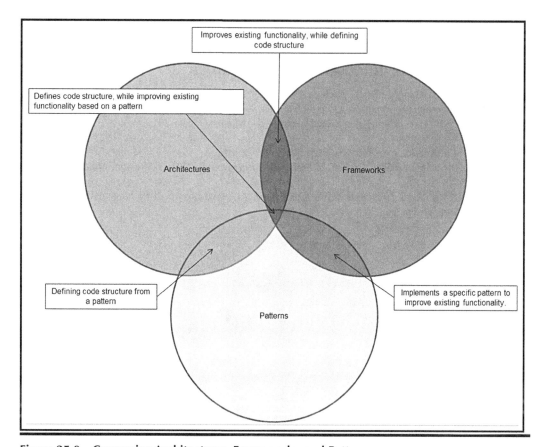

Figure 25.9 Comparing Architectures, Frameworks, and Patterns

25.6 Review Questions

1. Define software architecture. Identify and clarify four categories of software architectures.

2. Define software framework. Identify and clarify three categories of software frameworks.

3. Define software pattern. Identify and clarify four categories of software patterns.

4. What is the MVC stand for?

5. What interacts with the database in an MVC framework?

6. What interacts with the end user in an MVC framework?

7. While doing a difficult project for a class, your friend Bob designs an elegant solution for the problem using class diagrams. Bob has not implemented anything but can demonstrate its functionality on paper. Bob claims he has created a framework, is this true? Why or why not?

References and/or Recommended Readings

[Gamma et al. 1995] Gamma, Erich, Richard Helm, Ralph Johnson, and John Vlissides. 1995. *Design Patterns: Elements of Reusable Object-Oriented Software*. Boston, MA: Addison-Wesley.

[MacIntyre et. al. 2011] MacIntyre, Peter, Brian Danchilla, and Mladen Gogala. 2011. *Pro PHP Programming*. New York, NY: Apress.

[Pitt 2012] Pitt, Chris. 2012. *Pro Php Mvc*. New York, NY: Apress.

[Sommerville 2016] Sommerville, Ian. 2016. *Software Engineering*, 10th ed. Boston, MA: Pearson. See chapter 6.

[Wikipedia 2019] Wikipedia. 2019. Software Design Pattern. Accessed July 16, 2019. https://en.wikipedia.org/wiki/Software_design_pattern

Chapter 26

New Frontiers in Software Engineering

As the driving force behind computer science (CS) and information technology as well as the catalyst in many other disciplines, software engineering continues on its inevitable march forward. There are several amazing new frontiers on the horizon for the discipline. This chapter provides a survey of several new and exciting directions the field will be moving in. The chapter advances under the following subheadings:

- Empirical Software Engineering
- Data Science
- Bioinformatics
- Machine Learning
- Game Design
- Augmented and Virtual Reality
- Internet of Things (IoT)
- Cloud Computing
- Summary and Concluding Remarks

26.1 Empirical Software Engineering

As you have seen throughout the course, software engineering starts with observation of problems faced by human beings. This is followed by appropriate investigation/research, design of algorithmic solutions to these problems, automating the solutions, testing and refining the solutions, and reporting on the related activities and findings. The software engineering activities conform to the SDLC (review Chapter 1).

Empirical software engineering (ESE) emphasizes the observational aspect of the discipline. Here is a working definition: ESE concerns itself with observation of software engineering artifacts, as well as empirical validation of known software

DOI: 10.1201/9780367746025-32

engineering theories, tools, methodologies, and assumptions. Broad areas of focus include software evolution, software maintenance, and exploration of software repositories. For more information, see [Gueheneuc & Khomh 2018] and [Basili et al. 2006].

26.1.1 Rationale for Empirical Software Engineering

Recall that in Chapter 1 (Section 1.9), the fundamental challenge of software engineering—the software engineering dilemma—was introduced. Since the 1990s, software engineering has exploded in the diversity of different areas and applications. In many circles, the term software engineering is been incorrectly used to mean software construction. To make matters worse, the absence of appropriate software engineering standards often leads to the proliferation of software systems characterized by faulty design, substandard coding, and negligible or shoddy documentation (review Section 9.5). This unfortunate circumstance typically results in increased difficulty and cost of maintenance, as well as difficulty in accurately evaluating the work, or providing precise and meaningful feedback.

ESE confronts this problem by forcing a more scientific approach to the discipline. Data are collected and analyzed to evaluate the effectiveness or usefulness of a set of related software engineering artifacts for a given project. These data points include, but are not limited to case studies, surveys, error count, lines of code, rate of additional features added, amount of resources devoted to maintenance, quality evaluations, user satisfaction evaluations, etc.

26.1.2 Approaches to Empirical Software Engineering

ESE mandates that a software activity or resource must be proven to validate or improve an existing theory, methodology, tool, or assumption. Therefore, the burden-of-proof falls on the software engineers introducing the activity or resource. At first glance, this requirement may seem overly restrictive. However, remember that ESE techniques can be applied to any portion of the software engineering experience (i.e. the SDLC), not just the coding itself. This includes, but not limited to source control, bug reports, documentation, or other aspects of a project.

Once the development of the new resource is complete, the software engineer must then take steps to engage in evaluations that demonstrate the usefulness of the resource. Three common approaches to ESE are controlled experimentation, case studies, and surveys.

Controlled Experimentation: Running a controlled experiment is one of the strongest forms of evidence. However, the software engineer must be able to isolate what changes between the control and the experiment to determine its effect. Although this method typically provides the strongest evidence, it is not always possible to frame an experiment in this fashion. Sometimes project size, time, and different populations of people will affect the outcome. Likewise, there may be too many uncontrollable independent variables.

Case Studies: Another strategy that may be employed is the case study. The case study in software engineering is a procedure aimed at investigating and observing phenomena in a particular methodology. This experiment type is often more applicable in the general case but can take as much time to create as a controlled test.

Careful attention must be given in defining what use cases to analyze, determine how data should be collected, and finally collecting the data. Often, this methodology will produce both quantitative and qualitative results, thus forcing the researcher to analyze the data before reporting.

Surveys: Another common over-arching category of testing software engineering techniques is the survey. Here, the characteristics and effect of the software engineering resource will be evaluated across a large population of individuals. This approach relies on the data averaging out to reveal evidence. The primary problem with this mechanism is the large amount of subjectivity it has. Ideally, the questionnaire used for the survey would be carefully designed to minimize biases. Along with this problem, the population surveyed may deliver different results. For example, does a software engineer only survey experts in the field or do they include common users? Are the individuals being surveyed excited about the project or simply participating due to obligation? These and many more issues can affect a survey and must be considered.

26.2 Data Science

Data science is another burgeoning area of software engineering. Data science is a discipline that employs scientific methods, algorithms, and software engineering methodologies to extract and synthesize information from assorted (structured and unstructured) data sources. Being multidisciplinary in nature, data science combines skills from a number of areas including software engineering, programming, mathematics (forecasting and statistics), database systems, data warehousing, artificial intelligence (AI), and machine learning (ML). The ultimate objective is the provision of information that leads to improved decision making.

For a more detailed introduction to the data science terrain, see [Saltz & Stanton 2018]. Following is a brief discussion on selected areas that often come under the umbrella of data science: big data, data warehousing, and data mining.

26.2.1 Big Data

Big Data refers to the field where large data sets are stored for subsequent information extraction from them. It is important to note that big data deals with a much larger amount of data than a conventional application. So, what size constitutes 'big data'? This tends to be a moving target; the threshold typically occurs when a current system can no longer manage the volume of data required. However, the threshold usually falls somewhere between hundreds of gigabytes and hundreds of terabytes.

It should be noted that big data does not simply mean storing massive amounts of data. If this was the limit to the scope of the field, offline drives would be satisfactory. Big data implies storage as well as usage of the large data sets of interest. As you will see below, big data has also impacted the related field of ML with respect to inferring new information that may otherwise be missed.

Since more and more devices are being networked (see Section 26.7 on the Internet of Things) more and more devices can send information that is stored. In fact, in this information age, we are creating it has been estimated that we create about 2.5 quintillion bytes of data per day [Marr 2019]. Such massive datasets have their own set of challenges. Software systems and/or applications have to run multiple threads and processes in parallel in order to meet the demand for storage; and retrieval and information access must be very efficient. Because of this, conventional relational databases are not suited to handle such demands. Big data is one strategy for meeting this challenge.

26.2.2 Data Warehousing

A *data warehouse* is a typically read-only database that is constructed from multiple *source databases*, and designed for query and analysis rather than transaction processing (see [Foster 2016], Kimball & Ross 2013], and [Oracle 2015]). Data warehouses are designed to store summarized information drawn from their data sources and may conform to different architectural structures. The process of updating the data warehouse is called the *extract-transform-load* (ETL) process.

Data warehouses are used to store large volumes of information that may be subsequently accessed by various information retrieval systems and/or tools. Discussion of such resources is beyond the scope of this course; suffice it to say that an online search engine is an excellent example of a retrieval system.

26.2.3 Data Mining

Data mining (also called *knowledge discovery*) is the process of searching through and analyzing large volumes of (often) unstructured data with a view to identifying patterns and relationships, and presenting information in a manner that is consistent with user requirement. Data mining is often applied in the context of big data and/or data warehouses. To achieve its objective, data mining often employs statistical and/or descriptive mathematical models. Although data mining algorithms are not necessarily computationally heavy, they often place a heavy load on the data stores due to the high volume of read/write requests. Leading software engineering firms such as Oracle and IBM provide their data mining tools. However, there is an abundance of tools among smaller enterprises as well.

26.3 Bioinformatics

Another burgeoning area of CS that is driven by software engineering is *bioinformatics*. As the name suggests, bioinformatics relates to the collection and analysis of complex biological data. According to [Luscombe, Greenbaum, & Gerstein 2001], bioinformatics may be defined as "the application of computational techniques to understand and organize the information associated with biological macromolecules." Since bioinformatics covers everything from DNA sequencing to protein folding, the amount of information generated by this field is astounding.

Bioinformatics is an interdisciplinary field that is showing up in many institutions of higher learning. For obvious reasons, it is often sponsored and administered by the CS departments at these institutions. Bioinformatics draws from the following related fields:

- Computer Science
- Chemistry
- Biochemistry
- Biology
- Mathematics and Statistics
- Engineering

Bioinformatics relies on advances made in the CS-related areas of software engineering and big data. The field itself is broken down into the following five main areas:

- DNA sequence analysis
- Gene and protein expression
- Study of cellular organization
- Determining the three-dimensional structure of a protein
- Study of biological networks

Indeed, the field of bioinformatics is quite interesting as it advances at a rather quick pace; this trend is expected to continue. Additional references are provided in reference section (see [Jiang, Zhang, & Zhang 2013] and [Diniz & Canduri 2017]).

26.4 Machine Learning

Like big data, ML is a datacentric area of CS that involves the study of sophisticated algorithms and statistical models that enable software systems to carry out specific activities in the absence of explicit instructions. Rather than being instruction-driven, ML relies on patterns, known rules, and inferences to influence behavior (see [Nevala 2017] and [Wikipedia 2019a]).

ML may be construed as the logical progression from AI and big data. An ML system will analyze data to detect patterns similar to other known data. The ML system will then use those patterns to create an algorithm and act upon it. The *learning* aspect of ML refers to the act of developing an appropriate algorithm for a data set, and then implementing the algorithm.

26.4.1 Different Algorithms Used in Machine Learning

There are several different categories of ML algorithms that are used today. Here is a summary of the most common ones:

- **Supervised Learning:** The algorithm is trained on a data set created by humans to "learn" the correct response to any given input. The most notable example of this approach is the neural network.
- **Unsupervised Learning:** This represents a difficult expansion from the first category as this type of algorithm takes data that contains only inputs, must apply a model, and determine what the outputs should be. Data science is very important with unsupervised learning as the models used on the inputs can

create solutions that are not correct. An example of this type of ML would be clustering or cluster analysis.

- **Reinforcement Learning:** This category of ML attempts to train itself with input data to maximize some form of metric. Mathematically, the algorithm will provide more positive feedback to better-performing algorithms. This particular set of ML algorithms will often run many different tests at the same time. The vast majority will utterly fail and be culled from the tests. Those that do perform above a threshold will be tweaked and tested in subsequent iterations. The quintessential algorithm for reinforcement learning is the Monte Carlo.

- **Anomaly Detection:** This category of ML algorithms will look for unusual results or outliers in data. In a way, this objective is the opposite of supervised learning, where the goal is to train to place data within a known set of outputs. However, in this situation, the software engineer is looking to see if it does not fit the normal pattern. This category can also be used to analyze noise and deviations and is most commonly used in bank fraud detection. This category may use algorithms from supervised or unsupervised categories again with the different goal in mind. The K-nearest neighbor is a common algorithm for detecting outliers.

- **Rule-based** ML: In the 1980s, it was common for AI to simply be a set of rules that were handcrafted by the computer scientist. These were a very low level "if-something-then-do-something" style programming. Rule-based ML attempts to read data, classify it, and then derive its own reactive rule set.

Still, there are other ML algorithms that are emerging. This trend may likely continue as the relatively young field gains maturity and prominence.

26.4.2 Machine Learning Today

Nowadays, ML is used subtly but in a wide range of software applications that are used by the consuming public. Here are a few examples:

- Personal assistant applications ML to improve human-computer interaction. Examples include Cortana, Siri, Alexa, etc.
- Video surveillance use ML to help identify oddities.
- Search engines have long used ML to improve their search results.
- Social media ML to determine what people are paying attention to.
- Even email use ML to prevent spam or phishing messages from getting to the inbox.

So, while ML often does not get into the spotlight of CS, it is almost everywhere within the field. With modern processors and cheaper RAM algorithms that were not feasible to run in the 1980s, ML can now be incorporated into devices such as mobile phones. Many of the conveniences experienced from contemporary digital devices are probably directly linked to ML.

26.5 Game Design

Game design has become a very popular field of CS over the past two decades. Game design involves the combination of art and science to create a simulated software system for the purpose of facilitating entertainment, exercise, education, or other experimental purposes [Wikipedia 2019b]. Despite its name, and as the definition shows, game design addresses more than merely creating video games. It involves all forms of programming where graphical objects are interacting with each other.

In game design, the user (player) can often control some outcome of the program. This means game design principles can be used for other scenarios such as (but not limited to):

- Interactive simulation
- Actual game design
- Serious games
- Training programs
- Visual effects for amusement parks

26.5.1 The Peculiar Problem of Game Design

Game design, in general, is significantly different than conventional software engineering fields. Game design poses the questions:

- Does the game adequately simulate the intended problem?
- Is it fun?

In game design, the user experience is central to the software's mission, and there are more qualitative considerations than quantitative. Due to this phenomenon, it is impossible to create a set of requirements without having a prototype to interact with. This means that quite often, the final product is different from the initial conception (and sometimes significantly so). For this reason, game designers often have to decide on the relative importance of the art versus the science involved in each project.

From an end-user perspective, a game's program state is quite often significantly more complex than normal enterprise software state. Each object in the game will most likely have a positional and movement component. More than likely, each object will be changing its position per frame, and many of the objects have to interact with each other should they collide. This makes for a very difficult experience in modeling all possible states of interacting objects in a game. In contrast, consider writing an online business application. The controller element is not trying to destroy the database with a rocket launcher. Instead, they are interacting in a very precise and discrete manner that the software developer defines. This makes game design one of the more challenging programming problems to negotiate.

26.5.2 Challenges of Integrating Software Engineering with Game Design

Due to the aforementioned challenges, many game developers tend to skip sound software engineering principles on their games. While this practice does allow for greater fluidity and creative freedom, the resultant code will be virtually unusable for any sizable project. Therefore, it is crucial to apply sound software engineering principles to game design, whilst finding other appropriate strategies for dealing with the complexities of the endeavor.

In designing a game, there are several problems for which the software engineer will need to focus. Following are five such problem areas:

1. Software requirements should be clearly articulated at the beginning but with the assumption they are going to change.
2. Avoid use of prototype code on the final version unless rigorous software engineering techniques have been applied. This is a common problem that haunts many game developers.
3. The developer will need to re-evaluate the final product multiple times and as early as possible to avoid having to make costly changes to the code.
4. The design should incorporate heavy focus on code-reuse. This will lead to easier maintenance and greater scalability.
5. While using an iterative approach is not necessarily bad, the programmer will need to know when the game construction is finished; otherwise, the project will fall victim to scope creep.

With games and simulations becoming more relevant, future software engineers will need to be prepared to deal with these challenges. Conventionally, game design is considered arduous and difficult, especially in the area of debugging. However, this challenge presents an opportunity to push the field and provide new software engineering methodologies that adapt for these particular problems. The good news is, there is an abundance of good books on the subject matter. One such resource is [Koster 2005].

26.6 Augmented and Virtual Reality

Augmented and virtual realities are technologies that have been around for a while. However, continued advances keep them relevant for the foreseeable future.

26.6.1 What is Virtual Reality?

Virtual reality uses a headset to immerse a user/player in a completely artificial (virtual) world. The user will have no ability to use outside objects as a source of reference, therefore, this has to be done in a safe area. Research has shown that virtual reality possesses certain quality that other human-computer-interaction (HCI) devices do not have (see [Barnes 2019]).

Virtual reality (VR) has a captive audience. When a person uses a virtual headset, they observe only what has been placed in that virtual world. This experience prohibits users from being distracted. They cannot check their email or watch videos on YouTube. All they do is observe and interact in the virtual world that has been programmed. For this reason, VR is emerging as an excellent feature for learning. There are some drawbacks as well. Virtual reality headsets must have a high frame rate or they can induce motion sickness on the user. Another issue to be aware of is that some users may experience eye strain and/or headaches. Also, some people may become nauseous or experience exhaustion after an intense VR encounter.

26.6.2 What is Augmented Reality?

Augmented reality (AR) has some subtle differences from virtual reality, and the application of the technology is vastly different. Unlike, VR, augmented reality does not create a completely new virtual world. Rather, AR adds graphical artifacts to real-world objects, thus creating a new enhanced experience [Wikipedia 2019c]. Many games for the telephone have used AR; an example is Pokémon Go. Augmented reality does not require quite the high framerate as virtual reality does, but like VR, it can cause eyestrain. Another issue with AR is the fact that some devices are not capable of spatial awareness. This means if a graphical artifact moved behind a real-world object, the device would not have any means for occluding the virtual object. More modern AR devices such as the HoloLens and Magic Leap One have the capabilities of detecting their surroundings and allowing the virtual graphics to be occluded by real-world objects. This makes their ability to integrate into the real world much more seamless, thus creating a far more immersive experience.

26.6.3 Difference between Virtual Reality and Augmented Reality; Future Expectations

The main difference between VR and AR is that whereas VR creates a virtual object and often substitutes one's vision, AR enhances the experience with real objects.

While VR has some great potential, AR holds vast untapped potential for the foreseeable future. Many devices are becoming wireless so people will no longer need to be at a desk, they can take their device to the factory floor, to the meeting, warehouse, etc. Because the technology is so relatively new, there are very little conventions on how people should interact with it. This provides the opportunity for software engineers to be as creative with it as they desire.

26.7 Internet of Things

Today, internet connection is almost a necessity for life; the internet is ubiquitous and pervasive. Without it, a lot of devices simply do not work. As embedded systems become more powerful, more and more devices are including a Wi-Fi connection, so they can communicate with other personal computers and/or telephones. These devices can serve both as sensor and controllers for many conveniences around the home or workplace. The modern-day software engineer will therefore have to be prepared to write network applications for these devices.

26.7.1 What is the Internet of Things?

The concept for *Internet-of-Things* (IoT) has been around since 1982. However, it would not become widespread until about 2008. The term is used to refer to the vast global collection of interrelated devices, resources, and people, with the capacity to transfer information over a network. Technologies have advanced enough to the point where it is possible to network almost any electronic device. With advances in powerful mobile smartphones, people can install applications on their mobile phones (just like installation on a computer), and control things over the internet.

A second big revolution has occurred with Wi-Fi. Early Wi-Fi was unstable and fairly slow, especially considering the speed of category 6 cabling. However, as newer generations came out, the Wi-Fi signals have progressively become much more stable, essentially matching cable speed. This development, coupled with the ease of installation, has led to a general shift in attitude by the consumers to use Wi-Fi, making it easier to network devices.

Another area of significant change in embedded systems. Until around about 2005, embedded systems were mainly minimalistic. Consequently, extra concerns such as Wi-Fi or drivers did not enjoy high priority. However, hardware and memory are a lot cheaper than in the past, and it is no longer prohibitive to create hardware with a Wi-Fi connection. This new reality, coupled with accessibility to minimalistic Linux operating system, brings embedded systems much closer to personal computers (PCs) in terms of programming. Nowadays, embedded systems programmers can use a higher-level language (like C/C++ or Java) and use sockets provided by the operating system to deal with the networking portion. Since it has become so much easier to network, new devices are being made every day that will allow people to connect to them from their phones and computers. With these contributing elements, the IoT has emerged and thrived.

26.7.2 Adapting Software Engineering Techniques to Handle Networking from the Start

In the future, software engineers will need to be prepared to network their applications from the start. Conventionally, applications were written to be used offline, and then as an option to permit an online mode. Nowadays, the status quo is the reverse. Most applications require the internet or at least occasional access. Software engineers will be asked to create network APIs for different devices as well as create the socket code for them. They will need to understand sockets, web-servers, and web-services stack. Remember, it is no longer optional to deal with networking, it must be considered from the start.

26.8 Cloud Computing

Cloud computing (also called cloud technology) is a phenomenon that has become commonplace as a progression from a number of earlier groundswells in the related fields of computer networking and operating systems. Two of these developments are *cluster technology* and *grid computing*. Let's start this brief discussion by clarifying these three terms (for more details, see [Foster 2018]).

Cluster technology describes a system infrastructure that allows a conglomeration of computer systems to collaborate and operate as part of a larger whole. The computers within the cluster are referred to as nodes; each node conforms to a communication protocol established and administered by the cluster.

Grid computing is the term used to describe a federation of computer resources from multiple domains to form a coherent and integrated super system. The *grid* qualifies as a distributed system. Grid computing differs from conventional high-performance computing systems such as cluster computing in that grids tend to be more loosely coupled, heterogeneous, and geographically dispersed. Grid computing can therefore be construed as an upgrade of cluster computing.

Cloud computing is a metaphor used to describe the delivery of various computing requirements as a service to a mixed community of end-recipients. A *cloud* represents a complex abstraction that integrates various technologies and resources into a virtual information infrastructure. Cloud computing provides (typically centralized) services with a user's data, software, and computation on an application programming interface (API) over a network (typically the internet). End users access cloud services through a Web browser, a desktop application, or a mobile application.

Cloud computing consolidates on vast reservoirs of essential services (such as data storage and various software services) and making these services available to the consuming public at a much more affordable rate than would otherwise be possible. Due to the need for consolidation of processing power at a massive scale, along with the complexity involved in providing that consolidation, cloud computing is typically provided by leading software engineering companies (such as Google, Amazon, IBM, Oracle, Microsoft, Apple, etc.).

Cloud computing has brought with it a number of conveniences to the consuming public. Among the main ones are the following:

- **Operating System Services**: Thanks to cloud computing, students and enthusiasts can gain access to operating systems that would otherwise not be as attainable. For instance, in a recent course on Operating Systems Design, administered by the current author, students were able to use their cloud accounts to access the IBM System i operating systems, run simulations, and gain useful insights about the system. This was much cheaper than acquiring, installing, and maintaining the very expensive physical IBM System i infrastructure.
- **Database as a Service (DBaaS)**: Through DBaaS, individuals and/or small organizations have the option of creating, managing, and accessing complex databases without having expensive information infrastructure (hardware and/or software) on their local machines. This is done in a seamless manner, giving the illusion that the database is local when in fact it is being accessed remotely.
- **Software as a Service (SaaS)**: In the SaaS model, a conglomeration of software system(s) is hosted and maintained on a cloud facility, and licensed end users are able to access their relevant software system(s) through subscriptions. This seamless setup is much cheaper on the end-user side, avoiding expensive acquisition and maintenance by subscribing individuals/organizations.

- **Infrastructure as a Service (IaaS):** In IaaS, the cloud computing host provides streamlined network infrastructure services (such as servers, partitioning, virtualization, security, backup, etc.) to licensed subscribing clients.
- **Platform as a Service (PaaS):** This aspect of cloud computing facilitates development and deployment of applications without concern for or focus on the complexity of supporting platform or infrastructure. The objective is to promulgate the culture of platform independence for future software development initiatives.
- **Resource Repository:** Cloud facilities are excellent facilities for resource repositories where data and/or software are stored for the purpose of backup or sharing with others. For instance, when working on a software engineering project that involves multiple team members, having a cloud-based repository with assorted resources is extremely useful.

One gets the impression that cloud computing is one of the many noteworthy achievements of the software engineering industry that already has made, and will continue to have lasting impact on twenty-first century lifestyle. It is therefore reasonable to expect that cloud-based services will continue to gain preeminence in the foreseeable future.

26.9 Summary and Concluding Remarks

This chapter examined a number of burgeoning areas of software engineering. Let us summarize what has been covered:

Empirical software engineering (ESE) concerns itself with empirical observation of software engineering artifacts, as well as empirical validation of known software engineering theories, tools, methodologies, and assumptions. Broad areas of focus include software evolution, software maintenance, and exploration of software repositories. Three common approaches to ESE are controlled experimentation, case studies, and surveys.

Data science is a discipline that employs scientific methods, algorithms, and software engineering methodologies to extract and synthesize information from assorted (structured and unstructured) data sources. Being multidisciplinary in nature, data science combines skills from a number of areas including software engineering, programming, mathematics (forecasting and statistics), database systems, data warehousing, AI, and ML. The ultimate objective is the provision of information that leads to improved decision making.

Bioinformatics is an interdisciplinary field that is showing up in many institutions of higher learning. For obvious reasons, it is often sponsored and administered by the CS departments at these institutions. Bioinformatics draws from the following related fields: CS, chemistry, biochemistry, biology, mathematics/statistics, and engineering.

ML involves the study of sophisticated algorithms and statistical models that enable software systems to carry out specific activities in the absence of explicit instructions. The *learning* aspect of ML refers to the act of developing an appropriate algorithm for a data set, and then implementing the algorithm. Five prominent techniques used in ML are supervised learning, unsupervised learning, reinforcement learning, anomaly detection, and rule-based learning.

Game design involves the combination of art and science to create a simulated software system for the purpose of facilitating entertainment, exercise, education, or other experimental purposes. Game design offers great potential in areas of computer simulation, entertainment, and education.

Virtual reality uses a headset to immerse a user/player in a completely artificial (virtual) world. The user will have no ability to use outside objects as a source of reference, therefore, this has to be done in a safe area.

Augmented reality (AR) has some subtle differences from virtual reality, and the application of the technology is vastly different. Unlike, VR, augmented reality does not create a completely new virtual world. Rather, AR adds graphical artifacts to real-world objects, thus creating a new enhanced experience.

The term *internet of things* (IoT) is used to refer to the vast global collection of interrelated devices, resources, and people, with the capacity to transfer information over a network. Technologies have advanced enough to the point where it is possible to network almost any electronic device. With advances in powerful mobile smartphones, people can install applications on their mobile phones (just like installation on a computer), and control things over the internet.

Cloud computing is a metaphor used to describe the delivery of various computing requirements as a service to a mixed community of end-recipients. A *cloud* represents a complex abstraction that integrates various technologies and resources into a virtual information infrastructure. Cloud computing provides (typically centralized) services with a user's data, software, and computation on an application programming interface (API) over a network (typically the internet). End users access cloud services through a Web browser, a desktop application, or a mobile application. Cloud computing consolidates on vast reservoirs of essential services (such as data storage and various software services) and making these services available to the consuming public at a much more affordable rate than would otherwise be possible.

26.10 Review Questions

1. What is ESE? Explain its relevance. Briefly describe three common approaches that are employed in ESE.

2. Provide a working definition of *data science*. What are the disciplines that drive this field? Identify three prominent areas of data science and provide clarification of each.

3. Provide a working definition of *bioinformatics*. What disciplines does bioinformatics draw from? Identify three prominent areas of study in the field.

4. What is *ML*? State and clarify the five categories of ML algorithms covered in the chapter.

5. What is the significance of *game design* in contemporary software engineering?

6. Differentiate between *virtual reality* and *augmented reality*. What is the potential impact of each subfield on contemporary software engineering?

7. What do you understand by the term *internet of things*? Explain its relevance and implications for the future.

8. Define the term *cloud computing*. Identify and clarify four main areas of this field. How has cloud computing impacted contemporary software engineering and lifestyle in general? What are the implications for the future?

References and/or Recommended Readings

[Barnes 2019] Barnes, Stuart. 2019. Understanding Virtual Reality in Marketing: Nature, Implications and Potential. *Implications and Potential* (November 3, 2016).

[Basili et al. 2006] Basili, Victor, Deiter Rombach, Kurt Schneider, Barbara Kitchenham, et al. 2006. *Empirical Software Engineering Issues: Critical Assessment and Future Directions. International Workshop*, Dagstuhl Castle, Germany. New York: Springer.

[Diniz & Canduri 2017] Diniz, W. J. S. and F. Canduri. 2017. Bioinformatics: an overview and its applications. Accessed August 21, 2019. https://pdfs.semanticscholar.org/6d6f/611494102844abeaae12d37a29a353be94df.pdf

[Foster 2016] Foster, Elvis C. with Shripad Godbole. 2016. *Database Systems: A Pragmatic Approach*, 2nd ed. New York, NY: Apress.

[Foster 2018] Foster, Elvis C. 2018. Lecture Notes in Operating System Design. https://www.elcfos.com/lecture-series/index/entry/id/8/title/Operating-Systems-Design. See Lecture 10.

[Gueheneuc & Khomh 2018] Gueheneuc, Yann-Gael and Foutse Khomh. 2018. Empirical Software Engineering. Accessed August 21, 2019. http://www.ptidej.net/courses/log6306/fall16/readingnotes/1%20-%20Gueheneuc%20and%20Khomh%20-%20Empirical%20Software%20Engineering.pdf

[Jiang, Zhang, & Zhang 2013] Jiang, Rui, Xuegong Zhang, and Michael Q. Zhang. 2013. *Basics of Bioinformatics*. New York: Springer. Accessed August 21, 2019. https://courses.cs.ut.ee/MTAT.03.242/2017_fall/uploads/Main/Basics_of_Bioinformatics.pdf

[Kimball & Ross 2013] Kimball, Ralph & Margy Ross. 2013. *Data Warehouse Toolkit Classics: The Definitive Guide to Dimensional Modeling*, 3rd ed. New York: Wiley and Sons.

[Koster 2005] Koster, Ralph. 2005. *A Theory of Fun for Game Design*. Scottsdale, Arizona: Paraglyph Press.

[Luscombe, Greenbaum, & Gerstein 2001] Luscombe, N. M., D. Greenbaum, and M. Gerstein. 2001. What is Bioinformatics? A Proposed Definition and Overview of the Field. Accessed August 21, 2019. http://binf.gmu.edu/ivaisman/binf630/mim01-luscombe-what-is-bioinf.pdf

[Nevala 2017] Nevala, Kimberly. 2017. *The Machine Learning Primer*. Cary, NC: SAS Institute. Accessed August 22, 2019. https://www.sas.com/content/dam/SAS/en_us/doc/whitepaper1/machine-learning-primer-108796.pdf

[Marr 2019] Marr, Bernard. 2019. How Much Data Do We Create Every Day? The Mind-Blowing Stats Everyone Should Read. Accessed August 21, 2019. https://www.forbes.com/sites/bernardmarr/2018/05/21/how-much-data-do-we-create-every-day-the-mind-blowing-stats-everyone-should-read/#31ef322560ba

[Oracle 2015] Oracle Corporation. 2015. Oracle Database for Data Warehousing and Big Data. Accessed December 2015. http://www.oracle.com/technetwork/database/bi-datawarehousing/index.html.

[Saltz & Stanton 2018] Saltz, Jeffrey S. and Jeffrey M. Stanton. 2018. *An Introduction to Data Science*. Los Angeles, CA: Sage.

[Wikipedia 2019a] Wikipedia. 2019. Machine Learning. Accessed August 22, 2019. https://en.wikipedia.org/wiki/Machine_learning

[Wikipedia 2019b] Wikipedia. 2019. Dame Design. Accessed August 22, 2019. https://en.wikipedia.org/wiki/Game_design

[Wikipedia 2019c] Wikipedia. 2019. Augmented Reality. Accessed August 22, 2019. https://en.wikipedia.org/wiki/Augmented_reality

APPENDICES

G

This final division applies principles and methodologies from the previous chapters to a sample software engineering project: Included in the upcoming three chapters are excerpts from the initial system requirements, the requirements specification, and the design specification for an inventory management system.

The chapters are as follows:

- Appendix A: Project Proposal for a Generic Inventory Management System
- Appendix B: Requirements Specification for a Generic Inventory Management System
- Appendix C: Design Specification for a Generic of the Inventory Management System

DOI: 10.1201/9780367746025-33

APPENDICES

G

This final division applies principles and methodologies from the previous chapters to a sample software engineering project: Included in the upcoming three chapters are excerpts from the initial system requirements, the requirements specification, and the design specification for an inventory management system.

The chapters are as follows:

- Appendix A: Project Proposal for a Generic Inventory Management System
- Appendix B: Requirements Specification for a Generic Inventory Management System
- Appendix C: Design Specification for a Generic of the Inventory Management System

DOI: 10.1201/9780367746025-33

Appendix A: Project Proposal for a Generic Inventory Management System

This appendix provides excerpts from a sample project proposal (also called initial system requirement) for a generic Inventory Management System that may be suitable for a small or medium-sized organization. The document includes the following:

- Problem Definition
- Proposed Solution
- Scope of the System
- System Objectives
- Expected Benefits
- Overview of Storage Requirements
- Anticipated Outputs
- Feasibility Analysis Report
- Initial Project Schedule

A.1 Problem Definition

When operating a business, an inventory management system is one of the fundamental needs. Inventory management is critical to any business. The absence of such a system could result in several problems, some of which are stated below:

1. If a business does not know what items are in stock it might lose significant sale opportunities. On the other hand, if the inventory system shows items in stock that are actually not in stock then that is an embarrassing situation where a customer has to wait while the missing (nonexistent) item is being located. This lack of professionalism could most certainly drive customers away.
2. In addition to this, there is also the problem of knowing exactly how much inventory should be kept on hand. Keeping too much inventory on hand is costly, resulting in storage costs, insurance costs, and salaries of individuals to manage/locate inventory. On the other hand, not having sufficient stock on hand is also costly, in that the business could lose significant opportunities for additional sale.
3. Knowing exactly how much stock is on hand is crucial to knowing exactly when items need to be reordered.
4. Inability to properly manage the inventory could severely affect the productivity and profitability of the company.

A.2 Proposed Solution

This project seeks to address this problem by developing a generic Inventory Management System (IMS) that is tailored to the specific needs of any company.

This software system must be portable and platform independent (if possible). In light of this, the recommended software development tools include Java NetBeans on the front-end and MySQL as the back-end DBMS.

A.3 Scope of the System

The main components of the system will be as follows:

- A relational database in MySQL or another portable DBMS stores the essential data for subsequent analysis
- An Acquisitions Subsystem (AS) facilitates tracking of purchases, utilization, and sale of inventory items
- A Financial Management Subsystem (FMS) provides the financial implications of various transactions
- Point of Sale Subsystem (POSS) manages interfacing between the point of sale equipment (barcode scanners, and invoice printers, etc.) and the internal database

A.4 System Objectives

he Inventory Management System (IMS) will fulfill the following objectives:

- Accurate tracking of inventory
- Features to facilitate entry and update of inventory items and modify products and quantities
- Tracking of customers and suppliers contact information
- Tracking of outgoing purchase orders to suppliers
- Tracking of incoming purchase orders from customers
- Features to facilitate management of purchases and sale of goods
- Features to facilitate and handle return of goods purchased and sold
- Features to query various aspects of the company's inventory (e.g. lookup quantities for specific items) and the financial implications of this
- Facility to generate an inventory sheet
- Alerts for the reorder of inventory items
- Basic financial management capabilities which will give an overview of the financial status of the company (basic account receivable and accounts payable)

A.5 Expected Benefits

The IMS will bring a number of benefits to the organization:

- The system will help to keep accurate records of inventory. This will be an invaluable aid to sales as well as purchasing.
- The maintenance of accurate records should reduce the likelihood of inventory loss or misappropriation.
- The system will reduce the amount of human effort needed, and hence the overhead cost of the business.
- The system will contribute to improving the public image of the business—customers will appreciate being able to know instantly exactly what is available without having to waste time.
- The system should facilitate better management decisions, as well as improved productivity.
- The system will facilitate better management of the company's resources. For instance, instead of having money tied up in unnecessary inventory, the company can have a just-in-time (JIT) delivery system, and thereby free up funds that would otherwise be used in inventory storage, to be available for other aspects of the business.

A.6 Overview of Storage Requirements

It is anticipated that the system will contain the following main information entities:

- Inventory Master
- Item Categories

- Suppliers
- States or Provinces
- Customers
- Purchase Orders
- Purchase Invoices
- Purchase Returns
- Sale Orders
- Sale Invoices
- Sale Returns
- Chart of Accounts
- Accounting Balances
- Purchase Invoice Payment Plans
- Payments Made
- Financial Institutions of Interest
- Sale Invoice Payment Plans
- Payments Received
- Financial Transactions
- Investments

A.7 Anticipated Outputs

It will be possible to run queries and obtain reports on all information stored. As such, operations will be provided to extract and display information from the aforementioned information entities. Some of the more prominent outputs are as follows:

- Inventory Master Listings
- Sales Orders
- Sales Invoices
- Payments Received Log
- Accounts Receivable
- Returns History
- Purchase Orders
- Purchase Invoices
- Payments Made Log
- Accounts Payable
- Account Balances—Accounts Payable and Accounts Receivable
- Financial Transactions Log
- Investments Log

A.8 Feasibility Analysis Report

Possible alternate solutions examined are

- Alternative A: Manual system where inventory is kept track of by hand
- Alternative B: Buy an off-the-shelf solution and customize it

- Alternative C: Contract a software engineering firm to develop the system
- Alternative D: Develop system as a student project (for the purpose of illustration and academic research, this alternative will be taken)

A.8.1 Feasibility of Alternative A

The **technical feasibility** of this alternative may be summarized as follows:

- No expensive computer equipment would be required; only folders, files, and basic office equipment would be needed.
- No computer software would be needed.
- Not much knowledge and expertise is required since this kind of job would be basically filing forms and keeping track of inventory.
- No other technology would be required.

The **economic feasibility** of this alternative may be summarized as follows:

- The cost of this option is minimal (the wages of the employees who count stock and fill out the inventory forms).
- The cost of the office supplies needed for this system would be minimal.

The **operational feasibility** of this alternative may be summarized as follows:

- This method of having to manually count stock would be time-consuming and painstakingly difficult and it offers no guarantee of success.
- There would be no development time since this is a manual system.
- Implementation time would also be low—because this manual system is easy to learn not much training is required to be able to do the job.

The **risks** associated with this approach are extremely high. All the problems stated above would remain. This could potentially run the business into bankruptcy.

A.8.2 Feasibility of Alternative B

The **technical feasibility** of this alternative may be summarized as follows:

- The business no doubt already has the computer hardware that would be required to run the software.
- Customizing the software package could be tedious if the acquired software system is poorly designed.
- Maintaining the software could be problematic, especially if the system is poorly designed, and the documentation is scanty and/or inadequate.
- There is no guarantee that the purchased software system will be comprehensive enough.

In terms of **economic feasibility**, the initial investment may be affordable. However, estimating maintenance cost could be tricky.

The **operational feasibility** of this alternative may be summarized as follows:

- There would be no development time since the software is already made.
- The implementation time would be related to how well-designed the software system is. If it's well designed, installation time and time to learn the software should be minimal.
- User training would be required.

The **risks** associated with this approach are high, since there is no guarantee that software acquired will live up to expectations.

A.8.3 Feasibility of Alternative C

The **technical feasibility** of this alternative may be summarized as follows:

- Acquiring the required software development environment would be the responsibility of the contracted company.
- Knowledge and expertise might be required in learning the new software and how to use and apply it, but a well-developed software system that is custom-made for the business should be easy to learn.
- The developers themselves would have to spend some time learning about the operation and needs of the business in order to be able to develop a suitable software system.

The **operational feasibility** of this alternative may be summarized as follows:

- The initial investment would be very high because paying a software engineering company for development time can be very costly.
- Development time could be as much as 6 months to a year, but for an experienced software engineering firm, it shouldn't take more than a few months.
- The economic life of such custom-made software should be relatively long if the software is well made and well designed and if the needs of the business don't change drastically, in which case the system would either be not needed or wouldn't be adequate enough for the new needs.

The **operational feasibility** of this alternative may be summarized as follows:

- User training would be required.
- Implementation time should be short, provided that the system is adequately designed.

The **risks** associated with this approach could be high or low, depending on how the project is managed. There is no guarantee that software engineering firm will deliver a good job; however, precautions can be taken to ensure software quality.

A.8.4 Feasibility of Alternative D

The **technical feasibility** of this alternative may be summarized as follows:

- Acquisition of the required software development environment would be required.
- Required knowledge and expertise are readily available.

The **operational feasibility** of this alternative may be summarized as follows:

- Technology required is relatively inexpensive.
- The development time would be high, but the benefits gained from the experience would be huge.

The **operational feasibility** of this alternative may be summarized as follows:

- The implementation time would be long, mainly because the software would have to be developed by student effort.
- Minimal user training would be required.

The **risks** associated with this approach could be high or low, depending on how the project is administered. The student has total control over who works on the project, so depending on the quality of the project team risks may or may not be reduced.

A.8.5 Evaluation of Alternatives

The following table provides a comparison of the project alternatives, based on a number of feasibility factors. For each factor, a rank in the range {0 .. 20} is given for each alternative, as summarized in Figure A.1.

Feasibility Factors		Alt-A	Alt-B	Alt-C	Alt-D
Technical Feasibility		68	60	78	78
Availability of Hardware (bigger means better)	[Max 20]	20	15	20	20
Availability of Software (bigger means better)	[Max 20]	18	15	20	20
Availability of Expertise (bigger means better)	[Max 20]	14	16	20	20
Availability of Technology (bigger means better)	[Max 20]	16	12	18	18
Economic Feasibility		106	107	108	108
Engineering Cost (bigger means lower cost)	[Max 20]	18	17	17	17
Equipment Cost (bigger means lower cost)	[Max 20]	17	17	17	17
Operational Cost (bigger means lower cost)	[Max 20]	19	19	19	19
Facilities Cost (bigger means lower cost)	[Max 20]	19	19	19	19
Development Time (bigger means shorter time)	[Max 20]	18	18	19	19
Economic Life (bigger means longer time)	[Max 20]	16	18	18	18
Risk (bigger means lower risk)	[Max 20]	18	18	18	18
Operational Feasibility		50	52	56	56
User Attitude to Likely Changes (bigger means better)	[Max 20]	20	16	18	18
Likelihood of Organizational Policy Changes (bigger means fewer changes)	[Max 20]	16	19	18	18
Implementation Time (bigger means shorter time)	[Max 20]	14	17	20	20
Software Quality		16	110	176	176
Maintainability (bigger means better)	[Max 20]	00	10	16	16
Efficiency (bigger means better)	[Max 20]	04	10	16	16
User Friendliness (bigger means better)	[Max 20]	00	10	16	16
Documentation (bigger means better)	[Max 20]	04	10	16	16
Compatibility (bigger means better)	[Max 20]	00	10	16	16
Security (bigger means better)	[Max 20]	04	10	16	16
Reliability (bigger means better)	[Max 20]	02	10	16	16
Flexibility & Functionality (bigger means better)	[Max 20]	00	10	16	16
Adaptability (bigger means better)	[Max 20]	00	10	16	16
Growth Potential (bigger means better)	[Max 20]	02	10	16	16
Productivity (bigger means better)	[Max 20]	02	10	16	16
Overall Evaluation Score	[Max 500]	240	329	418	418

Note:
1. Feasibility factors were evaluated based on the foregoing discussion for each alternative.
2. Quality factors were evaluated based on estimates of the performance of each alternative.
3. Based on this analysis, alternative D has been selected as the most prudent.

Figure A.1 Feasibility Evaluation Grid Showing Comparison of System Alternatives

A.9 Initial Project Schedule

Figure A.2 shows an initial project schedule. This will be further refined as more details become available. The estimated duration (assuming three full-time software engineers) is 41 weeks.

Activity #	Activity Description	Weeks
1	Initial System Requirement	2
2	Requirements Specification	6
3	Design Specification	10
4	Database Creation	2
5	Software Development	14
6	Software Testing	2
7	Software Documentation	4
8	Software Installation & Delivery	1
	Total Estimated Duration	**41**
Assumption: Project team of three software engineers		

Figure A.2 Initial Project Schedule

Appendix B: Requirements Specification for a Generic Inventory Management System

This appendix provides excerpts from a sample requirements specification for a generic Inventory Management System (IMS) that may be suitable for a small or medium-sized organization. The document is a follow-up to the project proposal of Appendix A and includes the following:

- System Overview
- Storage Requirements
- Operational Requirements
- Business Rules
- Summary and Concluding Remarks

B.1 System Overview

B.1.1 Problem Definition

See Section A.1 of Appendix A. Typically, a refined problem statement is placed here.

B.1.2 Proposed Solution

See Section A.2 of Appendix A. Typically, a refined proposed solution is placed here.

B.1.3 System Architecture

The IMS project will have four main components as summarized in the two upcoming figures:

- Acquisitions Management Subsystem (AMS)
- Financial Management Subsystem (FMS)
- Point of Sale Subsystem (POSS)
- Database Backbone

Figure B.1 provides the object flow diagram (OFD) and Figure B.2 shows the information topology chart (ITC) for the IMS project. The OFD shows the main subsystems and how they interact; the ITC shows the main information entities and the related subsystems from which they will be managed.

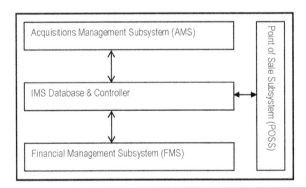

Figure B.1   IMS Object Flow Diagram

B.2 Storage Requirements

The main storage requirements of the system will be covered in this chapter. The conventions used are summarized below:

- Each information entity referenced is identified by a reference code and a descriptive name.
- For each entity, the attributes (data elements) to be stored are identified.

- The entities as presented will easily transition into a set of normalized relations in a normalized relational database.
- Data elements that will be implemented as foreign keys, in the normalized relational database, are identified by an asterisk (*) followed by a parenthesized comment, specifying what entity they reference.
- Data elements that will be used as primary key attributes (or part of the primary key) at database implementation time are also identified by parenthesized comments.
- For each entity, a comment describing the data to be stored is provided.

Inventory Management System (IMS)	
Acquisitions Management Subsystem (AMS)	E01: Item Definitions
	E02: Category Definitions
	E03: Supplier Definitions
	E04: State/Province Definitions
	E05: Location Definitions
	E06: Supplier-Item Matrix
	E07: Customer Definitions
	E08: Purchase Order Summary
	E09: Purchase Order Details
	E10: Purchase Invoice Summary
	E11: Purchase Invoice Details
	E12: Purchase Return Summary
	E13: Purchase Return Details
	E14: Sale Order Summary
	E15: Sale Order Details
	E16: Sale Invoice Summary
	E17: Sale Invoice Details
	E18: Sale Return Summary
	E19: Sale Return Details
Financial Management Subsystem (FMS)	E20: Chart of Accounts
	E21: Department Definitions
	E22: Account Balances
	E23: Purchase Payment Plans
	E24: Purchase Payments Log
	E25: Financial Institutions
	E26: Sale Payment Plans
	E27: Sale Payments Log
	E28: Employee Definitions
	E29: Payroll Log
	E30: Financial Transactions Log
	E31: Transaction Classifications
	E32: Investments Log
Point of Sale Subsystem (POSS)	E16: Sale Invoice Summary
	E17: Sale Invoice Details
	E18: Sale Return Summary
	E19: Sale Return Details
	E27: Sale Payments Log

Figure B.2   IMS Information Topology Chart

The Acquisitions Management Subsystem will be addressed first, followed by the Financial Management Subsystem.

B.2.1 Acquisitions Management Subsystem

Figure B.3 provides storage specifications for the main information entities comprising the Acquisitions Management Subsystem. This subsystem features entities E01 through E19.

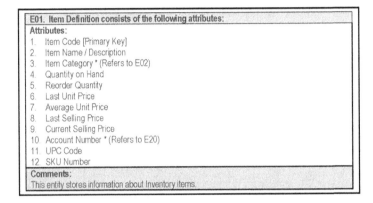

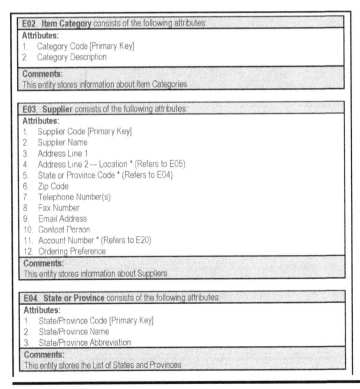

Figure B.3   Storage Specifications for the Acquisitions Management Subsystem

E05. Location consists of the following attributes:

Attributes:
1. Location Code [Primary Key]
2. Location Name
3. Location State or Province * (Refers to E04)
4. Location Abbreviation

Comments:
This entity stores information about locations (towns/cities)

E06. Supplier-Item Matrix consists of the following attributes:

Attributes:
1. Supplier Code * (Refers to E03) [Key 1]
2. Item Code * (Refers to E01) [Key 2]

Comments:
This entity stores information about which items come from which Supplier

---

E07. Customer consists of the following attributes:

Attributes:
1. Customer Code [Primary Key]
2. Customer Name
3. Address Line 1
4. Address Line 2 — Location * (Refers to E05)
5. State / Province Code * (Refers to E04)
6. Zip Code
7. Telephone Number(s)
8. Fax Number (s)
9. Email Address
10. Contact Person
11. Account Number * (Refers to E20)
12. Billing Preferences

Comments:
This entity stores information about Customers

E08. Purchase Order Summary consists of the following attributes:

Attributes:
1. Purchase Order Number [Primary Key]
2. Purchase Order Supplier Code * (Refers to E03)
3. Purchase Order Date
4. Purchase Order Status (Filled/ Partial/Outstanding)
5. Purchase Order Estimated Amount
6. Purchase Order Estimated Discount

Comments:
This entity stores information about Purchase Orders

E09. Purchase Order Detail consists of the following attributes:

Attributes:
1. Purchase Order Number * (Refers to E08) [Key 1]
2. Item Code * (Refers to E01) [Key 2]
3. Quantity Ordered
4. Order Unit Price

Comments:
This entity stores information about Purchase Order Details

Figure B.3   (Continued)

E10. Purchase Invoice consists of the following attributes:

Attributes:
1. Invoice Number [Alternate Key 1]
2. Invoice Supplier Code * (Refers to E03) [Alternate Key 2]
3. Invoice Date [Alternate Key 3]
4. Related Purchase Order Number * (Refers to E08)
5. Invoice Amount
6. Invoice Amount Outstanding
7. Discount
8. Tax
9. Comment
10. Purchase Invoice Reference Number [Primary Key]
11. Transaction Reference Number * (Refers to E30)

Comments:
This entity stores information about Purchase Invoices

E11. Purchase Invoice Detail consists of the following attributes:

Attributes:
1. Purchase Invoice Reference Number * (Refers to E10) [Key 1]
2. Item Number * (Refers to E01) [Key 2]
3. Item Quantity
4. Item Unit Price

Comments:
This entity stores information about Purchase Invoice Details

---

E12. Purchase Returns Summary consists of the following attributes:

Attributes:
1. Purchase Invoice Reference Number * (Refers to E10)
2. Return Date
3. Return Amount
4. Purchase Return ID [Primary Key]
5. Transaction Reference Number * (Refers to E30)

Comments:
This entity stores information about Purchase Returns

E13. Purchase-Returns Detail consists of the following attributes:

Attributes:
1. Purchase Return ID * (Refers to E12) [Key 1]
2. Item Code * (Refers to E01) [Key 2]
3. Quantity Returned
4. Return Unit Price
5. Comment

Comments:
This entity stores information about Purchase Return Details

Figure B.3   (Continued)

E14. Sale Order Summary consists of the following attributes:

Attributes:
1. Sales Order Number [Alternate Key 1]
2. Sales Order Customer Code * (Refers to E07) [Alternate Key 2]
3. Sales Order Date [Alternate Key 3]
4. Sales Order Status (Filled/ Partial/ Outstanding)
5. Sales Order Estimated Amount
6. Sales Order Estimated Discount
7. Sales Order Identification Number [Primary Key]

Comments:
This entity stores information about Sales Orders

E15. Sales Order Detail consists of the following attributes:

Attributes:
1. Sales Order Identification Number * (Refers to E14)
2. Item Code * (Refers to E01)
3. Quantity Ordered
4. Anticipated Unit Price

Comments:
This entity stores information about Sales Order Details

E16. Sales Invoice Summary consists of the following attributes:

Attributes:
1. Sales Invoice Number [Alternate Key 1]
2. Sales Invoice Customer Code * (Refers to E07) [Alternate Key 2]
3. Sales Invoice Date [Alternate Key 3]
4. Related Sale Order ID * (Refers to E14)
5. Sales Invoice Amount
6. Sales Invoice Amount Outstanding
7. Sales Invoice Discount
8. Sales Invoice Tax
9. Sales Invoice Comment
10. Sales Invoice Reference Number [Primary Key]
11. Transaction Reference Number * (Refers to E30)

Comments:
This entity stores information about Sales Invoices

E17. Sales Invoice Detail consists of the following attributes:

Attributes:
1. Sales Invoice Reference Number * (Refers to E16) [Key 1]
2. Item Code * (Refers to E01) [Key 2]
3. Item Quantity
4. Item Unit Price

Comments:
This entity stores information about Sales Invoice Details

E18. Sales Return Summary consists of the following attributes:

Attributes:
1. Sales Invoice Reference Number * (Refers to E16)
2. Return Date
3. Return Amount
4. Sale Return ID [Primary Key]
5. Transaction Reference Number * (Refers to E30)

Comments:
This entity stores information about Sales Returns

E19. Sales Return Detail consists of the following attributes:

Attributes:
1. Sale Return ID * (Refers to E18) [Key 1]
2. Item Code * (Refers to E01) [Key 2]
3. Quantity Returned
4. Return Unit Price
5. Comment

Comments:
This entity stores information about Sales Returns

Figure B.3   (Continued)

B.2.2 Financial Management Subsystem

Figure B.4 provides storage specifications for the Financial Management Subsystem. Entities E20 through E32 are featured in this subsystem.

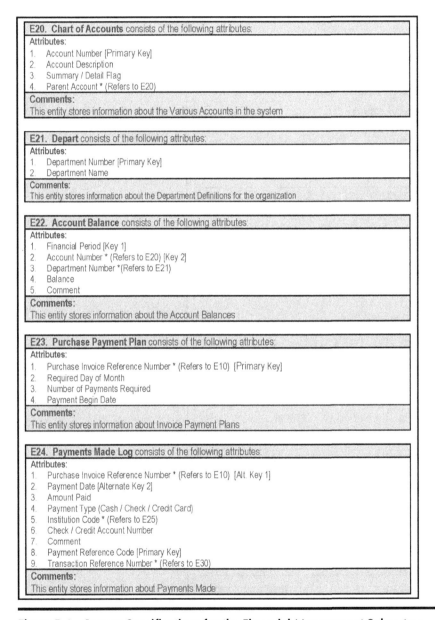

E20. Chart of Accounts consists of the following attributes:

Attributes:
1. Account Number [Primary Key]
2. Account Description
3. Summary / Detail Flag
4. Parent Account * (Refers to E20)

Comments:
This entity stores information about the Various Accounts in the system

E21. Depart consists of the following attributes:

Attributes:
1. Department Number [Primary Key]
2. Department Name

Comments:
This entity stores information about the Department Definitions for the organization

E22. Account Balance consists of the following attributes:

Attributes:
1. Financial Period [Key 1]
2. Account Number * (Refers to E20) [Key 2]
3. Department Number *(Refers to E21)
4. Balance
5. Comment

Comments:
This entity stores information about the Account Balances

E23. Purchase Payment Plan consists of the following attributes:

Attributes:
1. Purchase Invoice Reference Number * (Refers to E10) [Primary Key]
2. Required Day of Month
3. Number of Payments Required
4. Payment Begin Date

Comments:
This entity stores information about Invoice Payment Plans

E24. Payments Made Log consists of the following attributes:

Attributes:
1. Purchase Invoice Reference Number * (Refers to E10) [Alt. Key 1]
2. Payment Date [Alternate Key 2]
3. Amount Paid
4. Payment Type (Cash / Check / Credit Card)
5. Institution Code * (Refers to E25)
6. Check / Credit Account Number
7. Comment
8. Payment Reference Code [Primary Key]
9. Transaction Reference Number * (Refers to E30)

Comments:
This entity stores information about Payments Made

Figure B.4   Storage Specifications for the Financial Management Subsystem

E25. Financial Institution consists of the following attributes:

Attributes:
1. Institution Code [Primary Key]
2. Institution Name
3. Address Line 1
4. Address Line 2 — Location Code *(Refers to E05)
5. State / Province Code * (Refers to E04)
6. Zip Code
7. Telephone Number(s)
8. Fax Number(s)
9. Contact Person
10. Email Address

Comments:
This entity stores information about Financial Institutions

E26. Sales Payment Plan consists of the following attributes:

Attributes:
1. Sales Invoice Reference Number * (Refers to E16) [Key]
2. Required Day of Month
3. Number of Payments Required
4. Payment Begin Date

Comments:
This entity stores information about Sales Invoices Payment Plans

E27. Payments Received Log consists of the following attributes:

Attributes:
1. Sale Invoice Reference Number * (Refers to E16) [Alt. Key 1]
2. Payment Received Date [Alternate Key 2]
3. Amount Received
4. Payment Type (Cash / Check / Credit Card)
5. Institution Code * (Refers to E25)
6. Check / Credit Account Number
7. Comment
8. Payment Reference Code [Primary Key]
9. Transaction Reference Number * (Refers to E30)

Comments:
This entity stores information about Payments Received

E28. Employee consists of the following attributes:

Attributes:
1. Employee Identification Number [Primary Key]
2. Employee First Name
3. Employee Middle Name
4. Employee Last Name
5. Employee Address Line 1
6. Address Line 2 — Location Code * (Refers to E05)
7. State / Province Code * (Refers to E04)
8. Zip Code
9. Telephone Number(s)
10. Fax Number(s)
11. Email Address
12. Related Account Number * (Refers to E20)

Comments:
This entity stores information about Employees

E29. Payroll Log consists of the following attributes:

Attributes:
1. Payroll Log Reference Code [Primary Key]
2. Payroll Date
3. Employee Number *(Refers to E28)
4. Payment Amount
5. Insurance Amount
6. Investment Amount
7. Taxes
8. Comment

Comments:
This entity stores information on Payments made to Employees

Figure B.4 (Continued)

E30. Financial Transactions Log consists of the following attributes:

Attributes:
1. Transaction ID Number [Primary Key]
2. Transaction Date
3. Transaction Classification Code * (Refers to E31)
4. Account Number * (Refers to E20)
5. Department Number * (Refers to E21)
6. Debit Amount
7. Credit Amount
8. Accounting Period
9. Transaction Reference Number (see Note below)

Comments:
This entity stores information about Financial Transactions

Note: Transaction Reference Number ties back to one of the following entities:
PI: Purchase Invoice Transactions (E10)
PR1: Purchase Return Transactions (E12)
SI: Sales Invoice Transactions (E16)
SR: Sale Return Transactions (E18)
PP1: Cash Purchase Payments Transactions (E24)
PP2: Credit Purchase Payments Transactions (E24)
SP1: Cash Sales Payments Transactions (E27)
SP2: Credit Sales Payments Transactions (E27)
PR2: Payroll Log (E29)
INV: Investment Transactions (E32)

Comments:
These will be implemented via logical views.

E31. Transaction Classification consists of the following attributes:

Attributes:
1. Transaction Classification Code [Primary Key]
2. Transaction Classification Description

Note: Transaction Classifications include:
PI — Purchase Invoices
PR1 — Purchase Returns
PR2 — Payroll Log
SI — Sales Invoices
SR — Sales Returns
PP1 — Cash Purchase Payments
PP2 — Credit Purchase Payments
SP1 — Cash Sales Payments
SP2 — Credit Sales Payments
JED — Journal Entries Debit
JEC — Journal Entries Credit
INV — Investments

Comments:
This entity stores information about Transaction Classifications

E32. Investments Log consists of the following attributes:

Attributes:
1. Investment ID Code [Primary Key]
2. Investment Transaction Date
3. Investment Amount
4. Institution Code * (Refers to E25)
5. Institutional Account Number
6. Internal Account Number * (Refers to E20)
7. Comment
8. Transaction Reference Number * (Refers to E30)

Comments:
This entity stores information about Investments

Figure B.4   (Continued)

B.3 Operational Requirements

The information entities of the previous chapter will be implemented as a relational database using MySQL or some other appropriate DBMS. Superimposed on this database will be an OO GUI consisting of various operations that provide functionality and flexibility to the end-users.

The user functionality and flexibility will be provided based on the following approach:

- For each information entity, define an object type of the same name.
- For each object type, define a set of basic operations that will facilitate addition, modification, deletion, inquiry, and reporting of data related to that object type (note that each object type does not necessarily require all five basic operations).
- For selected object types, introduce flexibility and sophistication such as additional and/or more complex inquiry and/or report operations that facilitate end-users specifying selection criteria of their choice.
- Define additional utility operations that will support the user interface of the system.

Figure B.5 shows an initial list of operations that will be defined on each object type. An operation specification will be provided for each of these operations in the project's design specification. The reference code for each operation is indicated in square brackets.

Acquisitions Management Subsystem	
Object Type	**Operations**
E01 Inventory Item	**Add** Inventory Items [E1_A] **Modify** Inventory Items [E1_M] **Delete** Inventory Items [E1_Z] **Inquire** Inventory Items [E1_I] **Report** Inventory Items [E1_R]
E02 Item Category	**Add** Item Categories [E2_A] **Modify** Item Categories [E2_M] **Delete** Item Categories [E2_Z] **Inquire** Item Categories [E2_I] **Report** Item Categories [E2_R]
E03 Supplier	**Add** Suppliers [E3_A] **Modify** Suppliers [E3_M] **Delete** Suppliers [E3_Z] **Inquire** on Suppliers [E3_I] **Report** on Supplier [E3_R]
E04 State or Province	**Add** States or Provinces [E4_A] **Delete** States or Provinces [E4_Z] **Inquire** on States or Provinces [E4_I]
E05 Location Definitions	**Add** Location Definitions [E5_A] **Modify** Location Definitions [E5_M] **Delete** Location Definitions [E5_Z] **Inquire** on Location Definitions [E5_I] **Report** on Location Definitions [E5_R]

Figure B.5 IMS User Operations List *(Continued)*

Acquisitions Management Subsystem	
Object Type	Operations
E06 Supplier-Item Matrix	**Add** Supplier-Item Connections [E6_A] **Modify** Supplier-Item Connections [E6_M] **Delete** Supplier-Item Connections [E6_Z] **Inquire** Supplier-Item Connections [E6_I] **Report** Supplier-Item Connections [E6_R]
E07 Customer	**Add** Customers [E7_A] **Modify** Customers [E7_M] **Delete** Customers [E7_Z] **Inquire** Customers [E7_I]
E08 Purchase Order Summary	**Add** Purchase Orders [E8_A] **Modify** Purchase Orders [E8_M] **Delete** Purchase Orders [E8_Z] **Inquire** Purchase Orders [E8_I] **Report** Purchase Orders [E8_R]
E09 Purchase Order Detail	**Add** Purchase Order Details [E9_A] **Modify** Purchase Order Details [E9_M] **Delete** Purchase Order Details [E9_Z] **Inquire** Purchase Order Details [E9_I] **Report** Purchase Order Details [E9_R]
E10 Purchase Invoice	**Add** Purchase Invoices [E10_A] **Modify** Purchase Invoices [E10_M] **Delete** Purchase Invoices [E10_Z] **Inquire** Purchase Invoices [E10_I] **Report** Purchase Invoices [E10_R]
E11 Purchase Invoice Detail	**Add** Purchase Invoice Details [E11_A] **Modify** Purchase Invoice Details [E11_M] **Delete** Purchase Invoice Details [E11_Z] **Inquire** Purchase Invoice Details [E11_I] **Report** Purchase Invoice Details [E11_R]
E12 Purchase Returns Summary	**Add** Purchase Returns [E12_A] **Delete** Purchase Returns [E12_Z] **Inquire** Purchase Returns [E12_I] **Report** Purchase Returns [E12_R]
E13 Purchase Returns Detail	**Add** Purchase Returns Details [E13_A] **Modify** Purchase Returns Details [E13_M] **Delete** Purchase Returns Details [E13_Z] **Inquire** Purchase Returns Details [E13_I] **Report** Purchase Returns Details [E13_R]
E14 Sale Order Summary	**Add** Sale Orders [E14_A] **Modify** Sale Orders [E14_M] **Delete** Sale Orders [E14_Z] **Inquire** Sale Orders [E14_I] **Report** Sale Orders [E14_R]
E15 Sale Order Detail	**Add** Sale Order Details [E15_A] **Modify** Sale Order Details [E15_M] **Delete** Sale Order Details [E15_Z] **Inquire** Sale Order Details [E15_I] **Report** Sale Order Details [E15_R]

Figure B.5 (Continued)

Acquisitions Management Subsystem	
Object Type	Operations
E16 Sale Invoice Summary	**Add** Sale Invoices [E16_A] **Modify** Sale Invoices [E16_M] **Delete** Sale Invoices [E16_Z] **Inquire** Sale Invoices [E16_I] **Report** Sale Invoices [E16_R]
E17 Sale Invoice Detail	**Add** Sale Invoice Details [E17_A] **Modify** Sale Invoice Details [E17_M] **Delete** Sale Invoice Details [E17_Z] **Inquire** Sale Invoice Details [E17_I] **Report** Sale Invoice Details [E17_R]
E18 Sale Returns Summary	**Add** Sale Returns [E18_A] **Modify** Sale Returns [E18_M] **Delete** Sale Returns [E18_Z] **Inquire** Sale Returns [E18_I] **Report** Sale Returns [E18_R]
E19 Sale Return Detail	**Add** Sale Return Details [E19_A] **Delete** Sale Return Details [E19_Z] **Inquire** Sale Return Details [E19_I] **Report** Sale Return Details [E19_R]

Financial Management Subsystem	
Object Type	Operations
E20 Chart of Accounts	**Add** Accounts [E20_A] **Delete** Accounts [E20_Z] **Inquire** on Accounts [E20_I] **Report** on Accounts [E20_R]
E21 Department Definition	**Add** Department Definitions [E21_A] **Modify** Department Definitions [E21_M] **Delete** Department Definitions [E21_Z] **Inquire** Department Definitions [E21_I] **Report** Department Definitions [E21_R]
E22 Account Balances	**Add** Account Balances [E22_A] **Modify** Account Balances [E22_M] **Delete** Account Balances [E22_Z] **Inquire** Account Balances [E22_I] **Report** Account Balances [E22_R]
E23 Purchase Invoice Payment Plan	**Add** Purchase Invoice Payment Plans [E23_A] **Modify** Purchase Invoice Payment Plans [E23_M] **Delete** Purchase Invoice Payment Plans [E23_Z] **Inquire** Purchase Invoice Payment Plans [E23_I] **Report** Purchase Invoice Payment Plans [E23_R]
E24 Purchase Payments Log	**Add** Payments Made [E24_A] **Delete** Payments Made [E24_Z] **Inquire** Payments Made [E24_I] **Report** Payments Made [E24_R]
E25 Financial Institutions	**Add** Financial Institutions [E25_A] **Delete** Financial Institutions [E25_Z] **Inquire** Financial Institutions [E25_I] **Report** Financial Institutions [E25_R]

Figure B.5   (Continued)

Financial Management Subsystem	
Object Type	Operations
E26 Sales Invoice Payment Plan	**Add** Sale Payment Plans [E26_A] **Delete** Sale Payment Plans [E26_Z] **Inquire** Sale Payment Plans [E26_I] **Report** Sale Payment Plans [E26_R]
E27 Sale Payments Log	**Add** Payments Received [E27_A] **Delete** Payments Received [E27_Z] **Inquire** Payments Received [E27_I] **Report** Payments Received [E27_R]
E28 Employee Definition	**Add** Employee Definitions [E28_A] **Modify** Employee Definitions [E28_M] **Delete** Employee Definitions [E28_Z] **Inquire** Employee Definitions [E28_I] **Report** Employee Definitions [E28_R]
E29 Payroll Log	**Add** Payroll Entries [E29_A] **Modify** Payroll Entries [E29_M] **Delete** Payroll Entries [E29_Z] **Inquire** Payroll Entries [E29_I] **Report** Payroll Entries [E29_R]
E30 Financial Transactions	**Add** Financial Transactions [E30_A] **Modify** Financial Transactions [E30_M] **Delete** Financial Transactions [E30_Z] **Inquire** Financial Transactions [E30_I] **Report** Financial Transactions [E30_R]
E31 Transaction Classifications	**Add** Transaction Classifications [E31_A] **Delete** Transaction Classifications [E31_Z] **Inquire** Transaction Classifications [E31_I] **Report** Transaction Classifications [E31_R]
E32 Investment	**Add** Investments [E32_A] **Modify** Investments [E32_M] **Delete** Investments [E32_Z] **Inquire** Investments [E32_I] **Report** Investments [E32_R]

Figure B.5   (Continued)

B.4 System Rules

There are two main types of system rules that will be covered here:

- Data Integrity Rules
- Procedural and Derivation Rules

B.4.1 Data Integrity Rules

Data integrity rules include *referential integrity rules* and *data validation rules*. The **referential integrity rules** relate to the treatment of foreign keys. Specifically, each foreign key (i.e., referencing attribute) in a referencing entity must have values drawn from the entity that it references. The only exception to this rule is the instance where the foreign key value is null. However, there are instances in which foreign keys are allowed to have null values (for more clarification on database theory, see [Foster 2016]).

Data validation rules are guidelines that should characterize the data entering the software system. While these may be implemented at the database level, in the interest of user-friendliness, it is often expedient that these validation rules will be specified for related operations that facilitate the data entry.

Figure B.6 summarizes the main data integrity rules for the system; these are specified on an entity-by-entity basis. For more details on the relevance of these rules, please revisit Figures B.3 and B.4.

Entity	Integrity Rules
E01 Item Definitions	1. Primary key is Item Code. 2. Foreign keys reference E02 and E20.
E02 Category Definitions	Primary key is Category Code.
E03 Supplier Definitions	1. Primary key is Supplier Code. 2. Foreign keys reference E04, E05, and E20. 3. Validation checks on Telephone Number and Email suggested.
E04. State/Province Definitions	Primary key is Province Code.
E05 Location Definitions	1. Primary key is Location Code. 2. Foreign key references E04.
E06 Supplier-Item Matrix	1. Primary key is [Supplier Code & Item Code]. 2. Foreign keys reference E03 and E01.
E07 Customer Definitions	1. Primary key is [Customer Code. 2. Foreign keys reference E04, E05, and E20. 3. Validation checks on Telephone Number and Email suggested.
E08 Purchase Order Summary	1. Primary key is Purchase Order Number. 2. Foreign key references E03.
E09 Purchase Order Details	1. Primary key is [Purchase Order Number & Item Code]. 2. Foreign keys reference E08 and E01.
E10 Purchase Invoice Summary	1. Primary key is Purchase Invoice Reference Number. 2. Foreign keys reference E03, E08, and E30.
E11 Purchase Invoice Details	1. Primary key is [Purchase Invoice Reference Number & Item Code]. 2. Foreign keys reference E10 and E01.
E12 Purchase Return Summary	1. Primary key is Purchase Return ID. 2. Foreign keys reference E10 and E30.
E13 Purchase Return Details	1. Primary key is [Purchase Return ID & Item Code]. 2. Foreign keys reference E12 and E01.
E14 Sale Order Summary	1. Primary key is Sale Order Number. 2. Foreign key references E07.
E15 Sale Order Details	1. Primary key is [Sale Order Number & Item Code]. 2. Foreign keys reference E14 and E01.
E16 Sale Invoice Summary	1. Primary key is Sale Invoice Reference Number. 2. Foreign keys reference E07, E14, and E30.
E17 Sale Invoice Details	1. Primary key is [Sale Invoice Reference Number & Item Code]. 2. Foreign keys reference E16 and E01.
E18 Sale Return Summary	1. Primary key is Sale Return ID. 2. Foreign keys reference E16 and E30.
E19 Sale Return Details	1. Primary key is [Sale Return ID & Item Code]. 2. Foreign keys reference E18 and E01.
See figures A2-3 and A2-4 for more details. Also, all specified dates must be validated.	

Figure B.6 System Data Integrity Rules

Entity	Integrity Rules
E20 Chart of Accounts	1. Primary key is Account Number. 2. Foreign key reference E20.
E21 Department Definitions	Primary key is Department Number
E22 Account Balances	1. Primary key is [Account Number & Department Number]. 2. Foreign keys reference E20 and E21.
E23 Purchase Payment Plans	Primary key is Purchase Invoice Reference Number.
E24 Purchase Payments Log	1. Primary key is Payment Reference Code. 2. Foreign keys reference E10, E25, and E30.
E25 Financial Institutions	1. Primary key is Institution Code. 2. Foreign keys reference E04 and E05. 3. Telephone numbers and/or Email address must be valid.
E26 Sale Payment Plans	Primary key is Sale Invoice Reference Number.
E27 Sale Payments Log	1. Primary key is Payment Reference Code. 2. Foreign keys reference E16, E25, and E30.
E28 Employee Definitions	1. Primary key is Employee Identification Number. 2. Foreign keys reference E04, E05, and E20. 3. Telephone numbers and/or Email address must be valid.
E29 Payroll Log	1. Primary key is Payroll Reference Code. 2. Foreign key references E28.
E30 Financial Transactions Log	1. Primary key is Transaction ID Number. 2. Foreign keys reference E20, E21, and E31.
E31 Transaction Classifications	Primary key is Classification Code.
E32 Investments Log	1. Primary key is Investment ID Code. 2. Foreign keys reference E20, E25, and E30.
See figures A2-3 and A2-4 for more details. Also, all specified dates must be validated.	

Figure B.6   (Continued)

B.4.2 Procedural and Derivation Rules

Procedural and derivation rules relate to how the software system will actually work. The following derivation and procedural rules will be enforced:

1. Supplier and **Customer** entities (E03 and E07) will each store a record called **Miscellaneous** for **Cash Purchases** and **Cash Sales,** respectively.

2. **Cash Purchases** will impact the entities **Purchase Invoice Summary** (E10), **Purchase Invoice Detail** (E11), **Purchase Payments Log** (E24), **Item Definition** (E01), and **Financial Transactions Log** (E30) as follows:
   a. Write a record to **Purchase Invoice Summary** (related Purchase Order is null). If the **Supplier** is not listed in the **Supplier** entity (E03), use the default **Miscellaneous Supplier** record and put an appropriate remark in the invoice's **Comment-field**.
   b. Write corresponding detail records in **Purchase Invoice Detail**.
   c. Write a record in the **Purchase Payments Log**, for the relevant items.
   d. Adjust the **Quantity-on-Hand** in the **Item Definition** entity for the relevant items.
   e. Write corresponding record in the **Financial Transactions Log**.

3. **Credit Purchases** impact the entities **Purchase Order Summary** (E08), **Purchase Invoice Summary** (E10), **Purchase Invoice Detail** (E11), **Purchase Payments Log** (E24), **Item Definition** (E01), and **Financial Transactions Log** (E30) as follows:
   a. Write a record to **Purchase Invoice Summary**.
   b. Write corresponding records to **Purchase Invoice Detail**.
   c. Update the related **Purchase Order Summary** record.
   d. Adjust **Quantity-on-Hand** in **Item Definition** entity for the relevant items.
   e. Write **Accounts Payable** entry in **Financial Transactions Log**.

4. **Purchase Payments** involve the entities **Purchase Payments Log** (E24), **Purchase Invoice Summary** (E10), and **Financial Transactions Log** (E30) in the following way:
   a. Write a record to **Purchase Payments Log**.
   b. Update the **Invoice-Outstanding-Amount** in the **Purchase Invoice Summary**.
   c. Write a record in the **Financial Transactions Log**.

5. Each **Purchase Return** impacts the entities **Purchase Returns Summary** (E12), **Purchase Returns Detail** (E13), **Item Definition** (E01), **Purchase Invoice Summary** (E10), and **Financial Transactions Log** (E30) in the following way:
   a. Write a record to the **Purchase Returns Summary** entity.
   b. Write corresponding detail records to **Purchase Returns Detail**.
   c. Adjust the **Quantity-on-Hand** in the **Item Definition** entity, for the relevant items.
   d. Adjust **Outstanding-Amount** in **Purchase Invoice Summary**.
   e. Write corresponding record in **Financial Transactions Log**.

6. **Cash Sales** will impact the entities **Sale Invoice Summary** (E16), **Sale Invoice Detail** (E17), **Sale Payments Log** (E27), **Item Definition** (E01), and **Financial Transactions Log** (E30) as follows:
   a. Write a record to the **Sales Invoice Summary**. If the customer is not listed in the **Customer** entity, use the default **Miscellaneous Customer**, with an appropriate remark in the invoice's **Comment-field**.
   b. Write corresponding include records in the **Sale Invoice Detail**.
   c. Write a record in the **Sale Payments Log**.
   d. Adjust the **Quantity-on-Hand** in the **Item Definition** entity, for the relevant items.
   e. Issue a receipt and write record(s) in the **Financial Transactions Log**.

7. **Credit Sales Purchases** impact the entities **Sale Order Summary** (E14), **Sale Invoice Summary** (E16), **Sale Invoice Detail** (E17), **Sale Payments Log** (E27), **Item Definition** (E01), and **Financial Transactions Log** (E30) as follows:
   a. Write a record to the **Sales Invoice Summary** entity.
   b. Write corresponding detail records in **Sales Invoice Detail**.
   c. Update the related **Sale Order Summary** record.
   d. Adjust the **Quantity-on-Hand** in the **Item Definition** entity, for the relevant items.
   e. Write **Accounts Receivable** record in **Financial Transactions Log**.

8. **Sale Payments** involve the entities **Sale Payments Log** (E27), **Sale Invoice Summary** (E16), and **Financial Transactions Log** (E30) in the following way:
   a. Write a record to **Payments Received** entity.
   b. Update the **Invoice Outstanding Amount** in the **Sales Invoice Summary** entity.
   c. Issue a receipt.
   d. Write corresponding record in **Financial Transactions Log**.

9. Each **Sale Return** impacts the entities **Sale Returns Summary** (E18), **Sale Returns Detail** (E19), **Item Definition** (E01), **Sale Invoice Summary** (E16), and **Financial Transactions Log** (E30) in the following way:
   a. Write a record to the **Sales Returns Summary** log.
   b. Write corresponding details records to **Sales Returns Detail** log.
   c. Adjust the **Quantity-on-Hand** in the **Item Definition** entity, for the relevant items.
   d. Adjust the **Outstanding-Amount** for the related **Sale Invoice Summary**.
   e. Write corresponding record in **Financial Transactions Log**.

10. **Accounts Payable** is determined by the formula:

$$\text{Accounts Payble} = [\textbf{Purchase Invoices} \text{ with nonzero } \textbf{Outstanding Amounts} \text{ minus } \textbf{Purchase Returns} \text{ with nonzero } \textbf{Return Amounts}]$$

11. **Accounts Receivable** is determined by the formula:

$$\text{Accounts Receivable} = [\textbf{Sales Invoices} \text{ with nonzero } \textbf{Outstanding Amounts} \text{ minus } \textbf{Sales Returns} \text{ with nonzero } \textbf{Return Amounts}]$$

These rules will be extremely useful during system development. Some of them will be incorporated into the relevant operation specifications and included in the project's design specification.

B.5 Summary and Concluding Remarks

This document has provided a basic outline of the requirements for a generic Inventory Management System (IMS). It includes the following:

- An overview of the system, including problem definition, proposed solution, and system architecture.
- Storage requirements, including draft outlines for the basic information entities comprising the system.
- Operational requirements, including an initial list of user operations.
- System rules, including data integrity, derivation, and procedural rules.

The design specification will use these requirements to prepare a more detailed set of specifications from which the system will be developed.

Appendix C: Design Specification for a Generic Inventory Management System

This appendix provides excerpts from a sample design specification for a generic Inventory Management System (IMS) that may be suitable for a small or medium-sized organization. The document is a follow-up to the requirements specification of Appendix B and includes the following:

- System Overview
- Database Specification
- User Interface Specification
- Operations Specification
- Summary and Concluding Remarks

C.1 System Overview

C.1.1 Problem Definition

See Section B.1.1 of Appendix B. Typically, a refined problem statement is placed here.

C.1.2 Proposed Solution

See Section B.1.2 of Appendix B. Typically, a refined problem statement is placed here.

C.1.3 System Architecture

The System will have five main components:

- Acquisitions Management Subsystem (AMS)
- Financial Management Subsystem (FMS)
- System Controls Subsystem (SCS)
- Point of Sale Subsystem (POSS)
- Database Backbone

Figure C.1 shows the object flow diagram (OFD), while Figure C.2 shows the information topology chart (ITC) for the IMS project. Notice that on the ITC there are additional entities added under the System Controls Subsystem (SCS), compared to the diagram provided in Appendix B. This represents a refinement of the ITC provided in the requirements specification.

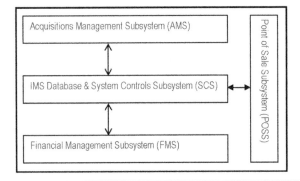

Figure C.1   IMS Object Flow Diagram

Inventory Management System (IMS)	
Acquisitions Management Subsystem (AMS)	E01: Item Definitions
	E02: Category Definitions
	E03: Supplier Definitions
	E04: State/Province Definitions
	E05: Location Definitions
	E06: Supplier-Item Matrix
	E07: Customer Definitions
	E08: Purchase Order Summary
	E09: Purchase Order Details
	E10: Purchase Invoice Summary
	E11: Purchase Invoice Details
	E12: Purchase Return Summary
	E13: Purchase Return Details
	E14: Sale Order Summary
	E15: Sale Order Details
	E16: Sale Invoice Summary
	E17: Sale Invoice Details
	E18: Sale Return Summary
	E19: Sale Return Details
Financial Management Subsystem (FMS)	E20: Chart of Accounts
	E21: Department Definitions
	E22: Account Balances
	E23: Purchase Payment Plans
	E24: Purchase Payments Log
	E25: Financial Institutions
	E26: Sale Payment Plans
	E27: Sale Payments Log
	E28: Employee Definitions
	E29: Payroll Log
	E30: Financial Transactions Log
	E31: Transaction Classifications
	E32: Investments Log
Point of Sale Subsystem (POSS)	E16: Sale Invoice Summary
	E17: Sale Invoice Details
	E18: Sale Return Summary
	E19: Sale Return Details
	E27: Sale Payments Log
System Controls Subsystem (SCS)	E33: Audit Additions Log
	E34: Audit Updates Log
	E35: Audit Deletions Log
	E36: Message File
	E37: System Users
	E38: Sessions Log

Figure C.2   IMS Information Topology Chart

C.2 Database Specification

The database specification will proceed under the following captions:

- Introduction
- Acquisitions Management Subsystem
- Financial Management Subsystem
- System Controls Subsystem

C.2.1 Introduction

The database specification of the system will be covered in this section. The methodology employed for database specification is the *object/entity specification grid* (O/ESG) developed by the current author. A Following is a summary of the conventions used:

- Each information entity referenced is identified by a reference code and a descriptive name.
- For each entity, the attributes (data elements) to be stored are identified.
- The entities as presented will easily transition into a set of normalized relations in a normalized relational database.
- Data elements that will be implemented as foreign keys, in the normalized relational database, are identified by comment in curly braces, specifying what entity they reference.
- For each attribute, the physical characteristics will be given (as described in the next section); the attributes implementation name will be indicated in square brackets; it will be indicated whether the attribute is a foreign key.
- Indexes (including primary key or candidate keys) to be defined on the entity are indicated.
- For each entity, a comment describing the data to be stored is provided. Additionally, the entity's implementation name is indicated in square brackets.
- Each operation defined on an entity will be given an implementation name, indicated in square brackets.

Naming of database objects will be very important for the following reasons:

- The database will host several objects. Without a proper naming convention, it will be extremely difficult to keep track of them.
- The naming convention will enable us to easily categorize database objects on sight.

Figure C.3 shows the object naming convention for the project; you will recall that this was first introduced in Chapter 12.

Object Name: SSSS_XXXXXXX_MMn where interpretations apply: ■ SSSS represents the system or subsystem abbreviation (2 – 4 bytes); ■ MMn represents the object mode or purpose (1-3 bytes); ■ XXXXXXXXX represents the descriptive name of the object (4-15 bytes).
For example, valid subsystem abbreviations for an Inventory Management System (IMS) are as follows: ■ AM: Acquisitions Management Subsystem ■ FM: Financial Management Subsystem ■ SC: System Controls Subsystem
Valid mode abbreviations include: ■ DM: Data Model ■ BR: A base relation (if relational DB model) ■ OT: An object type (if OO DB model) ■ LVn: A logical view (e.g. LV1, LV2, etc.) ■ NXn: An index to a base table or object type (e.g. NX1, NX2, etc.) ■ PK: Primary Key ■ FKn: Foreign Key (e.g. FK1, FK2, etc.) ■ ICn: Integrity Constraint (e.g.IC1, IC2, etc.) ■ AO: An ADD operation ■ MO: A MODIFY operation ■ ZO: A DELETE (Zap) operation ■ IO: An INQUIRE operation ■ FO: A FORECAST operation ■ RO: A REPORT operation ■ XO: A utility operation ■ DS: A database synonym or alias of a known database table ■ DC: A database constraint ■ DT: A database Trigger ■ DP: A database procedure or function ■ DK: A database package ■ MF: A Message file — a special purpose database table (file) to store the text (and other essential details) for diagnostic error and status messages
The descriptor used for a database base relation or object type is consistently used for other objects that directly relate to that object. For example, the objects related to the management of inventory items may be: ■ AM_ItemDef_BR — a base relation to store data on inventory items ■ AM_ItemDef_NX1 — an index on the base relation ■ AM_ItemDef_AO — an operation to ADD inventory items ■ AM_ItemDef_MO — an operation to MODIFY inventory items ■ AM_ItemDef_ZO — an operation to DELETE inventory items ■ AM_ItemDef_IO — an operation to INQUIRE on inventory items ■ AM_ItemDef_RO — an operation to REPORT on inventory items ■ AM_ItemDef_XO — a utility operation related to inventory items ■ AM_ItemDef_LV1 — a logical view of the base relation
Attribute implementation names are merely abbreviations of their more descriptive names, prefixed by an appropriate abbreviation of the entity.

Figure C.3   Proposed Object Naming Convention

C.2.2 *Acquisitions Management Subsystem*

The Acquisitions Management Subsystem (AMS) handles matters relating to the acquisition of resources for the organization. Following is the O/ESG for each information entity comprising this subsystem (Figure C.4).

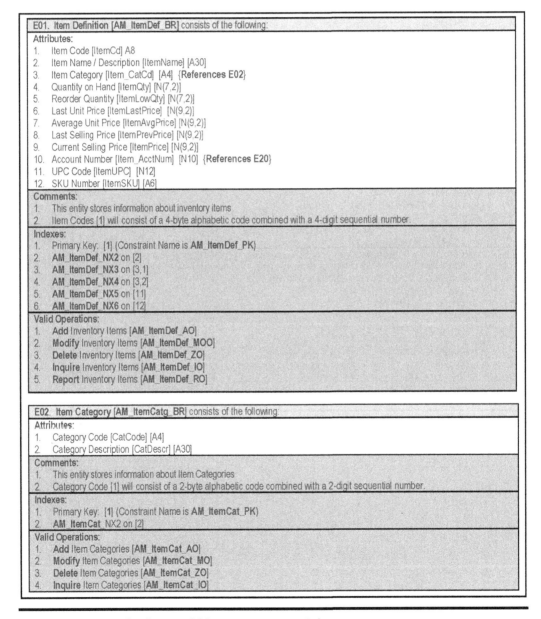

E01. Item Definition [AM_ItemDef_BR] consists of the following:

Attributes:
1. Item Code [ItemCd] A8
2. Item Name / Description [ItemName] [A30]
3. Item Category [Item_CatCd] [A4] {References E02}
4. Quantity on Hand [ItemQty] [N(7,2)]
5. Reorder Quantity [ItemLowQty] [N(7,2)]
6. Last Unit Price [ItemLastPrice] [N(9,2)]
7. Average Unit Price [ItemAvgPrice] [N(9,2)]
8. Last Selling Price [ItemPrevPrice] [N(9,2)]
9. Current Selling Price [ItemPrice] [N(9,2)]
10. Account Number [Item_AcctNum] [N10] {References E20}
11. UPC Code [ItemUPC] [N12]
12. SKU Number [ItemSKU] [A6]

Comments:
1. This entity stores information about inventory items
2. Item Codes [1] will consist of a 4-byte alphabetic code combined with a 4-digit sequential number.

Indexes:
1. Primary Key: [1] (Constraint Name is AM_ItemDef_PK)
2. AM_ItemDef_NX2 on [2]
3. AM_ItemDef_NX3 on [3,1]
4. AM_ItemDef_NX4 on [3,2]
5. AM_ItemDef_NX5 on [11]
6. AM_ItemDef_NX6 on [12]

Valid Operations:
1. **Add** Inventory Items [AM_ItemDef_AO]
2. **Modify** Inventory Items [AM_ItemDef_MOO]
3. **Delete** Inventory Items [AM_ItemDef_ZO]
4. **Inquire** Inventory Items [AM_ItemDef_IO]
5. **Report** Inventory Items [AM_ItemDef_RO]

E02. Item Category [AM_ItemCatg_BR] consists of the following:

Attributes:
1. Category Code [CatCode] [A4]
2. Category Description [CatDescr] [A30]

Comments:
1. This entity stores information about Item Categories
2. Category Code [1] will consist of a 2-byte alphabetic code combined with a 2-digit sequential number.

Indexes:
1. Primary Key: [1] (Constraint Name is AM_ItemCat_PK)
2. AM_ItemCat_NX2 on [2]

Valid Operations:
1. **Add** Item Categories [AM_ItemCat_AO]
2. **Modify** Item Categories [AM_ItemCat_MO]
3. **Delete** Item Categories [AM_ItemCat_ZO]
4. **Inquire** Item Categories [AM_ItemCat_IO]

Figure C.4   O/ESG for the Acquisitions Management Subsystem

(Continued)

```
E03. Supplier [AM_Supplier_BR] consists of the following:
Attributes:
1.   Supplier Code [SuppCode] [A8]
2.   Supplier Name [SuppName] [A30]
3.   Address Line 1 [SuppAddr1] [A30]
4.   Address Line 2 — Location [SuppLocCode] [A10] {References E05}
5.   State or Province Code [SuppStProvCd] {References E04} [A6]  // Redundant because E05 → E04
6.   Zip Code [SuppZip] [N8]
7.   Telephone Number(s) [SuppPhoneNo] [A16]
8.   Fax Number [SuppFaxNo] [A12]
9.   Email Address [SuppEmail] [A30]
10.  Contact Person [SuppContact] [A30]
11.  Account Number [SuppAccountNo] [N10] {References E20}
12.  Ordering Preference [SuppOrderPref] [A30]
Comments:
1.   This entity stores information about Suppliers
2.   Supplier Code [1] will consist of a 4-byte alphabetic code combined with a 4-digit sequential number, e.g. S01_299.
Indexes:
1.   Primary Key: [1] (Constraint Name is AM_Supplier_PK)
2.   AM_Supplier_NX2 on [2]
3.   AM_Supplier_NX3 on [11]
Valid Operations:
1.   Add Suppliers [AM_Supplier_AO]
2.   Modify Suppliers [AM_Supplier_MO]
3.   Delete Suppliers [AM_Supplier_ZO]
4.   Inquire on Suppliers [AM_Supplier_IO]
5.   Report on Suppliers [AM_Supplier_RO]


E04. State or Province [AM_StateProv_BR] consists of the following:
Attributes:
1.   State / Province Code [StateProvCode] [A6]
2.   State / Province Name [StateProvName] [A30]
Comments:
1.   This entity stores the List of States and/or Provinces.
2.   Province Code will consist of a 4-byte alphabetic code (possibly an acronym from the referenced country) combined with a 2-digit
     sequential number. Example: USA_01 .. USA_50 for the 50 states of the USA.
Indexes:
1.   Primary Key: [1] (Constraint Name is AM_StateProv_PK)
Valid Operations:
1.   Add States or Provinces [AM_StateProv_AO]
2.   Delete States or Provinces [AM_StateProv_ZO]
3.   Inquire States or Provinces [AM_StateProv_IO]
```

Figure C.4   (Continued)

E05. Location [AM_Location_BR] consists of the following:
Attributes:
1. Location Code [LocCode] [A10]
2. Location Name [LocName] [A30]
3. Location State or Province Code [LocStProvCd] [A6] {References E04}
4. Location Abbreviation [LocAbbr] [A4]
Comments:
1. This entity stores the List of Locations (cities and/or towns).
2. Location Code will consist of the concatenation of the related province code and 3-digit sequential number. Examples: USA_01_001 .. USA_01_189 for locations in state 01 of the USA.
Indexes:
1. Primary Key: [1] (Constraint Name is **AM_Location_PK**).
2. **AM_Location_NX2** on [2].
Valid Operations:
1. **Add** Locations [**AM_Location_AO**]
2. **Modify** Locations [**AM_Location_MO**]
3. **Delete** Locations [**AM_Location_ZO**]
4. **Inquire** on Locations [**AM_Location_IO**]

E06. Supplier-Item Matrix [AM_SuppItem_BR] consists of the following:
Attributes:
1. Supplier Code [SI_SuppCode] [A8] {References E03}
2. Item Code [SI_ItemCode] [A8] {References to E01}
Comments:
This entity stores information about which items come from which Supplier
Indexes:
1. Primary Key: [1, 2] (Constraint Name is **AM_SuppItem_PK**)
Valid Operations:
1. **Add** Supplier-Item Matrix entries [**AM_SuppItem_AO**]
2. **Delete** Supplier-Item Matrix entries [**AM_SuppItem_ZO**]
3. **Inquire** on Supplier-Item Matrix entries [**AM_SuppItem_IO**]
4. **Report** on Supplier-Item Matrix entries [**AM_SuppItem_RO**]

Figure C.4   (Continued)

E07. Customer [AM_Customer_BR] consists of the following:

Attributes:
1. Customer Code [CustCode] [A8]
2. Customer Name [CustName] [A30]
3. Address Line 1 [CustAddr1] [A30]
4. Address Line 2 [CustLocCode] [A10] {References E05}
5. State/Province Code [CustStProvCd] [A6] {References to E04} // Redundant because E05 → E04
6. Zip Code [CustZip] [N8]
7. Telephone Number(s) [CustPhoneNo] [A16]
8. Fax Number (s) [CustFaxNo] [A12]
9. Email Address [CustEmail] [A30]
10. Contact Person [CustContact] [A30]
11. Account Number [CustAcctNo] [N10] {References to E20}
12. Billing Preferences [CustBillingPref] [A30]

Comments:
1. This entity stores information about Customers.
2. Customer Code will consist of a 4-byte alphabetic code combined with a 4-digit sequential number, e.g. C01_199

Indexes:
1. Primary Key: [1] (Constraint Name is **AM_Customer_PK**)
2. **AM_Customer_**NX2 on [2]
3. **AM_Customer_**NX3 on [11]

Valid Operations:
1. **Add** Customers [**AM_Customer_AO**]
2. **Modify** Customers [**AM_Customer_MO**]
3. **Delete** Customers [**AM_Customer_ZO**]
4. **Inquire** on Customers [**AM_Customer_IO**]

E08. Purchase Order Summary [AM_POSum_BR] consists of the following:

Attributes:
1. Purchase Order Number [POS_RefNum] [N8]
2. Purchase Order Supplier Code [POS_SuppCode] [A8] {References E03}
3. Purchase Order Date [POS_Date] [N8]
4. Purchase Order Status [POS_Status] (Filled/ Partial/ Outstanding) [A30]
5. Purchase Order Estimated Amount [POS_EstAmount] [N(11,2)]
6. Purchase Order Estimated Discount [POS_EstDiscount] [N(11,2)]

Comments:
1. This entity stores information about Purchase Orders.
2. PO Code consists of 4-digit year concatenated with 4-digit sequential number (YYYYSSSS).
3. Date is of the format YYYYMMDD.

Indexes:
1. Primary Key: [1] (Constraint Name is **AM_POSum_PK**)
2. **AM_POSum_**NX2 on [2, 3]

Valid Operations:
1. **Add** Purchase Orders [**AM_POSum_AO**]
2. **Modify** Purchase Orders [**AM_POSum_MO**]
3. **Delete** Purchase Orders [**AM_POSum_ZO**]
4. **Inquire** on Purchase Orders [**AM_POSum_IO**]
5. **Report** on Purchase Orders [**AM_POSum_RO**]

Figure C.4 (Continued)

E09. Purchase Order Detail [AM_PODet_BR] consists of the following:

Attributes:
1. Purchase Order Number [POD_PONum] [N8] {References E08}
2. Item Code [POD_ItemCode] [A8] {References E01}
3. Item Quantity Ordered [POD_ItemQty] [N(7,2)]
4. Item Order Unit Price [POD_ItemUnitPrice] [N(9,2)]
5. Item Order Amount [POD_ItemAmount] [N(11,2)]

Comments:
1. This entity stores information about Purchase Order Details.
2. Order Amount is the product of quantity and unit price.

Indexes:
1. Primary Key: [1,2] (Constraint Name is **AM_PODet_PK**)

Valid Operations:
1. **Add** Purchase Order Details [AM_PODet_AO]
2. **Modify** Purchase Order Details [AM_PODet_MO]
3. **Delete** Purchase Order Details [AM_PODet_ZO]
4. **Inquire** on Purchase Order Details [AM_PODet_IO]
5. **Report** on Purchase Order Details [AM_PODet_RO]

E10. Purchase Invoice Summary [AM_PISum_BR] consists of the following:

Attributes:
1. Purchase Invoice Reference Number [PIS_RefNum] [N12]
2. Invoice Supplier-assigned Number [PIS_InvNum] [A8] // Issued by the supplier
3. Invoice Supplier Code [PIS_SuppCode] [A8] {References E03}
4. Invoice Date [PIS_InvDate] [N8]
5. Related Purchase Order Number [PIS_PONum] [N8] {References E08}
6. Invoice Amount [PIS_Amount] [N(11,2)]
7. Invoice Amount Outstanding [PIS_AmtOwed] [N(11,2)]
8. Discount [PIS_Discount] [N(11,2)]
9. Tax [PIS_InvTax] [N(9,2)]
10. Comment [PIS_Comment] [A30]
11. Transaction Reference Number [PIS_TransRefNum] [A16] {References E30}

Comments:
1. This entity stores information about Purchase Invoices.
2. PI Reference Number consists of the 4-digit year, 2-digit month and 6-digit sequential number: YYYYMMSSSSSS.
3. Transaction Reference Number has the literal "PI_" concatenated with the PI Reference Number: PI_YYYYMMSSSSSS.
4. Date is of the format YYYYMMDD.

Indexes:
1. Primary Key: [1] (Constraint Name is **AM_PISum_PK**)
2. **AM_PISum_NX2** on [3, 2, 4]
3. **AM_PISum_NX3** on [11]

Valid Operations:
1. **Add** Purchase Invoices [AM_PISum_AO]
2. **Modify** Purchase Invoices [AM_PISum_MO]
3. **Delete** Purchase Invoices [AM_PISum_ZO]
4. **Inquire** on Purchase Invoices [AM_PISum_IO]
5. **Report** on Purchase Invoices [AM_PISum_RO]

Figure C.4   (Continued)

E11. Purchase Invoice Detail [AM_PIDet_BR] consists of the following:

Attributes:
1. Purchase Invoice Reference [PID_RefNum] [N12] {References E10}
2. Item Number [PID_ItemCode] [A8] {References to E01}
3. Item Quantity [PID_ItemQty] [N(7,2)]
4. Item Unit Price [PID_ItemUnitPrice] [N(9,2)]
5. Item Amount [PID_ItemAmount] [N(11,2)]

Comments:
1. This entity stores information about Purchase Invoice Details.
2. Item Amount = [Item Quantity] * [Item Unit Price]

Indexes:
1. Primary Key: [1,2] (Constraint Name is AM_PIDet_PK)

Valid Operations:
1. **Add** Purchase Invoice Details [AM_PIDet_AO]
2. **Modify** Purchase Invoice Details [AM_PIDet_MO]
3. **Delete** Purchase Invoice Details [AM_PIDet_ZO]
4. **Inquire** on Purchase Invoice Details [AM_PIDet_IO]
5. **Report** on Purchase Invoice Details [AM_PIDet_RO]

E12. Purchase Returns Summary [AM_PRSum_BR] consists of the following:

Attributes:
1. Purchase Invoice Reference [PRS_RefNum] [N12] {References E10}
2. Return Date [PRS_RetDate] [N8]
3. Return Amount [PRS_Amount] [N(11,2)]
4. Purchase Return Identification [PRS_ReturnID] [N12]
5. Transaction Reference Number [PRS_TransRefNum] [A16] {References E30}

Comments:
1. This entity stores information about Purchase Returns.
2. PR Identification consists of 4-digit year, 2-digit month and 6-digit sequential number: YYYYMMSSSSSS.
3. Transaction Reference Number has literal "PR1_" concatenated with the PR Identification Number: PR1_YYYYMMSSSSSS.
4. Date is of the format YYYYMMDD.

Indexes:
1. Primary Key: [4] (Constraint Name is AM_PRSum_PK)
2. AM_PRSum_NX2 on [1,2]
3. AM_PRSum_NX3 on [5]

Valid Operations:
1. **Add** Purchase Returns [AM_PRSum_AO]
2. **Delete** Purchase Returns [AM_PRSum_ZO]
3. **Inquire** on Purchase Returns [AM_PRSum_IO]
4. **Report** on Purchase Returns [AM_PRSum_RO]

Figure C.4   (Continued)

E13. **Purchase-Returns Detail** [AM_PRDet_BR] consists of the following:

Attributes:
1. Purchase Return Identification [PRD_ReturnID] {**References E12**} [N12]
2. Item Code [PRD_ItemCode] [A8] {**References E01**}
3. Item Quantity Returned [PRD_ItemRetnQty] [N(7,2)]
4. Item Return Unit Price [PRD_ItemRetnPrice] [N(9,2)]
5. Item Return Amount [PRD_ItemRetnAmount] [N(11,2)]
6. Comment [PRD_Comment] [A40]

Comments:
1. This entity stores information about Purchase Return Details.
2. Item Amount = [Item Quantity] * [Item Unit Price].

Indexes:
Primary Key: [1,2] (Constraint Name **is AM_PRDet_PK**)

Valid Operations:
1. **Add** Purchase Returns Details [**AM_PRDet_AO**]
2. **Delete** Purchase Returns Details [**AM_PRDet_ZO**]
3. **Inquire** Purchase Returns Details [**AM_PRDet_IO**]
4. **Report** Purchase Returns Details [**AM_PRDet_RO**]

E14. **Sale Order Summary** [AM_SOSum_BR] consists of the following:

Attributes:
1. Sales Order Reference Number [SOS_RefNum] [N8]
2. Sales Order Customer-assigned Number [SOS_OrderNum] [A8]
3. Sales Order Customer Code [SOS_CustCode] [A8] {**References E07**}
4. Sales Order Date [SOS_Date] [N8]
5. Sales Order Status [SOS_Status] (Filled/ Partial/ Outstanding) [A1]
6. Sales Order Estimated Amount [SOS_EstAmount] [N(11,2)]
7. Sales Order Estimated Discount [SOS_EstDiscount] [N(11,2)]

Comments:
1. This entity stores information about Sales Orders.
2. Sale Order Reference Coding system consists of 4-digit year concatenated with 4-digit sequence number: YYYYSSSS.
3. Date is of the format YYYYMMDD.

Indexes:
1. Primary Key: [1] (Constraint Name is **AM_SOSum_PK**)
2. **AM_SOSum_NX2** on [3,2,4]

Valid Operations:
1. **Add** Sale Orders [**AM_SOSum_AO**]
2. **Modify** Sale Orders [**AM_SOSum_MO**]
3. **Delete** Sale Orders [**AM_SOSum_ZO**]
4. **Inquire** Sale Orders [**AM_SOSum_IO**]
5. **Report** Sale Orders [**AM_SOSum_RO**]

Figure C.4   (Continued)

E15. **Sales Order Detail [AM_SODet_BR]** consists of the following:

Attributes:
1. Sales Order Reference Number [SOD_RefNum] [N8] {**References E14**}
2. Item Code [SOD_ItemCode] [A8] {**References to E01**}
3. Quantity Ordered [SOD_ItemQty] [N(7,2)]
4. Anticipated Unit Price [SOD_ItemUnitPrice] [N(9,2)]
5. Estimated Amount [SOD_ItemjAmount] [N(11,2)]

Comments:
1. This entity stores information about Sales Order Details.
2. Item Amount = [Item Quantity] * [Item Unit Price].

Indexes:
1. Primary Key: [1,2] (Constraint Name is **AM_SODet_PK**)

Valid Operations:
1. **Add** Sale Order Details [**AM_SODet_AO**]
2. **Modify** Sale Order Details [**AM_SODet_MO**]
3. **Delete** Sale Order Details [**AM_SODet_ZO**]
4. **Inquire** Sale Order Details [**AM_SODet_IO**]
5. **Report** Sale Order Details [**AM_SODet_RO**]

E16. **Sales Invoice Summary [AM_SISum_BR]** consists of the following:

Attributes:
1. Sales Invoice Reference Number [SIS_RefNum] [N12]
2. Sales Invoice External Number [SIS_InvNum] [A8]
3. Sales Invoice Customer Code [SIS_CustCode] [A8] {**References to E07**}
4. Sales Invoice Date [SIS_InvDate] [N8]
5. Related Sale Order Number [SIS_SORefNum] [N8] {**References E14**}
6. Sales Invoice Amount [SIS_Amount] [N(11,2)]
7. Sales Invoice Amount Outstanding [SIS_AmtOwed] [N(11,2)]
8. Sales Invoice Discount [SIS_Discount] [N(11,2)]
9. Sales Invoice Tax [SIS_InvTax] [N(9,2)]
10. Transaction Reference Number [SIS_TransRefNum] [A16] {**References E30**}
11. Sales Invoice Comment [SIS_Comment] [A30]

Comments:
1. This entity stores information about Sales Invoices.
2. Sales Invoice Reference consists of the 4-digit year, 2-digit month, and 6-digit sequential number: YYYYMMSSSSSS.
3. Transaction Reference Number has the literal "SI_" concatenated with the SI Reference Number: SI_YYYYMMSSSSSS.
4. Date is of the format YYYYMMDD.

Indexes:
1. Primary Key: [1] (Constraint Name is **AM_SISum_PK**)
2. AM_SISum_NX2 on [2,3,4]
3. AM_SISum_NX3 on [10]

Valid Operations:
1. **Add** Sale Invoices [**AM_SISum_AO**]
2. **Modify** Sale Invoices [**AM_SISum_MO**]
3. **Delete** Sale Invoices [**AM_SISum_ZO**]
4. **Inquire** Sale Invoices [**AM_SISum_IO**]
5. **Report** Sale Invoices [**AM_SISum_RO**]

Figure C.4   (Continued)

E17. Sales Invoice Detail [AM_SIDet_BR] consists of the following:

Attributes:
1. Sales Invoice Reference Number [SID_RefNum] [N12] {**References E16**}
2. Item Code [SID_ItemCode] [A8] {**References E01**}
3. Item Quantity [SID_ItemQty] [N(7,2)]
4. Item Unit Price [SID_ItemUnitPrice] [N(9,2)]
5. Item Amount [SID_ItemAmount] [N(11,2)]

Comments:
1. This entity stores information about Sales Invoice Details.
2. Item Amount = [Item Quantity] * [Item Unit Price].

Indexes:
1. Primary Key: [1,2] (Constraint Name is **AM_SIDet_PK**)

Valid Operations:
1. **Add** Sale Invoice Detail [**AM_SIDet_AO**]
2. **Modify** Sale Invoice Details [**AM_SIDet_MO**]
3. **Delete** Sale Invoice Details [**AM_SIDet_ZO**]
4. **Inquire** Sale Invoice Details [**AM_SIDet_IO**]
5. **Report** Sale Invoice Details [**AM_SIDet_RO**]

E18. Sale Return Summary [AM_SRSum_BR] consists of the following:

Attributes:
1. Sales Invoice Return Identification Number [SRS_ReturnID] [N12]
2. Sale Invoice Reference Number [SRS_RefNum] [N12] {**References E16**}
3. Return Date [SRS_RetDate] [N8]
4. Return Amount [SRS_Amount] [N(11,2)]
5. Transaction Reference Number [SRS_TransRefNum] [A16] {**References E30**}

Comments:
1. This entity stores information about Sale Returns.
2. Sale Return ID consists of the 4-digit year, 2-digit month and 6-digit sequential number: YYYYMMSSSSSS.
3. Transaction Reference Number has literal "SR_" concatenated with the SI Identification Number: SR_YYYYMMSSSSSS.
4. Date is of the format YYYYMMDD.

Indexes:
1. Primary Key: [1] (Constraint Name is **AM_SRSum_PK**)
2. **AM_SRSum_NX2** on [2,3]
3. **AM_SRSum_NX3** on [5]

Valid Operations:
1. **Add** Sale Returns [**AM_SRSum_AO**]
2. **Modify** Sale Returns [**AM_SRSum_MO**]
3. **Delete** Sale Returns [**AM_SRSum_ZO**]
4. **Inquire** Sale Returns [**AM_SRSum_IO**]
5. **Report** Sale Returns [**AM_SRSum_RO**]

E19. Sale Return Detail [AM_SRDet_BR] consists of the following:

Attributes:
1. Sale Return Identification [SRD_ReturnID] [N12] {**References E18**}
2. Item Code [SRD_ItemCode] [A8] {**References E01**}
3. Quantity Returned [SRD_ItemRetnQty] [N(7,2)]
4. Return Unit Price [SRD_ItemRetnPrice] [N(9,2)]
5. Item Return Amount [SRD_ItemRetnAmount] [N(11,2)]
6. Comment [SRD_Comment] [A30]

Comments:
1. This entity stores information about Sales Returns.
2. Item Amount = [Item Quantity] * [Item Unit Price].

Indexes:
1. Primary Key: [1,2] (Constraint Name is **AM_SRDet_PK**)

Valid Operations:
1. **Add** Sale Return Detail [**AM_SRDet_AO**]
2. **Delete** Sale Return Detail [**AM_SRDet_ZO**]
3. **Inquire** Sale Return Detail [**AM_SRDet_IO**]
4. **Report** Sale Return Detail [**AM_SRDet_RO**]

Figure C.4   (Continued)

C.2.3 *Financial Management Subsystem*

The Financial Management Subsystem (FMS) addresses matters that have financial implications. Following is the O/ESG for each information entity comprising this subsystem (Figure C.5).

E20. Chart of Accounts [FM_ChartAccts_BR] consists of the following:

Attributes:
1. Account Number [AcctNum] [N10]
2. Account Description [AcctDesc] [A30]
3. Summary/Detail Flag [AcctFlag] [A1]
4. Parent Account [AcctParent] [N10] {References E20}

Comments:
1. This entity stores information about the various Accounts in the system.
2. Account Number consists of 10-byte numeric code where the first 4 bytes represents account category and the last 6 bytes represent the account: CCCCAAAAAA.

Indexes:
1. Primary Key: [1] (Constraint Name is FM_ChartAccts_PK)
2. FM_ChartAccts_NX2 on [2]

Valid Operations:
1. Add Accounts [FM_ChartAccts_AO]
2. Delete Accounts [FM_ChartAccts_ZO]
3. Inquire Accounts [FM_ChartAccts_IO]
4. Report Accounts [FM_ChartAccts_RO]

E21. Department [FM_Department_BR] consists of the following:

Attributes:
1. Department Number [DeptNum] [N4]
2. Department Name [DeptName] [A30]

Comments:
1. This entity stores information about the various Departments in the system.
2. Department Number consists of 4-byte numeric code.

Indexes:
1. Primary Key: [1] (Constraint Name is FM_Department_PK)
2. FM_Department_NX2 on [2]

Valid Operations:
1. Add Departments [FM_Department_AO]
2. Delete Departments [FM_Department_ZO]
3. Inquire Departments [FM_Department_IO]
4. Report Departments [FM_Department_RO]

Figure C.5 O/ESG for the Financial Management Subsystem

(Continued)

E22. Account Balances Log [FM_AcctBal_BR] consists of the following:

Attributes:
1. Financial Period [AB_Period] [N6]
2. Account Number [AB_AccountNum] [N10] {References E20}
3. Department Number [AB_DeptNum] [N4] {References E21}
4. Balance [AB_Balance] [N(11,2)]
5. Comment [AB_Comment] [A30]

Comments:
1. This entity stores periodic information about the Account Balances for each Department.
2. Period is of the format YYYYMM, e.g. 201204.

Indexes:
1. Primary Key: [1,2,3] (Constraint Name is **FM_AcctBal_PK**)

Valid Operations:
1. **Add** Account Balances [**FM_AcctBal_AO**]
2. **Modify** Account Balances [**FM_AcctBal_MO**]
3. **Delete** Account Balances [**FM_AcctBal_ZO**]
4. **Inquire** Account Balances [**FM_AcctBal_IO**]
5. **Report** Account Balances **FM_AcctBal_RO**]

E23. Purchase Invoice Payment Plan [FM_PIPayPlan_BR] consists of the following:

Attributes:
1. Purchase Invoice Reference Number [N12] [PIPP_InvRefNum] {**References to E10**}
2. Required Day of Month [PIPP_PayDay] [N2]
3. Number of Payments Required [PIPP_NumPaymts] [N3]
4. Payment Begin Date [PIPP_BeginDate] [N8]

Comments:
1. This entity stores information about Purchase Invoice Payment Plans.
2. Date is of the format YYYYMMDD.

Indexes:
1. Primary Key: [1] (Constraint Name is **FM_PIPayPlan_PK**)

Valid Operations:
1. **Add** Purchase Invoice Payment Plans [**FM_PIPayPlan_AO**]
2. **Modify** Purchase Invoice Payment Plans [**FM_PIPayPlan_MO**]
3. **Delete** Purchase Invoice Payment Plans [**FM_PIPayPlan_ZO**]
4. **Inquire** Purchase Invoice Payment Plans [**FM_PIPayPlan_IO**]
5. **Report** Purchase Invoice Payment Plans [**FM_PIPayPlan_RO**]

Figure C.5   (Continued)

E24. Purchase Payments Log [FM_PurchPay_BR] consists of the following:

Attributes:
1. Payment Reference Number [PP_RefNum] [N12]
2. Purchase Invoice Reference Number [PP_InvRevNum] [N12] {References E10}
3. Payment Date [PP_PayDate] [N8]
4. Amount Paid [PP_Amount] [N(11,2)]
5. Payment Type [PP_PayType] (Cash/Check/Credit Card) [A4]
6. Institution Code [PP_InstCode] [A8] {References E25}
7. Check/Credit Account Number [PP_InstAcctNum] [N16]
8. Transaction Reference Number [PP_TransRefNum] [A16] {References E30}
9. Comment [PP_Comment] A30

Comments:
1. This entity stores information about Purchase Payments Made.
2. Payment Reference involves a 4-digit year, 2-digit month and 6-digit sequence number: YYYYMMSSSSSS.
3. Transaction Reference Number consists of the literals "PP1_" or "PP2_" concatenated with the Purchase Payment Reference Number: PP1_YYYYMMSSSSSS or PP2_YYYYMMSSSSSS.
4. Date is of the format YYYYMMDD.

Indexes:
1. Primary Key: [1] (Constraint Name is FM_PurchPay_PK)
2. FM_PurchPay_NX2 on [2,3]
3. FM_PurchPay_NX3 on [8]

Valid Operations:
1. Add Purchase Payments [FM_PurchPay_AO]
2. Delete Purchase Payments [FM_PurchPay_ZO]
3. Inquire Purchase Payments [FM_PurchPay_IO]
4. Report Purchase Payments [FM_PurchPay_RO]

E25. Financial Institution [FM_FinInst_BR] consists of the following:

Attributes:
1. Institution Code [InstCode] [A8]
2. Institution Name [InstName] [A30]
3. Address Line 1 [InstAddr1] [A30]
4. Address Line 2 — Location Code [InstLocCode] [A10] {References E05}
5. State/Province Code [InstStProvCode] [A6] {References E04} // Redundant since E05 → E04
6. Zip Code [InstZip] [A8]
7. Telephone Number(s) [InstPhoneNo] [A16]
8. Fax Number(s) [InstFaxNo] [A10]
9. Contact Person [InstContact] [A30]
10. Email Address [InstEmail] [A30]

Comments:
1. This entity stores information about Financial Institutions.
2. Institution Code will consist of a 4-byte alphabetic code combined with a 4-digit sequential number, e.g. I01_0145.

Indexes:
1. Primary Key: [1] (Constraint Name is FM_FinInst_PK)
2. FM_FinInst_NX2 on [2]
3. FM_FinInst_NX3 on [7]

Valid Operations:
1. Add Financial Institutions [FM_FinInst_AO]
2. Modify Financial Institutions [FM_FinInst_MO]
3. Delete Financial Institutions [FM_FinInst_ZO]
4. Inquire Financial Institutions [FM_FinInst_IO]
5. Report Financial Institutions [FM_FinInst_RO]

Figure C.5 (Continued)

E26. Sale Invoice Payment Plan FM_SIPayPlan_BR] consists of the following:

Attributes:
1. Sales Invoice Reference Number [SIPP_InvRefNum] [N12] {**References E16**}
2. Required Day of Month [SIPP_PayDay] [N2]
3. Number of Payments Required SIPP_NumPaymts] [N3]
4. Payment Begin Date [SIPP_BeginDate] [N8]

Comments:
This entity stores information about Sale Invoices Payment Plans.

Indexes:
1. Primary Key: [1] (Constraint Name is **FM_SIPayPlan_PK**)

Valid Operations:
1. **Add** Sale Invoice Payment Plans [**FM_SIPayPlan_AO**]
2. **Delete** Sale Invoice Payment Plans [**FM_SIPayPlan_ZO**]
3. **Inquire** Sale Invoice Payment Plans [**FM_SIPayPlan_IO**]
4. **Report** Sale Invoice Payment Plans [**FM_SIPayPlan_RO**]

E27. Sale Payments Log [FM_SalePay_BR] consists of the following:

Attributes:
1. Sale Payment Reference Number [SP_RefNum] [N12]
2. Sales Invoice Reference Number [SP_InvRefNum] [N12] {**References E16**}
3. Payment Received Date [SP_PayDate] [N8]
4. Amount Received [SP_Amount] [N(11,2)]
5. Payment Type [SP_PayType] (Cash/Check/Credit Card) [A4]
6. Institution Code [SP_InstCode] [A8] {**References E25**}
7. Check/Credit Institution Account Number [SP_InstAcctNum] [N16]
8. Transaction Reference Number [SP_TransRefNum] [A16] {**References E30**}
9. Comment [SP_Comment] [A30]

Comments:
1. This entity stores information about Sale Payments Received.
2. Sale Payment Reference consists of the 4-digit year, 2-digit month and 6-digit sequence number: YYYYMMSSSSSS.
3. Transaction Reference Number consists of the literals "SP1_" or "SP2_" concatenated with the Sale Payment Reference Number: SP1_YYYYMMSSSSSS or SP2_YYYYMMSSSSSS.
4. Date is of the format YYYYMMDD.

Indexes:
1. Primary Key: [1] (Constraint Name is **FM_SalePay_PK**)
2. **FM_SalePay_NX2** on [2, 3]
3. **FM_SalePay_NX3** on [8]

Valid Operations:
1. **Add** Payments Received [**FM_SalePay_AO**]
2. **Delete** Payments Received [**FM_SalePay_ZO**]
3. **Inquire** Payments Received [**FM_SalePay_IO**]
4. **Report** Payments Received [**FM_SalePay_RO**]

Figure C.5   (Continued)

```
E28.  Employee [FM_Employee_BR] consists of the following:
Attributes:
1.    Employee Identification Number [EmpNum]  [N8]
2.    Employee First Name [EmpFName] [A15]
3.    Employee Middle Name(s) [EmpMName] [A25]
4.    Employee Last Name [EmpLName] [A15]
5.    Employee Address Line 1 [EmpAddr1] [A30]
6.    Employee Address Line 2 — Location Code [EmpLocCode] [A10] {References E05}
7.    Employee State/Province Code [EmpStProvCode] [A6] {References E04} // Redundant since E05 → E04
8.    Employee Zip Code [IEmpZip] [A8]
9.    Employee Telephone Number [EmpPhone1] [A12]
10.   Employee Alternate Telephone Number [EmpPhone2] [A12]
11.   Employee Email Address [EmpEmail] [A30]
12.   Employee Emergency Contact Person [EmpContact] [A30]
13.   Employee Emergency Contact Telephone [EmpEmgPhone] [A12]
14.   Employee Related Account Number [EmpAcctNum] [N10] {References E20}
Comments:
1.    This entity stores information about Employees.
2.    Employee Number consists of a 4-digit year combined with a 4-digit sequence number, e.g. 20100199.
Indexes:
1.    Primary Key: [1] {Constraint Name is [FM_Employee_PK]
2.    FM_Employee_NX2 on [4,2,3]
3.    FM_Employee_NX3 on [9]
4.    FM_Employee_NX4 on [11]
Valid Operations:
1.    Add Employees [FM_Employee_AO]
2.    Modify Employees [FM_Employee_MO]
3.    Delete Employees [FM_Employee_ZO]
4.    Inquire Employees [FM_Employee_IO]
5.    Report Employees [FM_Employee_RO]
```

Figure C.5   (Continued)

E29. Payroll Log [FM_PayrollLog_BR] consists of the following:

Attributes:

1. Payroll Log Identification Code [PL_RefCode] [A17]
2. Payroll Date [PL_PayDate] [N8]
3. Employee ID Number [PL_EmpNum] [N8] {**References E28**}
4. Payment Gross Amount [PL_PayGross] [N(11,2)]
5. Insurance Payment [PL_InsOrgAmount] [N(7,2)]
6. Insurance Employee Contribution [PL_InsEmpAmount] [N(7,2)]
7. Investment Payment [PL_InvOrgAmount] [N(7,2)]
8. Investment Employee Contribution [PL_InvEmpAmount] [N(7,2)]
9. Federal Tax Deduction [PL_FedTax] [N(7,2)]
10. Federal State/Prov Deduction [PL_ProvTax] [N(7,2)]
11. Other Deductions [PL_OtherDeduct] [N(7,2)]
12. Payment Net Amount to Employee [PL_PayNet] [N(11,2)]
13. Total Organizational Expense for this Employment [PL_TotalOutlay] [N(11,2)]
14. Related Accounting Period [PL_AcctPeriod] [N6]
15. Payroll Reference Number [PL_RefNum] [N12]
16. Transaction Reference Number [PL_TranRefNum] [A16] {**References E30**}
17. Comment [PL_Comment] [A30]

Comments:

1. This entity stores information about Payroll disbursements every payroll period.
2. Payroll Log Identification Code consists of the Payroll Date concatenated with the Employee Number, e.g. 20190630_20040145.
3. Date is of the format YYYYMMDD.
4. Accounting Period is of the format YYYYMM.
5. Payroll Reference Number consists of the Accounting Period concatenated with a 6-digit sequence number: YYYYMMSSSSSS.
6. Transaction Reference Number has literal "PR2_" concatenated with the Payroll Reference Number: PR2_YYYYMMSSSSSS.

Indexes:

1. Primary Key: [1] (Constraint Name is [**FM_PayrollLog_PK**)
2. **FM_PayrollLog_NX2** on [2,3]
3. **FM_PayrollLog_NX3** on [15]
4. **FM_PayrollLog_NX4** on [16]

Valid Operations:

1. **Add** Employees [**FM_PayrollLog_AO**]
2. **Modify** Employees [**FM_PayrollLog_MO**]
3. **Delete** Employees [**FM_PayrollLog_ZO**]
4. **Inquire** Employees [**FM_PayrollLog_IO**]
5. **Report** Employees [**FM_PayrollLog_RO**]

Figure C.5   (Continued)

E30. Financial Transactions [FM_FinTran_BR] consists of the following:

Attributes:
1. Transaction Date [FT_Date] [N8]
2. Transaction Classification Code [FT_ClassCode] [A3] {**References to E31**}
3. Account Number [FT_AcctNum] [N10] {**References E20**}
4. Department Number [FT_DeptNum] [N4] {**References E21**}
5. Debit Amount [FT_Debit] [N(11,2)]
6. Credit Amount [FT_Credit] [N(11,2)]
7. Accounting Period [FT_Period] [N6]
8. Transaction Reference Number [FT_TransRefNum] [A16] // See Note below
9. Transaction Comment [FT_Comment] [A40]

Comments:
1. This entity stores information about Financial Transactions.
2. Transaction Reference Number is of the form CCC_YYYYMMSSSSSS where CCC represents the classification code.
3. Accounting Period is of the form YYYYMM.

Indexes:
1. Primary Key: [8] (Constraint Name is **FM_FinTran_PK**)
2. **FM_FinTran_NX2** on [4,3]
3. **FM_FinTran_NX3** on [2,4,3]

Valid Operations:
1. **Add** Financial Transactions [**FM_FinTran_AO**]
2. **Delete** Financial Transactions [**FM_FinTran_ZO**]
3. **Inquire** Financial Transactions [**FM_FinTran_IO**]
4. **Report** Financial Transactions [**FM_FinTran_RO**]

Note: Based on the classification code, each Transaction Reference ties back to one of the following entities:

PI: Purchase Invoice Transactions (**E10**)
PR1: Purchase Return Transactions (**E12**)
SI: Sale Invoice Transactions (**E16**)
SR: Sale Return Transactions (**E18**)
PP1: Cash Purchase Payments Transactions (**E24**)
PP2: Credit Purchase Payments Transactions (**E24**)
SP1: Cash Sale Payments Transactions (**E27**)
SP2: Credit Sale Payments Transactions (**E27**)
INV: Investment Transactions (**E32**)
PR2: Payroll Log (**E29**)

Comments: These will be implemented via logical views.

Figure C.5   (Continued)

E31. Transaction Classification [FM_TranClass_BR] consists of the following:

Attributes:
1. Transaction Classification Code [TransCode] [A3]
2. Transaction Classification Description [TransDesc] [A30]

Comments:
This entity stores information about Transaction Classifications.

Indexes:
1. Primary Key: [1] (Constraint Name is **FM_TranClass_PK**)

Valid Operations:
1. **Add** Transaction Classifications [**FM_TranClass_AO**]
2. **Delete** Transaction Classifications [**FM_TranClass_ZO**]
3. **Inquire** Transaction Classifications [**FM_TranClass_IO**]
4. **Report** Transaction Classifications [**FM_TranClass_RO**]

Note: Transaction Classifications include:

PI ≡ Purchase Invoices;	**PR1** ≡ Purchase Returns;
SI ≡ Sales Invoices;	**SR** ≡ Sales Returns;
PP1 ≡ Cash Purchase Payments;	**PP2** ≡ Credit Purchase Payments;
SP1 ≡ Cash Sales Payments;	**SP2** ≡ Credit Sales Payments;
JED ≡ Journal Entries Debit;	**JEC** ≡ Journal Entries Credit;
INV ≡ Investments;	**PR2** ≡ Payroll Entry.

E32. Investments Log [FM_Invest_BR] consists of the following:

Attributes:
1. Investment Reference Number [Inv_RefNum] [N12]
2. Investment Transaction Date [Inv_Date] [N8]
3. Investment Amount [Inv_Amount] [N(11,2)]
4. Institution Code [Inv_InstCode] [A8] {**References E25**}
5. Institutional Account Number [Inv_AcctNum] [N16]
6. Internal Account Number [Inv_IntAcctNum] [N10] {**References E20**}
7. Transaction Reference Number [Inv_TransRefNum] [A16] {**References E30**}
8. Comment [Inv_Comment] [A30]

Comments:
1. This entity stores information about Investments activities.
2. Investment Reference Number involves a 4-digit year, 2-digit month and 6-digit sequence number: YYYYMMSSSSSS.
3. Transaction Reference Number consists of the literals "INV_" concatenated with the Investment Reference Number: INV_YYYYMMSSSSSS.

Indexes:
1. Primary Key: [1] (Constraint Name is **FM_Invest_PK**)
2. **FM_Invest_NX2** on [2,4]
3. **FM_Invest_NX3** on [7]

Valid Operations:
1. **Add** Investment Actions [**FM_Invest_AO**]
2. **Modify** Investment Actions [**FM_Invest_MO**]
3. **Delete** Investment Actions [**FM_Invest_ZO**]
4. **Inquire** Investment Actions [**FM_Invest_IO**]
5. **Report** Investment Actions [**FM_Invest_RO**]

Figure C.5   (Continued)

C.2.4 Systems Control Subsystem

The System Controls Subsystem (SCS) concerns itself with covert but essential matters that facilitate the smooth operation of the system. Following is the O/ESG for each information entity comprising this subsystem (Figure C.6):

E33. Audit File for Addition [SC_AudAdd_BR] includes the following:

Attributes:
1. Session ID [AA_AddID] [N12] {References E38}
2. Add Code [AA_AddCode] [N4]
3. Add Detail 1 [AA_AddDet1] [A75]
4. Add Detail 2 [AA_AddDet2] [A75]
5. Add Detail 3 [AA_AddDet3] [A75]
6. Add Detail 4 [AA_AddDet4] [A75]
7. Add File [AA_AddFile] [A12]

Comments:
1. This entity logs all additions of data to the system.
2. Session ID identifies session and will be in the form of a date and a sequence number: YYYYMMDDSSSS.
3. Add Code will be in the form of a sequence number.

Indexes:
1. Primary Key: [1,2] (Constraint Name is SC_AudAdd_PK)

Valid Operations:
1. Add Audit File Additions [SC_AudAdd_AO]
2. Delete Audit File Additions [SC_AudAdd_ZO]
3. Inquire on Audit File Additions [SC_AudAdd_IO]
4. Report Audit File Additions [SC_AudAdd_RO]

E34. Audit File for Update [SC_AudUpd_BR] includes the following:

Attributes:
1. Session ID [AU_UpdateID] [N12] {References E38}
2. Update Code [AU_UpdateCode] [N4]
3. Before Detail 1 [AU_UpdateBefDet1] [A75]
4. Before Detail 2 [AU_UpdateBefDet2] [A75]
5. Before Detail 3 [AU_UpdateBefDet3] [A75]
6. Before Detail 4 [AU_UpdateBefDet4] [A75]
7. After Detail 1 [AU_UpdateAftDet1] [A75]
8. After Detail 2 [AU_UpdateAftDet2] [A75]
9. After Detail 3 [AU_UpdateAftDet3] [A75]
10. After Detail 4 [AU_UpdateAftDet4] [A75]
11. Update File [AU_UpdateFile] [A12]

Comments:
1. This entity logs all update of data in the system.
2. Session ID identifies session and will be in the form of a date and a sequence number: YYYYMMDDSSSS.
3. Update Code will be in the form of a sequence number.

Indexes:
1. Primary Key: [1,2] (Constraint Name is SC_AudUpd_PK)

Valid Operations:
1. Add Audit File Updates [SC_AudUpd_AO]
2. Delete Audit File Updates [SC_AudUpd_ZO]
3. Inquire on Audit File Updates [SC_AudUpd_IO]
4. Report Audit File Updates [SC_AudUpd_RO]

Figure C.6   O/ESG for the System Controls Subsystem

(Continued)

E35. Audit File for Deletion [SC_AudDel_BR] includes the following:

Attributes:
1. Session ID [AD_DelID] [N12] {References E38}
2. Delete Code [AD_DelCode] [N4]
3. Delete Detail 1 [AD_DelDetail1] [A75]
4. Delete Detail 2 [AD_DelDetail2] [A75]
5. Delete Detail 3 [AD_DelDetail3] [A75]
6. Delete Detail 4 [AD_DelDetail4] [A75]
7. Delete Comment [AD_DelComment] [A75]
8. Delete File [AD_DelFile] [A12]

Comments:
1. This table logs all deletion of data from the system.
2. Session ID identifies session and will be in the form of a date and a sequence number: YYYYMMDDSSSS.
3. Delete Code be in the form of a sequence number.

Indexes:
1. Primary Key: [1,2] {Constraint Name is **SCAudDelPK**}

Valid Operations:
1. Add Audit File Deletions [**SC_AudDel_AO**]
2. Delete Audit File Deletions [**SC_AudDel_ZO**]
3. Inquire on Audit File Deletions [**SC_AudDel_IO**]
4. Report Audit File Deletions [**SC_AudDel_RO**]

E36. System Message [SC_Message_MF] includes the following:

Attributes:
1. Message ID [MsgID] [A7]
2. Message Description [MsgDesc] [A75]

Comments:
1. This entity is used for storing defined system messages.
2. Message ID will be of the form SSEEEnn [SS = System or subsystem (AM, FM, SC), EEE = Entity Reference (E01...E31), nn = Sequence #]. Example. AME0101...AME0112.

Indexes:
1. Primary Key: [1] {Constraint Name is **SC_Message_PK**}

Valid Operations:
1. Add Messages **SC_Message_AO**]
2. Delete Messages [**SC_Message_ZO**]
3. Retrieve Messages [**SC_Message_RO**]

E37. System User [SC_SysUser_BR] includes the following:

Attributes:
1. User Code [UserCode] [N8]
2. User Full Name [UserName] [A35]
3. User Login Name [UserLogin] [A20]
4. User Password [UserPass] [A16]

Comments:
1. This entity defines users who will be directly using the system.
2. The User Code will be of the form YYYYSSSS (Y = Year, S = Sequence #).

Indexes:
1. Primary Key: [1] {Constraint Name is **SC_SysUser_PK**}
2. **SC_SysUser_NX2** on [2]
3. **SC_SysUser_NX3** on [3]

Valid Operations:
1. Add System Users [**SC_SysUser_AO**]
2. Modify System Users [**SC_SysUser_MO**]
3. Delete System Users [**SC_SysUser_ZO**]
4. Inquire on System Users [**SC_SysUser_IO**]
5. Report on System Users [**SC_SysUser_RO**]

Figure C.6   (Continued)

```
E38. Sessions Log [SC_SessLog_BR] includes the following:
Attributes:
1.  Session ID [SessLogID] [N12]
2.  User Code [SessLogUser] [N8] {References E37}
3.  Login Time [SessLoginTime] [N6]
4.  Logout Time [SessLogoutTime] [N6]
5.  Normal/Abnormal Flag [SessLogNormFlag] [A1]
Comments:
1.  This entity defines sessions of users using the system.
2.  Session ID identifies session and will be in the form of a date and a sequence number: YYYYMMDDSSSS.
3.  The Normal/Abnormal Flag indicates whether the session ended normally or abnormally (i.e. N for Normal or A for Abnormal).
Indexes:
1.  Primary Key: [1] (Constraint Name is SC_SessLog_PK)
Valid Operations:
1.  Add Sessions [SC_SessLog_AO]
2.  Delete Sessions [SC_SessLog_ZO]
3.  Inquire on Sessions [SC_SessLog_IO]
4.  Report on Sessions [SC_SessLog_RO]


E39. Session Mark [SC_SessMark_BR] includes the following:
Attributes:
1.  Session Mark [SessionMark] [N4]
2.  Session Date [SessionDate] [N8]
Comments:
This table is used to facilitate auditing by session. Each day, the Session Mark is initialized before users can access the system.
Indexes:
1.  Primary Key: [2] (Constraint Name is SC_SessMark_PK)
Valid Operations:
1.  Initialize System [SC_Initialize_XO]
```

Figure C.6   (Continued)

C.3 User Interface Specification

The IMS will employ a graphical user interface (GUI) that is easy for end-users to understand and use. This chapter provides a brief overview.

C.3.1 User Interface Topology

The user interface design is based on Schneiderman's *object-action interface* (OAI) model for user interfaces. The real benefit of this approach is that it is consistent with the way people tend to think: People do not think about the functional intricacies of their daily activities; rather, they think about objects and what they desire to do with them. Because of the natural fit to the typical thought process on the job, user learning will be enhanced.

The menu system will be hierarchical, as represented in the user interface topology chart (UITC) of Figure C.7. The user interface will be a GUI with the following features:

- At the highest level, the main menu will consist of three options representing the four subsystems.
- For each subsystem, the menu options will point to each information entity managed in that subsystem.
- Any option taken from a subsystem menu will invoke a pop-up menu with the operations relevant to that particular information entity (object type).

Inventory Management System Main Menu
1. Acquisitions Management Subsystem (AMS)
2. Financial Management Subsystem (FMS)
3. System Controls Subsystem (SCS)
4. Point of Sale Subsystem (POSS)

1. Acquisitions Management Subsystem (AMS)
1.1 Item Definitions
1.1.1 Add Item Definitions
1.1.2 Modify Item Definitions
1.1.3 Delete Item Definitions
1.1.4 Inquire/Report on Item Definitions

1.2 Item Category Definitions
1.2.1 Add Item Category Definitions
1.2.2 Modify Item Category Definitions
1.2.3 Delete Item Category Definitions
1.2.4 Inquire/Report on Item Category Definitions

1.3 Supplier Definitions
1.3.1 Add Supplier Definitions
1.3.2 Modify Supplier Definitions
1.3.3 Delete Supplier Definitions
1.3.4 Inquire/Report on Supplier Definitions

1.4 State/Province Definitions
1.4.1 Add State/Province Definitions
1.4.2 Modify State/Province Definitions
1.4.3 Delete State/Province Definitions
1.4.4 Inquire/Report on State/Province Definitions

1.5 Location Definitions
1.5.1 Add Location Definitions
1.5.2 Modify Location Definitions
1.5.3 Delete Location Definitions
1.5.4 Inquire/Report on Location Definitions

1.6 Supplier-Item Matrix
1.6.1 Add Supplier-Item Combinations
1.6.2 Delete Supplier-Item Combinations
1.6.3 Inquire/Report on Supplier-Item Combinations

1.7 Customer Definitions
1.7.1 Add Customer Definitions
1.7.2 Modify Customer Definitions
1.7.3 Delete Customer Definitions
1.7.4 Inquire/Report on Customer Definitions

1. Acquisitions Management Subsystem (AMS)
1.8 Purchase Order Summaries
1.8.1 Add Purchase Order Summaries
1.8.2 Modify Purchase Order Summaries
1.8.3 Delete Purchase Orders
1.8.4 Inquire/Report on Purchase Order Summaries

1.9 Purchase Order Details
1.9.1 Add Purchase Order Details
1.9.2 Modify Purchase Order Details
1.9.3 Delete Purchase Order Details
1.9.4 Inquire/Report on Purchase Order Details

1.10 Purchase Invoice Summaries
1.10.1 Add Purchase Invoice Summaries
1.10.2 Modify Purchase Invoice Summaries
1.10.3 Delete Purchase Invoices
1.10.4 Inquire/Report on Purchase Invoice Summaries

1.11 Purchase Invoice Details
1.11.1 Add Purchase Invoice Details
1.11.2 Modify Purchase Invoice Details
1.11.3 Delete Purchase Invoice Details
1.11.4 Inquire/Report on Purchase Invoice Details

1.12 Purchase Return Summaries
1.12.1 Add Purchase Return Summaries
1.12.2 Delete Purchase Returns
1.12.3 Inquire/Report on Purchase Return Summaries

1.13 Purchase Return Details
1.13.1 Add Purchase Return Details
1.13.2 Modify Purchase Return Details
1.13.3 Delete Purchase Return Details
1.13.4 Inquire/Report on Purchase Return Details

1.14 Sale Order Summaries
1.14.1 Add Sale Order Summaries
1.14.2 Modify Sale Order Summaries
1.14.3 Delete Sale Orders
1.14.4 Inquire/Report on Sale Order Summaries

1.15 Sale Order Details
1.15.1 Add Sale Order Details
1.15.2 Modify Sale Order Details
1.15.3 Delete Sale Order Details
1.15.4 Inquire/Report on Sale Order Details

Figure C.7   IMS User Interface Topology Chart

2. Acquisitions Management Subsystem (AMS)
1.16   Sale Invoice Summaries
1.16.1 Add Sale Invoice Summaries
1.16.2 Modify Sale Invoice Summaries
1.16.3 Delete Sale Invoices
1.16.4 Inquire/Report on Sale Invoice Summaries
1.17   Sale Invoice Details
1.17.1 Add Sale Invoice Details
1.17.2 Modify Sale Invoice Details
1.17.3 Delete Sale Invoice Details
1.17.4 Inquire/Report on Sale Invoice Details
1.18   Sale Return Summaries
1.18.1 Add Sale Return Summaries
1.18.2 Delete Sale Returns
1.18.3 Inquire/Report on Sale Return Summaries
1.19   Sale Return Details
1.19.1 Add Sale Return Details
1.19.2 Modify Sale Return Details
1.19.3 Delete Sale Return Details
1.19.4 Inquire/Report on Sale Return Details

3. Financial Management Subsystem (FMS)
2.1   Account Definitions
2.1.1 Add Account Definitions
2.1.2 Modify Account Definitions
2.1.3 Delete Account Definitions
2.1.4 Inquire/Report on Account Definitions
2.2   Department Definitions
2.2.1 Add Department Definitions
2.2.2 Modify Department Definitions
2.2.3 Delete Department Definitions
2.2.4 Inquire/Report on Department Definitions
2.3   Account Balances
2.3.1 Add Account Balances
2.3.2 Modify Account Balances
2.3.3 Delete Account Balances
2.3.4 Inquire/Report on Account Balances
2.4   Purchase Payment Plans
2.4.1 Add Purchase Payment Plans
2.4.2 Modify Purchase Payment Plans
2.4.3 Delete Purchase Payment Plans
2.4.4 Inquire/Report on Purchase Payment Plans

2. Financial Management Subsystem (FMS)
2.5   Purchase Payment Logs
2.5.1 Add Purchase Payment Logs
2.5.2 Delete Purchase Payment Logs
2.5.3 Inquire/Report on Purchase Payment Logs
2.6   Financial Institution Definitions
2.6.1 Add Financial Institution Definitions
2.6.2 Modify Financial Institution Definitions
2.6.3 Delete Financial Institution Definitions
2.6.4 Inquire/Report on Financial Institution Definitions
2.7   Sale Payment Plans
2.7.1 Add Sale Payment Plans
2.7.2 Modify Sale Payment Plans
2.7.3 Delete Sale Payment Plans
2.7.4 Inquire/Report on Sale Payment Plans
2.8   Sale Payment Logs
2.8.1 Add Sale Payment Logs
2.8.2 Delete Sale Payment Logs
2.8.3 Inquire/Report on Sale Payment Logs
2.9   Employee Definitions
2.9.1 Add Employee Definitions
2.9.2 Modify Employee Definitions
2.9.3 Delete Employee Definitions
2.9.4 Inquire/Report on Employee Definitions
2.10   Payroll Logs
2.10.1 Add Payroll Logs
2.10.2 Delete Payroll Logs
2.10.3 Inquire/Report on Payroll Logs
2.11   Financial Transaction Logs
2.11.1 Add Financial Transaction Logs
2.11.2 Modify Financial Transaction Logs
2.11.3 Delete Financial Transaction Logs
2.11.4 Inquire/Report on Financial Transaction Logs
2.12   Financial Classification Definitions
2.12.1 Add Financial Classification Definitions
2.12.2 Delete Financial Classification Definitions
2.12.3 Inquire/Report on Financial Classifications
2.13   Financial Investment Logs
2.13.1 Add Financial Investment Logs
2.13.2 Modify Financial Investment Logs
2.13.3 Delete Financial Investment Logs
2.13.4 Inquire/Report on Financial Investment Logs

Figure C.7   (Continued)

3. Point of Sale Subsystem (POSS)
{Selected operations from the AMS}
3.1 Sale Invoice Summaries
3.1.1 Add Sale Invoice Summaries
3.1.2 Modify Sale Invoice Summaries
3.1.3 Delete Sale Invoices
3.1.4 Inquire/Report on Sale Invoice Summaries
3.2 Sale Invoice Details
3.2.1 Add Sale Invoice Details
3.2.2 Modify Sale Invoice Details
3.2.3 Delete Sale Invoice Details
3.2.4 Inquire/Report on Sale Invoice Details
3.3 Sale Return Summaries
3.3.1 Add Sale Return Summaries
3.3.2 Delete Sale Returns
3.3.3 Inquire/Report on Sale Return Summaries
3.4 Sale Return Details
3.4.1 Add Sale Return Details
3.4.2 Modify Sale Return Details
3.4.3 Delete Sale Return Details
3.4.4 Inquire/Report on Sale Return Details
3.5 Sale Payment Logs
3.5.1 Add Sale Payment Logs
3.5.2 Delete Sale Payment Logs
3.5.3 Inquire/Report on Sale Payment Logs

4. System Controls Subsystem (SCS)
4.1 Insertion Audit Logs
4.1.1 Add Insertion Audit Logs
4.1.2 Delete Insertion Audit Logs
4.1.3 Inquire/Report on Insertion Audit Logs
4.2 Modification Audit Logs
4.2.1 Add Modification Audit Logs
4.2.2 Delete Modification Audit Logs
4.2.3 Inquire/Report on Modification Audit Logs
4.3 Removal Audit Logs
4.3.1 Add Removal Audit Logs
4.3.2 Delete Removal Audit Logs
4.3.3 Inquire/Report on Removal Audit Logs
4.4 System User Definitions
4.4.1 Add System User Definitions
4.4.2 Modify System User Definitions
4.4.3 Delete System User Definitions
4.4.4 Inquire/Report on System User Definitions
4.5 User Session Logs
4.5.1 Add User Session Logs
4.5.2 Delete User Session Logs
4.5.3 Inquire/Report on User Session Logs
4.5.4 Initialize System for the Day

Figure C.7 (Continued)

This user interface can be easily implemented using an OO RAD tool such as Delphi or NetBeans. Moreover, it can be implemented in one of two ways:

- **Static Approach:** The options may be hard-coded into four menu operations—a main menu and one for each subsystem. For each option taken by the end-user, the appropriate operation would be invoked. This approach is simple but not very flexible. Whenever options are to be changed or new options added, the appropriate menu operation(s) have to be modified.
- **Dynamic Approach:** The options may be loaded into database tables created for that purpose. Each entry would be the option description along with the system name of the actual operation to be invoked. When the user selects an option, the corresponding operation for that option is invoked. The approach provides more flexibility to managers of the system: Menu changes (modifications or option additions) will not necessitate changes to the menu operations. Obviously, this approach requires a bit more intelligent programming effort.

C.3.2 Utility Operations

The menu system will be controlled by the following utility operations:

- **IMS_Login_XO**: This operation presents the user with a basic system login that allows authorized users to access the system. Upon successful login, the operations **SC_Initialize_XO**, **SC_SessLog_AO**, and **IMS_Menu0_XO** are called. Operation specifications for **SC_Initialize_XO** and **SC_SessLog_AO** appear in Section C.4.6.
- **IMS_Menu0_XO**: This operation is invoked by **IMS_Login_XO** after successful login. The subsystems AMS, FMS, POSS, and SCS are presented to the user.
- **IMS_Menu1_XO**: This operation is invoked by **IMS_Menu0_XO** after successful login. The options of the ACS are presented.
- **IMS_Menu2_XO**: This operation is invoked by **IMS_Menu0_XO** after successful login. The options of the FMS are presented.
- **IMS_Menu3_XO**: This operation is invoked by **IMS_Menu0_XO** after successful login. The options of the POSS are presented.
- **IMS_Menu4_XO**: This operation is invoked by **IMS_Menu0_XO** after successful login. The options of the SCS are presented.

C.3.3 Message Specification

All user messages will be stored in a system-wide message file (as a database table) called **SC_Message_MF** (see entity E36 of Figure C.6). Each message will be assigned a unique code. Messages will be displayed in the form of pop-up messages from related operations.

C.3.4 Help Specification

The system will host a hypermedia-based help system. Users will access the hypermedia help by clicking appropriate links until they get to the desired help they seek. Moreover, the help system will be organized according to the UITC, so that help will be provided operation-by-operation. As the system matures, the help system can be improved to include context-sensitivity.

C.4 Operations Specification

In this section, operation specifications for selected operations in the system are provided. The algorithms needed for some of the operations (for different entities) are similar. Therefore, in the interest of brevity, instead of repeating the same pseudo-code for these similar operations, the following generic operation outlines will be referenced. The section proceeds as follows:

- System Rules
- Procedural and Derivation Rules
- Generic Pseudo-codes

- Acquisitions Management Subsystem
- Financial Management Subsystem
- System Controls Subsystem

C.4.1 System Rules

System rules are special guidelines that characterize the smooth operation of the software system. They typically include *data integrity rules*, *derivation rules*, and *procedural rules*. Additionally, generic operations serve to simplify and minimize duplication relating to specifying the requirements for each system operation.

Data integrity rules include *referential integrity rules* and *data validation rules*. The **referential integrity rules** have been implied in the database design of Section C.2. In particular, each foreign key (i.e. referencing attribute) in a referencing entity must have values drawn from the entity that it references. The only exception to this rule is the instance where the foreign key value is null.

Data validation rules are guidelines that should characterize the data entering the software system. While these may be implemented at the database level, in the interest of user-friendliness, these validation rules will be specified for related operations that facilitate the data entry. For instance, you will observe that ADD/MODIFY operations for Province Definitions (entity E04 of Figure C.4) will not allow blank province names.

Figure C.8 summarizes the main data integrity rules for the system; these are specified on an entity-by-entity bases.

Entity	Integrity Rules
E01 Item Definitions	1. Primary key is Item Code. 2. Foreign keys reference E02 and E20.
E02 Category Definitions	Primary key is Category Code.
E03 Supplier Definitions	1. Primary key is Supplier Code. 2. Foreign keys reference E04, E05, and E20. 3. Validation checks on Telephone Number and Email suggested.
E04 State/Province Definitions	Primary key is Province Code.
E05 Location Definitions	1. Primary key is Location Code. 2. Foreign key references E04.
E06 Supplier-Item Matrix	1. Primary key is [Supplier Code & Item Code]. 2. Foreign keys reference E03 and E01.
E07 Customer Definitions	1. Primary key is [Customer Code. 2. Foreign keys reference E04, E05, and E20. 3. Validation checks on Telephone Number and Email suggested.
E08 Purchase Order Summary	1. Primary key is Purchase Order Number. 2. Foreign key references E03.
E09 Purchase Order Details	1. Primary key is [Purchase Order Number & Item Code]. 2. Foreign keys reference E08 and E01.
E10 Purchase Invoice Summary	1. Primary key is Purchase Invoice Reference Number. 2. Foreign keys reference E03, E08, and E30.
E11 Purchase Invoice Details	1. Primary key is [Purchase Invoice Reference Number & Item Code]. 2. Foreign keys reference E10 and E01.
E12 Purchase Return Summary	1. Primary key is Purchase Return ID. 2. Foreign keys reference E10 and E30.
E13 Purchase Return Details	1. Primary key is [Purchase Return ID & Item Code]. 2. Foreign keys reference E12 and E01.
E14 Sale Order Summary	1. Primary key is Sale Order Number. 2. Foreign key references E07.
E15 Sale Order Details	1. Primary key is [Sale Order Number & Item Code]. 2. Foreign keys reference E14 and E01.
E16 Sale Invoice Summary	1. Primary key is Sale Invoice Reference Number. 2. Foreign keys reference E07, E14, and E30.
E17 Sale Invoice Details	1. Primary key is [Sale Invoice Reference Number & Item Code]. 2. Foreign keys reference E16 and E01.
E18 Sale Return Summary	1. Primary key is Sale Return ID. 2. Foreign keys reference E16 and E30.
E19 Sale Return Details	1. Primary key is [Sale Return ID & Item Code]. 2. Foreign keys reference E18 and E01.
See section A3.2 more details on each entity. Also, all specified dates must be validated.	

Figure C.8   System Data Integrity Rules

Entity	Integrity Rules
E20 Chart of Accounts	1. Primary key is Account Number. 2. Foreign key reference E20.
E21 Department Definitions	Primary key is Department Number.
E22 Account Balances	1. Primary key is [Account Number & Department Number]. 2. Foreign keys reference E20 and E21.
E23 Purchase Payment Plans	Primary key is Purchase Invoice Reference Number.
E24 Purchase Payments Log	1. Primary key is Payment Reference Code. 2. Foreign keys reference E10, E25, and E30.
E25 Financial Institutions	1. Primary key is Institution Code. 2. Foreign keys reference E04 and E05. 3. Telephone numbers and/or Email address must be valid.
E26 Sale Payment Plans	Primary key is Sale Invoice Reference Number.
E27 Sale Payments Log	1. Primary key is Payment Reference Code. 2. Foreign keys reference E16, E25, and E30.
E28 Employee Definitions	1. Primary key is Employee Identification Number. 2. Foreign keys reference E04, E05, and E20. 3. Telephone numbers and/or Email address must be valid.
E29 Payroll Log	1. Primary key is Payroll Reference Code. 2. Foreign key references E28.
E30 Financial Transactions Log	1. Primary key is Transaction ID Number. 2. Foreign keys reference E20, E21, and E31.
E31 Transaction Classifications	Primary key is Classification Code.
E32 Investments Log	1. Primary key is Investment ID Code. 2. Foreign keys reference E20, E25, and E30.
See section A3.2 more details on each entity. Also, all specified dates must be validated.	

Figure C.8 (Continued)

C.4.2 Procedural and Derivation Rules

Procedural and derivation rules relate to how the software system will actually work. The following derivation and procedural rules will be enforced:

1. Supplier and **Customer** entities (E03 and E07) will each store a record called **Miscellaneous** for **Cash Purchases** and **Cash Sales** respectively.

2. Cash **Purchases** will impact the entities **Purchase Invoice Summary** (E10), **Purchase Invoice Detail** (E11), **Purchase Payments Log** (E24), **Item Definition** (E01), and **Financial Transactions Log** (E30) as follows:
   a. Write a record to **Purchase Invoice Summary** (related Purchase Order is null). If the **Supplier** is not listed in the **Supplier** entity (E03), use the default **Miscellaneous Supplier** record and put an appropriate remark in the invoice's **Comment-field**.
   b. Write corresponding detail records in **Purchase Invoice Detail**.
   c. Write a record in the **Purchase Payments Log**, for the relevant items.
   d. Adjust the **Quantity-on-Hand** in the **Item Definition** entity for the relevant items.
   e. Write corresponding record in the **Financial Transactions Log**.

3. **Credit Purchases** impact the entities **Purchase Order Summary** (E08), **Purchase Invoice Summary** (E10), **Purchase Invoice Detail** (E11), **Purchase Payments Log** (E24), **Item Definition** (E01), and **Financial Transactions Log** (E30) as follows:
   a. Write a record to **Purchase Invoice Summary**.
   b. Write corresponding records to **Purchase Invoice Detail**.
   c. Update the related **Purchase Order Summary** record.
   d. Adjust **Quantity-on-Hand** in **Item Definition** entity for the relevant items.
   e. Write **Accounts Payable** entry in **Financial Transactions Log**.

4. **Purchase Payments** involve the entities **Purchase Payments Log** (E24), **Purchase Invoice Summary** (E10), and **Financial Transactions Log** (E30) in the following way:
   a. Write a record to **Purchase Payments Log**.
   b. Update the **Invoice-Outstanding-Amount** in the **Purchase Invoice Summary**.
   c. Write a record in the **Financial Transactions Log**.

5. Each **Purchase Return** impacts the entities **Purchase Returns Summary** (E12), **Purchase Returns Detail** (E13), **Item Definition** (E01), **Purchase Invoice Summary** (E10), and **Financial Transactions Log** (E30) in the following way:
   a. Write a record to the **Purchase Returns Summary** entity.
   b. Write corresponding detail records to **Purchase Returns Detail**.
   c. Adjust the **Quantity-on-Hand** in the **Item Definition** entity, for the relevant items.
   d. Adjust **Outstanding-Amount** in **Purchase Invoice Summary**.
   e. Write corresponding record in **Financial Transactions Log**.

6. **Cash Sales** will impact the entities **Sale Invoice Summary** (E16), **Sale Invoice Detail** (E17), **Sale Payments Log** (E27), **Item Definition** (E01), and **Financial Transactions Log** (E30) as follows:
   a. Write a record to the **Sales Invoice Summary**. If the customer is not listed in the **Customer** entity, use the default **Miscellaneous Customer**, with an appropriate remark in the invoice's **Comment-field**.
   b. Write corresponding include records in the **Sale Invoice Detail**.
   c. Write a record in the **Sale Payments Log**.
   d. Adjust the **Quantity-on-Hand** in the **Item Definition** entity, for the relevant items.
   e. Issue a receipt and write record(s) in the **Financial Transactions Log**.

7. **Credit Sales Purchases** impact the entities **Sale Order Summary** (E14), **Sale Invoice Summary** (E16), **Sale Invoice Detail** (E17), **Sale Payments Log** (E27), **Item Definition** (E01), and **Financial Transactions Log** (E30) as follows:
   a. Write a record to the **Sales Invoice Summary** entity.
   b. Write corresponding detail records in **Sales Invoice Detail**.
   c. Update the related **Sale Order Summary** record.
   d. Adjust the **Quantity-on-Hand** in the **Item Definition** entity, for the relevant items.
   e. Write **Accounts Receivable** record in **Financial Transactions Log**.

8. **Sale Payments** involve the entities **Sale Payments Log** (E27), **Sale Invoice Summary** (E16), and **Financial Transactions Log** (E30) in the following way:
   a. Write a record to **Payments Received** entity.
   b. Update the **Invoice Outstanding Amount** in the **Sales Invoice Summary** entity.
   c. Issue a receipt.
   d. Write corresponding record in **Financial Transactions Log**.

9. Each **Sale Return** impacts the entities **Sale Returns Summary** (E18), **Sale Returns Detail** (E19), **Item Definition** (E01), **Sale Invoice Summary** (E16), and **Financial Transactions Log** (E30) in the following way:
   a. Write a record to the **Sales Returns Summary** log.

    b. Write corresponding details records to **Sales Returns Detail** log.
    c. Adjust the **Quantity-on-Hand** in the **Item Definition** entity, for the relevant items.
    d. Adjust the **Outstanding-Amount** for the related **Sale Invoice Summary**.
    e. Write corresponding record in **Financial Transactions Log**.

10. **Accounts Payable** is determined by the formula:

$$\text{Accounts Payable} = [\text{\textbf{Purchase Invoices} with nonzero \textbf{Outstanding Amounts}}$$
$$\text{minus \textbf{Purchase Returns} with nonzero \textbf{Return Amounts}}]$$

11. **Accounts Receivable** is determined by the formula:

$$\text{Accounts Receivable} = [\text{\textbf{Sales Invoices} with nonzero \textbf{Outstanding Amounts}}$$
$$\text{minus \textbf{Sales Returns} with nonzero \textbf{Return Amounts}}]$$

These rules will be extremely useful during system development. Some of them will be incorporated into the relevant operation specifications and included in the project's design specification.

C.4.3 Generic Pseudo-codes

Generic pseudo-codes are provided for the ADD operation, the MODIFY operation, and the DELETE operation in Figures C.9, C.10, and C.11 respectively.

```
START
WHILE (User wishes to continue)
      Accept Key Field(s);
      Check Record Absence or Existence in the primary file;
      IF     (Record Absent)
            Accept Non-key Fields;
            Validate Non-key Fields based on Validation Rules;
            WHILE(Any Error Exists),
                  Re-display Non-key Fields for possible Update;
                  Display appropriate error message(s);
                  Validate Non-key Fields based on Validation Rules;
            END-WHILE;
            Re-display full Record for confirmation;
            IF     (Confirmation Obtained)
                  Write New Record to the primary file;
                  Write New Record to file SC_AudAdd_BR via operation SC_AudAdd_AO;
            ENDIF;
            ELSE  Inform the User that nothing was saved;  END-ELSE;
      ENDIF;
      ELSE  Display Message ('Record already exists'); END-ELSE;
      Check if User wishes to quit and set an exit flag if necessary;
END-WHILE;
Generate Edit-List;
STOP
```

Figure C.9   Generic ADD Pseudo-code

```
START
WHILE (User wishes to continue)
        Accept Key Field(s);
        Check Record Absence or Existence in the primary file;
        IF      (Record Present)
                Retrieve Record and update Audit Log Fields (with before-values);
                Display Non-key Fields for possible Update;
                Validate Non-key Fields based on Validation Rules;
                WHILE(Any Error Exists),
                        Re-display Non-key Fields for possible Update;
                        Display appropriate error message(s);
                        Validate Non-key Fields based on Validation Rules;
                END-WHILE;
                Re-display full Record for confirmation;
                IF      (Confirmation Obtained)
                        Update Audit Log Fields (with current-values);
                        Write New Record to file SC_AudUpd_BR via operation SC_AudUpd_AO;
                        Update Record in the primary file;
                ENDIF;
                ELSE  Inform the User that nothing was saved; END-ELSE;
        ENDIF;
        ELSE  Display Message ('Record does not exists'); END-ELSE;
        Check if User wishes to quit and set an exit flag if necessary;
END-WHILE;
Generate Edit-List;
STOP
```

Figure C.10   Generic MODIFY Pseudo-code

```
START
WHILE (User wishes to continue)
        Accept Key Field(s);
        Check Record Absence or Existence in the primary file;
        IF      (Record Present)
                Retrieve Record;
                Display full Record for confirmation;
                IF      (Deletion Confirmation Obtained)
                        Update Audit Log Fields (with current-values);
                        Write New Record to file SC_AudDel_BR via operation SC_AudDel_AO;
                        Delete Record from the primary file;
                ENDIF;
                ELSE  Inform the User that nothing was saved; END-ELSE;
        ENDIF;
        ELSE  Display Message ('Record does not exists'); END-ELSE;
        Check if User wishes to quit and set an exit flag if necessary;
END-WHILE;
Generate Edit-List;
STOP
```

Figure C.11   Generic DELETE Pseudo-code

C.4.4 Acquisitions Management Subsystem

The operation specifications for the first two operations in the Acquisitions Management Subsystem (AMS) are provided in this section via Figures C.12 and C.13. These operations relate to management of the entities specified in Section C.2. The methodology employed in specifying the operational requirements is the *extended operation specification* (EOS) as discussed in Chapter 15 .

Operation Biography:
System: Inventory Management System
Subsystem: Acquisitions Management
Operation Name: **AM_ItemDef_AO / AM_ItemDef_MO / AM_ItemDef_ZO**
Operation Description: Facilitates addition of items to the Item Master table.
Operation Category: Mandatory
Complexity Rank: 8 of 10
Spec. Author: E. Foster
Date: 7-10-2016

Inputs:
New Item Form
AM _ItemDef_BR — Inventory Item Definition (E01)
AM ItemCatg_BR — Item Categories Definition (E02)
FM_ChartAccts_BR — Chart of Accounts (E20)

Outputs:
AM _ItemDef_BR — Inventory Item Definition (E01)

Validation Rules for Inserting Data:
1. Item Code must not previously exist
2. Category Code must already exist in **AM_ItemCat_BR**
3. Blank Item Name not allowed
4. Account Number must already exist in **FM_ChartAccts_BR**

Validation Rules for Modifying Data:
1. Item Code must previously exist
2. Category Code must already exist in **AM_ItemCat_BR**
3. Blank Item Name not allowed
4. Account Number must already exist in **FM_ChartAccts_BR**

Special Notes:
When a new Item is added or purchased, the Quantity on Hand and the Quantity Owned are adjusted.

Operation Outline: See Generic ADD / MODIFY / DELETE pseudo-code.

Figure C.12   Operation Specification for Managing Inventory Items Data

---

Operation Biography:
System: Inventory Management System
Subsystem: Acquisitions Management
Operation Name: **AM_ItemDef_IO / AM_ItemDef_RO**
Operation Description: Facilitates inquiry/report on Inventory Items.
Operation Category: Mandatory   Complexity Rank: 10 of 10
Specification Author: E. Foster   Specification Date: 7-10-2016

Inputs:
AM_ItemDef_LV1 — logical view that combines Item Definition (E01), Item Category (E02), and Chart of Accounts (E20)

Outputs: Monitor / Printer

Validation Rules: None

Special Notes:
It will be possible to query Items via any of the following access paths:
1.   By Category Name & Item Name
2.   By Category Code & Item Code
3.   Item Code or Item Name
4.   By Account Number
5.   UPC Code or SKU number

// Main Operation Outline:
START:
        While User Wishes to Continue
                Present the User with the options mentioned above;
                Depending on the User's choice, Invoke the relevant sub-operation for that purpose;
        End-While;
STOP.

// Outline for Option 1:
START
While User Wishes to Continue
        Prompt user for Category Name and Item Name;
        Starting at that point in **AM_ItemDef_LV1**, Load a Virtual Data Collection Object with all records until End-of-File,
                ordering by the information by [Category Name & Item Name];
        Display the Virtual Data Collection Object;
End-While;
STOP

// Outline for Option 2:
START
While User Wishes to Continue
        Prompt user for Category Code and Item Code;
        Starting at that point in **AM_ItemDef_LV1**, Load a Virtual Data Collection Object with all records until End-of-File,
                ordering by the information by [Category Code & Item Code];
        Display the Virtual Data Collection Object;
End-While;
STOP

{The rest of the sub-operations will be similar}

Figure C.13   Operation Specification for Inquiry/Report on Inventory Items Data

C.4.5 *Financial Management Subsystem*

The operation specifications for the first two operations in the Financial Management Subsystem (FMS) will be provided in this section — see Figures C.14 and C.15..

```
Operation Biography:
System:                 Inventory Management System
Subsystem:              Acquisitions Management
Operation Name:         FM_ChartAccts_AO / FM_ChartAccts_MO / FM_ChartAccts_ZO
Operation Description:  Facilitates addition of items to the Item Master table.
Operation Category:     Mandatory
Complexity Rank:        6 of 10
Spec. Author:           E. Foster
Date:                   7-10-2016

Inputs:
New Item Form
FM _ChartAccts_BR — Chart of Accounts (E20)

Outputs:
FM _ChartAccts_BR — Chart of Accounts (E20)

Validation Rules for Inserting Data:
1.  Account Code must not previously exist
2.  Blank Item Name not allowed

Validation Rules for Modifying Data:
1.  Account Code must previously exist
2.  Blank Item Name not allowed

Special Notes:
Each detail account belongs to a summary account.

Operation Outline: See Generic ADD / MODIFY / DELETE pseudo-code.
```

Figure C.14   Operation Specification for Managing Chart of Accounts Items

```
Operation Biography:
System:                 Inventory Management System
Subsystem:              Financial Management
Operation Name:         FM_ChartAccts_IO / FM_ChartAccts_RO
Operation Description:  Facilitates inquiry/report on Chart of Accounts
Operation Category:     Important
Complexity Rank:        8 of 10
S Spec. Author:         E. Foster
Date:                   7-10-2016

Inputs:
FM _ChartAccts_LV1 — Logical view of Chart of Accounts (E20) that connects E20 with itself

Outputs:
Monitor / Printer

Validation Rules:    None

Special Notes: It will be possible to query Chart of Accounts by Account Number or Account Name

Operation Outline:
START:
While User Wishes to Continue
        Prompt user for Account Number or Account Name;
        Prompt user for preference (Accounts by Account Number or Account Name);
        If By Account Number
                Starting at that point in FM _ChartAccts_LV1 Load a Virtual Data Collection Object with all records until
                        End-of-File, ordering by [Account Number];
                Display the Virtual Data Collection Object;
        End-If;
        If By Account Name
                Starting at that point in FM _ChartAccts_LV1, Load a Virtual Data Collection Object with all records until
                        End-of-File, ordering by [Account Name];
                Display the Virtual Data Collection Object;
        End-If;
End-While;
STOP
```

Figure C.15   Operation Specification for Inquiry/Report on Chart of Accounts Items

C.4.6 *System Controls Subsystem*

The operation specifications for operations in the System Controls Subsystem (SCS) are provided in this section. Since these operations are different from others in the system, most of the related operation specifications are presented (Figures C.16 to C.29).

Operation Biography:
System:	Inventory Management System
Subsystem:	System Controls
Operation Name:	**SC_AudAdd_AO**
Operation Description:	Facilitates addition of records to the Audit Table for Data Additions.
Operation Category:	Enhancement
Complexity Rank:	6 of 10
Specification Author:	E. Foster
Specification Date:	7-10-2016

Inputs:
SC_AudAdd_BR — Audit Table for Data Additions (E33)

Outputs:
SC_AudAdd_BR — Audit Table for Data Additions (E33)

Validation Rules: None

Special Notes:
1. This operation requires a different logic from the generic ADD operation.
2. This operation will be called with arguments for the attributes (columns) of **SC_AudAdd_BR**.
3. The source table is the table written to by the operation that called this operation.

Operation Outline:
START
  Use the arguments received to load **SC_AudAdd_BR** record fields;
  Write the new record to **SC_AudAdd_BR**;
STOP

Figure C.16   Operation Specification for Adding to the Audit Table for Data Additions

```
Operation Biography:
System:                 Inventory Management System
Subsystem:              System Controls
Operation Name:         SC_AudAdd_ZO
Operation Description:  Facilitates deletion of records from the Audit Table for Data Additions.
Operation Category:     Enhancement
Complexity Rank:        6 of 10
Specification Author:   E. Foster
Specification Date:     7-10-2016


Inputs:
SC_AudAdd_BR — Audit Table for Data Additions (E33)


Outputs:
SC_AudAdd_BR — Audit Table for Data Additions (E33)


Validation Rules:  None


Special Notes:
1.   This operation requires a different logic from the generic DELETE operation.
2.   The user will specify a starting Session ID, and an ending Session ID. Records between both sessions (inclusive) will be
     deleted.


Operation Outline:
START:
 Prompt user for Starting Session and Ending Session;
 Delete all records from SC_AudAdd_BR that satisfy the condition (Start Session <= AA_AddID <= Stop Session);
STOP
```

Figure C.17   Operation Specification for Deletion from the Audit Table for Data Additions

```
Operation Biography:
System:                 Inventory Management System
Subsystem:              System Controls
Operation Name:         SC_AudAdd_IO / SC_AudAdd_RO
Operation Description:  Facilitates inquiry/report on the Audit Table for Data Additions.
Operation Category:     Enhancement
Complexity Rank:        8 of 10
Specification Author:   E. Foster
Specification Date:     7-10-2016


Inputs:
SC_AudAdd_LV — Logical view that joins Audit Table for Data Additions (E33) with Sessions Log (E38) and
System User (E37)


Outputs:
Monitor / Printer


Validation Rules:    None


Special Notes:
It will be possible to query audited data additions By Session ID.


Operation Outline:
START:
While User Wishes to Continue
     Prompt user for Session ID;
     Starting at that point in SC_AudAdd_LV, Load a Virtual Data Collection Object with all records until End-of-File,
          Ordering by [Session ID];
     Display the Virtual Data Collection Object;
End-While;
STOP
```

Figure C.18   Operation Specification for Inquiry/Report on the Audit Table for Data Additions

Operation Biography:
System: Inventory Management System
Subsystem: System Controls
Operation Name: **SC_AudUpd_AO**
Operation Description: Facilitates addition of records to the Audit Table for Data Modifications.
Operation Category: Enhancement
Complexity Rank: 6 of 10
Specification Author: E. Foster
Specification Date: 7-10-2016

Inputs:
SC_AudUpd_BR — Audit Table for Data Modifications (E34)

Outputs:
SC_AudUpd_BR — Audit Table for Data Modifications (E34)

Validation Rules: None

Special Notes:
1. This operation will be called with arguments for the attributes (columns) of **SC_AudUpd_BR**.
2. The source table is the table written to by the operation that called this operation.

Operation Outline:
START
  Use the arguments received to load **SC_AudUpd_BR** record fields;
  Write the new record to **SC_AudUpd_BR**;
STOP

Figure C.19   Operation Specification for Adding to the Audit Table for Data Modifications

Operation Biography:
System: Inventory Management System
Subsystem: System Controls
Operation Name: **SC_AudUpd_ZO**
Operation Description: Facilitates deletion of records from the Audit Table for Data Modifications.
Operation Category: Enhancement
Complexity Rank: 6 of 10
Specification Author: E. Foster
Specification Date: 7-10-2016

Inputs:
SC_AudUpd_BR — Audit Table for Data Modifications (E34)

Outputs:
SC_AudUpd_BR — Audit Table for Data Modifications (E34)

Validation Rules: None

Special Notes:
The user will specify a starting Session ID, and an ending Session ID. Records between both sessions (inclusive) will be deleted.

Operation Outline:
START:
  Prompt user for Starting Session and Ending Session;
  Delete all records from **SC_AudUpd_BR** that satisfy the condition (Start Session <= **AU_UpdateID** <= Stop Session);
STOP

Figure C.20   Operation Specification for Deletion from Audit Table for Data Modifications

Operation Biography:
System: Inventory Management System
Subsystem: System Controls
Operation Name: **SC_AudUpd_IO / SC_AudUpd_RO**
Operation Description: Facilitates inquiry/report on the Audit Table for Data Modifications.
Operation Category: Enhancement
Complexity Rank: 8 of 10
Specification Author: E. Foster
Specification Date: 7-10-2016

Inputs:
SC_AudUpd_LV — Logical view that joins Audit Table for Data Modifications (E34) with Sessions Log (E38) and System User (E37)

Outputs:
Monitor / Printer

Validation Rules: None

Special Notes:
It will be possible to query audited data modifications By Session ID.

Operation Outline:
START:
While User Wishes to Continue
         Prompt user for Session ID;
         Starting at that point in **SC_AudUpd_LV**, Load a Virtual Data Collection Object with all records until End-of-File,
                 ordering by [Session ID];
         Display the Virtual Data Collection Object;
End-While;
STOP

Figure C.21   Operation Specification for Inquiry/Report on Audit Table for Data Modifications

Operation Biography:
System: Inventory Management System
Subsystem: System Controls
Operation Name: **SC_AudDel_AO**
Operation Description: Facilitates addition of records to the Audit Table for Data Deletions.
Operation Category: Enhancement
Complexity Rank: 6 of 10
Specification Author: E. Foster
Specification Date: 7-10-2016

Inputs:
SC_AudDel_BR — Audit Table for Data Deletions (E35)

Outputs:
SC_AudDel_BR — Audit Table for Data Deletions (E35)

Validation Rules: None

Special Notes:
1.   This operation will be called with arguments for the attributes (columns) of **SC_AudDel_BR**.
2.   The source table is the table written to by the operation that called this operation.

Operation Outline:
START
  Use the arguments received to load **SC_AudDel_BR** record fields;
  Write the new record to **SC_AudDel_BR**;
STOP

Figure C.22   Operation Specification for Addition to Audit Table for Data Deletions

Operation Biography:
System: Inventory Management System
Subsystem: System Controls
Operation Name: **SC_AudDel_ZO**
Operation Description: Facilitates deletion of records from the Audit Table for Data Deletions.
Operation Category: Enhancement
Complexity Rank: 6 of 10
Specification Author: E. Foster
Specification Date: 7-10-2016

Inputs:
SC_AudDel_BR — Audit Table for Data Deletions (E35)

Outputs:
SC_AudDel_BR — Audit Table for Data Deletions (E35)

Validation Rules: None

Special Notes:
The user will specify a starting Session ID, and an ending Session ID. Records between both sessions (inclusive) will be deleted.

Operation Outline:
START:
  Prompt user for Starting Session and Ending Session;
  Delete all records from **SC_AudDel_BR** that satisfy the condition (Start Session <= **AD_DelID** <= Stop Session);
STOP

Figure C.23   Operation Specification for Deletion from Audit Table for Data Deletions

Operation Biography:
System: Inventory Management System
Subsystem: System Controls
Operation Name: **SC_AudDel_IO / SC_AudDel_RO**
Operation Description: Facilitates inquiry/report on the Audit Table for Data Deletions.
Operation Category: Enhancement
Complexity Rank: 8 of 10
Specification Author: E. Foster
Specification Date: 7-10-2016

Inputs:
SC_AudDel_LV — Logical view that joins Audit Table for Data Deletions (E35) with Sessions Log (E38) and System User (E37)

Outputs:
Monitor / Printer

Validation Rules: None

Special Notes:
It will be possible to query audited data deletions By Session ID.

Operation Outline:
START: /* Inquire */
While User Wishes to Continue
        Prompt user for Session ID;
        Starting at that point in **SC_AudDel_LV**, Load a Virtual Data Collection Object with all records until End-of-File,
              Ordering by [Session ID;
        Display the Virtual Data Collection Object;
End-While;
STOP

Figure C.24   Operation Specification for Inquiry/Report on Audit Table for Data Deletions

```
Operation Biography:
System:                   Inventory Management System
Subsystem:                System Controls
Operation Name:           SC_SessLog_AO
Operation Description:    Facilitates addition of records to the Session Log file.
Operation Category:       Enhancement        Complexity Rank:     6 of 10
Specification Author:     E. Foster           Specification Date:  7-10-2016

Inputs:
System Login-in Screen
SC_SessLog_BR — Sessions Log (E38)
SC_SysUser_BR — System User (E37)

Outputs:
SC_SessLog_BR — Sessions Log (E38)
SC_SessMark_BR — Sessions Mark (E39)

Validation Rules:
1.  Session ID must not previously exist
2.  This operation is called by IMS_Login_XO after a successful user login with incoming parameter thisUserCode

Special Notes:
1. A Session record is added when a user successfully logs on.
2. The table SCSessMark is accessed and its record incremented with each successful log on.

Operation Outline:
START
  Retrieve system values for currentTime and currentDate;
  Use currentDate to access SC_SessMark_BR and increment its SessionMark by 1;

  // Set Session attributes and write SessionLog
  SCSessLogID := SC_SessMark_BR.SCSessionDate concatenated to SC_SessMark_BR.SCSessionMark;
  SCSessLogUser := thisUserCode;
  SCSessLoginTime := currentTime;
  Write new record in SC_SessLog_BR;
STOP
```

Figure C.25   Operation Specification for Adding Session Log Entry

```
Operation Biography:
System:                   Inventory Management System
Subsystem:                System Controls
Operation Name:           SC_SessLog_ZO
Operation Description:    Facilitates deletion of records from the Sessions Log.
Operation Category:       Enhancement
Complexity Rank:          6 of 10
Specification Author:     E. Foster
Specification Date:       7-10-2016

Inputs:
SC_SessLog_BR — Sessions Log (E34)

Outputs:
SC_SessLog_BR — Sessions Log (E34)

Validation Rules:  None

Special Notes:
1.  Only authorized personnel must have access to this operation.
2.  The user will specify a starting Session ID, and an ending Session ID. Records between both sessions (inclusive) will be
    deleted.

Operation Outline:
START:
  Prompt user for Starting Session and Ending Session;
  Delete all records from SC_SessLog_BR that satisfy the condition (Start Session <= SCSessLogID <= Stop Session);
STOP
```

Figure C.26   Operation Specification for Deleting Sessions from the Sessions Log

```
Operation Biography:
System:                   Inventory Management System
Subsystem:                System Controls
Operation Name:           SC_SessLog_IO / SC_SessLog_RO
Operation Description:    Facilitates inquiry/report on Sessions
Operation Category:       Enhancement
Complexity Rank:          8 of 10
Specification Author:     E. Foster
Specification Date:       7-10-2016


Inputs:
SC_SessLog_LV — Logical view that joins Sessions Log (E38) with the System User (E37)


Outputs:
Monitor / Printer


Validation Rules:    None


Special Notes:
It will be possible to query session log via any of the following access paths:
1.   By Session ID and User Code
2.   By User Full Name or By User Short Name


// Outline
START
While User Wishes to Continue
       Present the User with the options mentioned above.
       If (By Session ID and User Code)
             Prompt user for Session ID and User Code;
             Starting at that point in SC_SessLog_LV, Load a Virtual Data Collection Object with all records
                   until End-of-File, ordering by [Session ID & User Code];
             Display the Virtual Data Collection Object;
       End-If;
       If (By User Name)
             Prompt for User Name;
             Starting at that point in SC_SessLog_LV, Load a Virtual Data Collection Object with all records
                   until End-of-File, ordering by [User Name];
             Display the Virtual Data Collection Object;
       End-If
End-While;
STOP
```

Figure C.27   Operation Specification for Inquiry/Report on Sessions

```
Operation Biography:
System:                   Inventory Management System
Subsystem:                System Controls
Operation Name:           SC_Initialize_XO
Operation Description:    Initializes the system each day of usage, and sets the initial Session Mark for that day
Operation Category:       Mandatory
Complexity Rank:          3 of 10
Specification Author:     E. Foster
Specification Date:       7-10-2016


Inputs:  SC_SessMark_BR — Session Mark (E39)


Outputs:  SC_SessMark_BR — Session Mark (E39)


Validation Rules:    None


Special Notes: None


Operation Outline:
START
    Retrieve currentDate  in form YYYYMMDD;
    initStart := 0;
    // Starter := currentDate + initStart;
    Use currentDate to check SC_SessMark_BR for a record;
    If (there is no record in SC_SessMark_BR with SessionMark >= Start)
        SessionMark := initStart;
        SessionDate := currentDate;
        Write a new record to SC_SessMark_BR
    End-If
STOP
```

Figure C.28   Operation Specification for System Initialization

```
Operation Biography:
System:                    Inventory Management System
Subsystem:                 System Controls
Operation Description:     Manages the IMS menu system and processes user's choices.
Operation Category:        Mandatory
Complexity Rank:           8 of 10
Specification Author:      E. Foster
Specification Date:        7-10-2016

Special Notes:

1.  The menu system will be controlled by the following utility operations:
■   IMS_Login_XO: This operation presents the user with a basic system login that allows authorized users to access
    the system. Upon successful login, the operations SC_Initialize_XO, SC_SessLog_AO, and IMS_Menu0_XO are
    called. IMS_Menu0_XO: This operation is invoked by IMS_Login_XO after successful login. The subsystems
    AMS, FMS, POSS, and SCS are presented to the user.
■   IMS_Menu1_XO: This operation is invoked by IMS_Menu0_XO after successful login. The options of the ACS are
    presented.
■   IMS_Menu2_XO: This operation is invoked by IMS_Menu0_XO after successful login. The options of the FMS are
    presented.
■   IMS_Menu3_XO: This operation is invoked by IMS_Menu0_XO after successful login. The options of the POSS
    are presented.
■   IMS_Menu4_XO: This operation is invoked by IMS_Menu0_XO after successful login. The options of the SCS are
    presented.

2.  The UITC depicted in figure A3-7 provides an overview for the menu structure for the system.
```

Figure C.29   Guidelines for Managing the System Login and Menus Infrastructure

C.5 Summary and Concluding Remarks

This design specification has outlined the blueprint for an Inventory Management System that will be robust, and adaptable to any organization. It includes the following:

- An overview of the system, including problem definition, proposed solution, and system architecture. The overview includes an information topology chart (ITC) that identifies the key information entities in four-component subsystems— Acquisitions Management Subsystem (AMS), Financial Management Subsystem (FMS), Point of Sale Subsystem (POSS), and System Controls Subsystem (SCS).
- Database specification that includes a set of design conventions, naming conventions, and an O/ESG for each of the information entities comprising the system in all three sub-systems.
- User interface specification that outlines menu hierarchy via a user interface topology chart (UITC). Also included are a message specification that outlines how system messages will be handled, and a help specification that outlines how the help system will be implemented.
- Operations specification that includes extended operation specifications (EOSs) for critical operations comprising the system in all four sub-systems.

Possible future enhancements include the following:

- **Point-of-Sale Subsystem:** The preliminary system prototype would basically provide a front-end that allows point-of-sale (POS) machines to be used. It would be necessary to have the POS machines programmed to interface with the system in a seamless manner.
- **Just-in-Time Subsystem:** This would require automating the re-order process. The idea is to trigger an automatic purchase order to a supplier for goods, once the re-order limit for that item has been reached.
- **Refine the System Controls Subsystem:** More features can be added to the System Controls Subsystem. For instance, a feature could be added that dynamically generates different menus for different users, depending on the features they have been authorized to access.

Despite these potential enhancements, this is a solid start. Now have fun with the development and implementation!

INDEX

Page numbers in *Italics* refer to figures.

Printed in the United States
by Baker & Taylor Publisher Services